Texts in Mathematics

Volume 5

A Treatise on the Binomial Theorem

Volume 1
An Introduction to Discrete Dynamical Systems and their General Solutions
F. Oliveira-Pinto

Volume 2
Adventures in Formalism
Craig Smoryński

Volume 3
Chapters in Mathematics. From π to Pell
Craig Smoryński

Volume 4
Chapters in Probability
Craig Smoryński

Volume 5
A Treatise on the Binomial Theorem
Craig Smoryński

Texts in Mathematics Series Editor
Dov Gabbay   dov.gabbay@kcl.ac.uk

A Treatise on the Binomial Theorem

Craig Smoryński

©   Individual author and College Publications 2012. All rights reserved.

ISBN 978-1-84890-085-1

College Publications
Scientific Director: Dov Gabbay
Managing Director: Jane Spurr
Department of Computer Science
King's College London, Strand, London WC2R 2LS, UK

http://www.collegepublications.co.uk

Cover designed by Laraine Welch
Printed by Lightning Source, Milton Keynes, UK

All rights reserved. No part of this publication may be reproduced, stored in a retrieval system or transmitted in any form, or by any means, electronic, mechanical, photocopying, recording or otherwise without prior permission, in writing, from the publisher.

Contents

Preface   vii

Chapter 1. The Finite Binomial Theorem   1
1. Preparation: $(a+b)^2 = a^2 + 2ab + b^2$ 1
2. Extracting Square Roots   6
3. Finding Square Roots One Decimal at a Time   10
4. Third, Fourth, and Fifth Roots   14
5. The Full Finite Binomial Theorem   23
6. Combinatorics in Early India   30
7. The Chinese Application: Horner's Method   36
8. Pascal at Long Last   51

Chapter 2. Newton's Binomial Theorem   63
1. Newton's Discovery   63
2. Discussion of Newton's Remarks   68
3. Application of Newton's Theorem   71
4. A Brief Calculus Primer, I: Area   78
5. A Brief Calculus Primer, II: Slopes and Tangents   81
6. A Brief Calculus Primer, III: Fundamental Theorem of Calculus   90
7. A Brief Calculus Primer, IV: Series   94
8. Power Series   104
9. Taylor's Theorem   118
10. Newton's Method and the Mean Value Theorem   128

Chapter 3. The Binomial Theorem Proven   147
1. Prefatory Remarks   147
2. Landen's Attempted Proof   148
3. Euler's Attempted Proof   151
4. Hutton's Attempted Proof   153
5. Peacock's Attempted Metaphysical Proof   161
6. D'Alembert and The Limit Concept   169
7. The New Maths of the 19th Century   191
8. Bolzano's Proof   195
9. Cauchy's Proof   222
10. The Exponential and Logarithmic Functions   246
11. Abel's Proof   250
12. Taylor's Theorem   279

Appendix A. Using the Calculator		287
1. Horner's Method on the Calculator		287
2. Exact Calculations on the *TI-83*		299
3. Extended Calculations on the *TI-83*		319
4. Extended Calculations on the *TI-89*		329
Index		337

Preface

In explaining to Dr. Watson the brilliance of his archenemy Professor James Moriarty, Sherlock Holmes said, in part,

> At the age of twenty-one he wrote a treatise upon the Binomial Theorem, which has had a European vogue. On the strength of it, he won the Mathematical Chair at one of our smaller Universities, and had, to all appearances, a most brilliant career before him.

In quoting this passage in his edited life of Sherlock Holmes, J.R. Hamilton added the footnote

> The Binomial Theorem is beautiful and useful, but it is difficult to see how Moriarty could have thrown any new light upon it.[1]

New light can come in many forms. A new proof, a new result, or even an up to date survey of what was known at the time. What was the time, what was known by then, and what is known now?

According to Watson's chronicles, Moriarty died in 1891. And Sydney Paget's portrait of the professor shows a man no longer young, perhaps in his 60s? He couldn't have been much older or the comparatively younger Holmes's battle with him at the Reichenbach Falls would have been completely one-sided. So a 21-year old Moriarty's treatise on the Binomial Theorem would have appeared in the 1850s, 1860s if Moriarty were only in his 50s when he succumbed. Most of the important results on the Binomial Theorem had been well-established by then, but there was room for a bit more. As an amateur historian, I cannot tout my ignorance of any contemporary expositions of the full story of the Binomial Theorem as the nonexistence of such at the time of Moriarty's publication, but the time may have been ripe for such.

Questioning the contents of a treatise on the Binomial Theorem by a fictional character may perhaps be of interest only to a mathematically literate Holmes enthusiast who wishes to complete the canon by authoring Moriarty's treatise on the Binomial Theorem as well, perhaps, as his *The Dynamics of an Asteroid*. Such are, indeed, tempting propositions and my encounters with the Binomial Theorem in some of my recent books have tempted me with the former project. My complete and utter lack of ability to reproduce the style of another writer, in this case that of the typical Victorian author, and the amateur nature of my grasp of mathematical history preclude me from doing such. But I believe

---

[1] J.R. Hamilton (ed.), *My Life With Sherlock Holmes: Conversations in Baker Street by John H. Watson, M.D.*, John Murray, London, 1968, p. 67.

Sherlock Holmes is not only big in literature, but has been the subject of numerous films, including the parody *The Adventures of Sherlock Holmes' Smarter Brother*. In this film, Moriarty is twice seen with a blackboard behind him bearing mathematical formulæ. In one scene, the board contains elementary arithmetic mistakes. Here we see him and his cohort in front of a board on which the Finite Binomial Theorem is written. Unfortunately, the two gentlemen obstruct the full statement of the Theorem and it is impossible to determine whether the statement is correct or, given that actor Leo McKern (right) portrays Moriarty as a mathematical illiterate, another load of rubbish.

At the time I write this, a new Sherlock Holmes movie, *Sherlock Holmes: A Game of Shadows*, features the Binomial Theorem more directly in the form of a secret code based in part on the binomial coefficients devised by two Oxford mathematicians, Alain Goriely and Derek Moulton, who describe the code in "The mathematics behind *Sherlock Holmes: A Game of Shadows*", *Siam News*, volume 45 number 3, April 2012, pp. 1, 8. The article includes an annotated reproduction of Moriarty's densely covered blackboard.

I have learned enough about the history of the Binomial Theorem to produce a serviceable survey and such is the work that lies before you now. It is, of course, subject to J.R. Hamilton's doubts about its ability to throw any new light upon the subject, and I can make no such claim. But it does gather the information in a way that I hope will be both accessible and interesting.

I recognise four phases in the development of the Binomial Theorem. First is the discovery of the finite form of the Theorem for positive integral exponents, along with the applications thereof. Newton's generalisation for rational exponents resulting in an infinite series ushered in the second phase, an era of powerful applications and questionable foundations. A third phase initiated by Jakob Bernoulli coincides with the development of the theory of probability and consists of applications of the finite form of the theorem to this theory. Finally, the growing concern with the foundations of mathematics and the question of rigour resulted in the fourth phase, the many attempts to prove Newton's Binomial Theorem.

The first, second, and fourth of these phases offer a coherent narrative that can be told in roughly chronological order. The third interrupts the flow and, as I have nothing to say about it that was not covered in my book on probability[2], I have decided against its inclusion here. The present book thus consists of three chapters on the finite form of the Binomial Theorem, Newton's generalisation, and the eventual proof of Newton's Theorem.

The initial importance of the Binomial Theorem, both in the case of its finite algebraic expansion for positive integral coefficients and that of Newton's infinite series expansion for rational coefficients, is its application to the extraction of roots. This being the case, such approximations form a secondary thread in the first two chapters of this book and I have allowed myself to follow this thread a short distance away from our primary binomial concern. In the main body of the text this occurs primarily in Chapter 2, section 10, where I discuss Newton's Method. And in Chapter 1, the discussion of the implementation of various procedures on the calculator grew so long that I moved the material to an appendix so as not to interrupt the flow of the exposition. I suppose there is no harm in passing over this material, especially, for one not interested in playing with the calculator, the programming tasks of the Appendix.

The reader who chooses to follow these computational paths will not find all the calculation too painful if he uses a calculator or a computer. Even the best comfy chair has only so much lap space and I find the calculator most handy and thus give instructions for it. In doing so I standardise on a couple of calculators from Texas Instruments because of their popularity in American schools and my familiarity with them. Specifically, I use the *TI-83 Plus* (roughly equivalent to the *TI-84*)[3] and the *TI-89 Titanium*. My assumption is that only an expert would have the latter and, thus, I am not as expansive in my explanations of its use, which is limited to tasks not easily performed on the *TI-83 Plus*. I do presuppose some familiarity with the calculator and make no attempt to replace its manual in explaining the application of the calculator. Expressions and instructions to be entered into the calculator, as well as representation of its output, are given in **sans serif**. Programs are presented pretty much as they would appear in the program editor. The exceptions are that i. I try to present long instructions on single lines instead of wrapping them as on the calculator screen, and ii. when the nesting of loops and conditionals gets too extreme I use indents to increase readability.

I also stray from the Binomial Theorem in Chapter 2 in those sections the titles of which begin with the phrase "A Brief Calculus Primer". Ostensibly this reduces the prerequisites of the book by introducing and explaining the Calculus needed. Given that the real prerequisite is a certain mathematical

---

[2]Craig Smoryński, *Chapters in Probability*, College Publications, London, 2012.

[3]The *TI-84* has some additional functions that the *TI-83* does not. Thus, the instructions given here for the *TI-83* should apply directly to the currently more common *TI-84*. The exceptions would be those remarks on the performance of the calculator when executing various programs, the *TI-84* presumably being slightly faster.

maturity of the kind traditionally acquired in the Calculus course, I could more realistically suggest its inclusion is for the benefit of the reader whose Calculus is a bit rusty or who has perhaps had a watered down Calculus course. But the real justification for including such material here is to provide the backdrop for the Binomial Theorem, the history of which is intimately tied to that of the Calculus.

Prerequisites for understanding this book are more general than specific. A good understanding of algebra and the ability to follow algebraic manipulations is vital. And there is the aforementioned familiarity with the calculator. Chapter 1 makes no more assumptions than this. Chapter 2 uses the Calculus, but not in a way that requires full mastery of the subject. Concepts and results of the Calculus are introduced and explained as needed. Whether this is adequate for the reader with no background in the Calculus remains to be seen. The real prerequisite is not the knowledge of specific results of the Calculus nor the calculational mastery of the subject acquired through the excessive drill of the modern Calculus course, but the mathematical maturity traditionally derived from a solid course in the subject. And this is especially true in Chapter 3, where the theory of the Calculus takes centre stage. The coverage is certainly not as complete as that in the old Advanced Calculus course or the modern Real Analysis course, being limited to those concepts necessary to explain the attempts to prove the Binomial Theorem.

I suppose, had I been a bit more thorough, this could have turned into an introduction to the theory of the Calculus, i.e., a book on Advanced Calculus in which binomial concerns serve to introduce the fundamental concepts. I have, however, come to the Binomial Theorem from a different direction. The Binomial Theorem popped up several times when I was writing *Adventures in Formalism*[4] and I slowly came to realise that its story proves a good case study for that book. Chapter 1 of the present book provides the background, Chapter 2 and the unsuccessful proof attempts discussed in Chapter 3 present Newton's generalisation as a major example of the Type I formalism discussed in that book, while the Calculus primer of Chapter 2 and the introduction of deeper analytic concepts for the successful proofs of Chapter 3 offer detailed examples of Type II formalism in action. Lest I paint too convincing a picture of the present volume as a companion volume to my earlier effort, I hasten to add that the present work stands on its own, no prior familiarity with my book on formalism being necessary for understanding the work you have before you.

This book, like my previous efforts on formalism and probability, is a hybrid—half mathematical textbook and half history. It belongs to that genre of historical writing called *Problemgeschichte*—problem history—best exemplified at the advanced level by the algebraic works of Harold Edwards and at the elementary level by the works of Herbert Meschkowski. My books are written at an intermediate level, offering more mathematical meat than Meschkowski, but not requiring the stronger mathematical background that Edwards does. It remains to be seen how successful I have been in my endeavours.

---

[4]College Publications, London, 2012.

The inclusion of history in mathematical exposition has a great deal to offer. Foremost among the benefits is the motivation it offers for the introduction of new concepts through the presentation of the contexts out of which these concepts naturally emerged. The chief disadvantage of including history in a mathematics textbook is the page count. The much maligned Definition-Theorem-Proof style of exposition is most efficient as regards paper usage, and is ideal for readers who have reached a certain level of mathematical maturity and have already developed their own perspectives on the field.

One thing my book on probability makes abundantly clear is that, to fit the history of the mathematics into a reasonable amount of space, one must somehow limit the mathematics covered. The field of probability, even of elementary probability, is too vast for a single-volume history-based textbook. I had no space to elaborate and even stinted a bit historically. For example, in my all too brief mention of Markov chains and their invention in 1906 by Andrei Andreevich Markov, I did not discuss the context of their introduction, but applied them instead to problems that had already been introduced. My conclusion is that the historical approach to teaching mathematics is more suited to microtopics, say individual theorems, than to full theories.

The Binomial Theorem is an ideal candidate for such an historically based microtopical discussion, as I hope the present volume shows. While I do omit some things (Multinomial Theorem, binomial probability, subtle differences in algorithms for finding roots of numbers, "more of the same" failed attempts at proving the Binomial Theorem, and the full Binomial Theorem in the complex plane), the ensuing treatment of the Theorem and its rôle in the development of rigour is fairly complete both mathematically and historically. And I think it offers a coherent and inherently interesting account.

In writing the following I have received material assistance from several individuals and would like to offer my thanks to my old friends Eckart Menzler-Trott, Fred Thulin, and Jimmie Johnson, as well as to my more recent contact Henrik Kragh Sørensen.

I would also like to express my gratitude to the anonymous referee who, contrary to my usual experience with referees, made a number of useful comments that I have taken to heart and addressed in this final version.

CHAPTER 1

The Finite Binomial Theorem

Crudely stated, the Finite Binomial Theorem states the result of multiplying a binomial $a + b$ by itself repeatedly:
$$(a + b)(a + b) = a^2 + 2ab + b^2$$
$$(a + b)(a + b)(a + b) = a^3 + 3a^2b + 3ab^2 + b^3$$
$$(a + b)(a + b)(a + b)(a + b) = a^4 + 4a^3b + 6a^2b^2 + 4ab^3 + b^4,$$
etc. The Theorem is, of course, more than just the collection of the results of such computations; it describes the common features of these results: for n factors, the product simplifies to a sum of $n + 1$ terms of the forms
$$\binom{n}{k} a^k b^{n-k}$$
for $k = 0, 1, \ldots, n$, where the numbers $\binom{n}{k}$, called *binomial coefficients*, depend only on n, k and not on a, b. Moreover, these coefficients are characterised in one of several computational manners, each more efficient than actually carrying out the multiplications and collecting like terms. The result is often credited to Blaise Pascal who rediscovered and publicised it in the 17th century. However, the result was already known in Europe for over a century at the time of his rediscovery, and before that it had been known to various extents for some centuries in India, the Islamic World, and China. And knowledge of the instance,
$$(a + b)(a + b) = a^2 + 2ab + b^2, \tag{1}$$
is considerably older, going back to the Babylonians.

1. Preparation: $(a + b)^2 = a^2 + 2ab + b^2$

Formula (1) underlies the solution to the *quadratic equation*, first solved by the Babylonians. The general quadratic equation in one variable x is the equation
$$Ax^2 + Bx + C = 0, \tag{2}$$
where $A \neq 0$. To solve it, one first divides by A to get
$$x^2 + \frac{B}{A}x + \frac{C}{A} = 0.$$
Then one isolates the constant term:
$$x^2 + \frac{B}{A}x = -\frac{C}{A}.$$

One now wants to add a constant D to each side of the equation so as to make the left side a perfect square:

$$x^2 + \frac{B}{A}x + D = (x+b)^2$$
$$= x^2 + 2bx + b^2,$$

by (1). Clearly one can take $D = b^2$, where $2b = B/A$, whence $b^2 = B^2/(4A^2)$:

$$x^2 + \frac{B}{A}x + \frac{B^2}{4A^2} = \frac{-C}{A} + \frac{B^2}{4A^2}$$

$$\left(x + \frac{B}{2A}\right)^2 = \frac{-4AC + B^2}{4A^2}$$

whence

$$x + \frac{B}{2A} = \frac{\pm\sqrt{B^2 - 4AC}}{2A}$$

and

$$x = \frac{-B \pm \sqrt{B^2 - 4AC}}{2A}. \tag{3}$$

Formula (3) is known as the *quadratic formula*, was known in several ancient cultures, and is taught in one's first course in Algebra. The derivation is also taught as the method of *completing the square* and is generally more useful to know. I find in my experience, however, that students are more resistant to learning the method than to memorising the formula. There is some legitimacy to their preference.

Consider, e.g.,

$$2x^2 - 7x + 6 = 0. \tag{4}$$

To solve by completing the square, we first divide by $A = 2$:

$$x^2 - \frac{7}{2}x + 3 = 0.$$

Transpose the 3:

$$x^2 - \frac{7}{2}x = -3.$$

Divide $7/2$ by 2, square it, and add the result to both sides:

$$x^2 - \frac{7}{2}x + \frac{49}{16} = -3 + \frac{49}{16}$$

$$\left(x - \frac{7}{4}\right)^2 = \frac{-48 + 49}{16} = \frac{1}{16}.$$

Then take the square root,

$$x - \frac{7}{4} = \pm\frac{1}{4},$$

and add $7/4$ to both sides:

$$x = \frac{7}{4} \pm \frac{1}{4} = \frac{6}{4}, \frac{8}{4} = \frac{3}{2}, 2.$$

The solutions to (4) are thus $x = 3/2$ and $x = 2$.

Using the quadratic formula (3), the student simply plugs in the values $A = 2, B = -7, C = 6$:

$$x = \frac{-(-7) \pm \sqrt{49 - 4 \cdot 2 \cdot 6}}{2 \cdot 2}$$
$$= \frac{7 \pm \sqrt{49 - 48}}{4} = \frac{7 \pm 1}{4} = \frac{6}{4}, \frac{8}{4} = \frac{3}{2}, 2,$$

as before, but with much less work.

The quadratic formula is used so often in mathematics that the efficiency of remembering and applying it cannot be denied. However, there are two ways of memorising it. One is to derive it repeatedly until one remembers the outcome without trying; the other is to commit the formula to memory by rote. The danger of the latter approach is that a slight slip, such as a forgotten minus sign in front of the B or forgetting the A in the denominator of (3) will result in a major error.

Completing the square has other uses in elementary maths. The graph of a quadratic function,

$$y = Ax^2 + Bx + C \tag{5}$$

is a parabola which opens vertically—upwards if A is positive and downwards if A is negative. The lowest (highest) point on an upwards (respectively, downwards) opening parabola is called its *vertex* and can be found by the method of completing the square:

$$\frac{y}{A} = x^2 + \frac{B}{A}x + \frac{C}{A}$$

$$\frac{y}{A} - \frac{C}{A} = x^2 + \frac{B}{A}x$$

$$\frac{y}{A} - \frac{C}{A} + \frac{B^2}{4A^2} = x^2 + \frac{B}{A}x + \frac{B^2}{4A^2} = \left(x + \frac{B}{2A}\right)^2$$

$$\frac{y}{A} = \left(x + \frac{B}{2A}\right)^2 + \frac{C}{A} - \frac{B^2}{4A^2}$$

$$y = A\left(x + \frac{B}{2A}\right)^2 + \frac{4AC - B^2}{4A}.$$

This is minimised (maximised) for A positive (respectively, negative) when

$$x + \frac{B}{2A} = 0, \quad \text{i.e., } x = -\frac{B}{2A}, \tag{6}$$

and for such x one has

$$y = \frac{4AC - B^2}{4A}. \tag{7}$$

Again, a numerical example may illustrate this more readily. Let

$$y = 2x^2 - 7x + 6. \tag{8}$$

Then
$$\frac{y}{2} = x^2 - \frac{7}{2}x + 3$$
$$\frac{y}{2} - 3 = x^2 - \frac{7}{2}x$$
$$\frac{y}{2} - 3 + \frac{49}{16} = x^2 - \frac{7}{2}x + \frac{49}{16}$$
$$\frac{y}{2} - \frac{48}{16} + \frac{49}{16} = \left(x - \frac{7}{4}\right)^2$$
$$\frac{y}{2} = \left(x - \frac{7}{4}\right)^2 - \frac{1}{16}$$
$$y = 2\left(x - \frac{7}{4}\right)^2 - \frac{1}{8}, \tag{9}$$

and the vertex of the parabola is at $(7/4, -1/8)$.

In graphing (8), one would also want to find the x-intercepts and the y-intercept. The latter is easy: plug $x = 0$ into (8) to get $y = 6$. For the x-intercepts, one could set (8) or (9) equal to 0 and solve. Starting with the former, we could simply apply the quadratic formula as before to get $x = 3/2$ and $x = 2$. The bilateral symmetry of the parabola tells us that the vertex lies halfway between these points,

$$x = \frac{1}{2}\left(\frac{3}{2} + 2\right) = \frac{1}{2}\left(\frac{3}{2} + \frac{4}{2}\right) = \frac{7}{4}.$$

The y-coordinate can then be obtained without deriving (9) by plugging $x = 7/4$ into (8):

$$y = 2\left(\frac{7}{4}\right)^2 - 7 \cdot \frac{7}{4} + 6 = \frac{2 \cdot 49}{16} - \frac{49}{4} + 6$$
$$= \frac{49}{8} - \frac{2 \cdot 49}{8} + \frac{6 \cdot 8}{8} = \frac{-49}{8} + \frac{48}{8} = \frac{-1}{8}.$$

The reader with a graphing calculator might want to verify these assertions directly. For the Texas Instruments calculators I suggest a window with Xmin = $-.35$, Xmax = 4.35, Ymin = -1, and Ymax = 6.5. The graph is vertically squashed, but all the crucial points appear when one traces the curve via the TRACE button.

This ability to find the vertex of a parabola allows one to solve a limited set of maximum/minimum problems that ordinarily would require the use of the Calculus. A perennial favourite is the following:

1.1.1. Example. *A farmer has 100 feet of fencing with which to make a rectangular enclosure for his chickens. What is the maximum area he can enclose?*

Referring to FIGURE 1, we see we wish to maximise the value of wl subject to the constraint that $2w + 2l = 100$, where w, l denote the width and length,

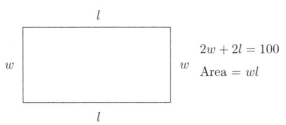

FIGURE 1.

respectively, of the enclosed area. Because the perimeter is fixed, we can express the length in terms of the width,
$$l = \frac{100 - 2w}{2} = 50 - w,$$
and thus the area is a quadratic function of the width:
$$\text{Area} = wl = w(50 - w) = 50w - w^2.$$
Using y for the area and x for w, we are looking to maximise the quadratic function
$$y = -x^2 + 50x.$$
Here $A = -1$ is negative, whence we are dealing with a downward opening parabola, the vertex of which is its highest point. Plugging $A = -1, B = 50, C = 0$ into (6) yields
$$x = -\frac{B}{2A} = -\frac{50}{-2} = 25,$$
and (7) yields
$$y = \frac{4AC - B^2}{4A} = \frac{0 - 50^2}{-4} = 625.$$
Recalling $w = x$ and $l = 50 - w$, we see that $w = l = 25$. Thus the farmer's enclosure is a square with sides 25 feet in length and an area of 625 square feet. [In the present example, since we already have the factorisation $wl = w(50-w)$, it is quicker to note that the x-intercepts being 0 and 50, the maximum occurs at their midpoint $w = 25$. One then calculates $25(50 - 25) = 25 \cdot 25 = 625$.]

This example is usually followed by the

1.1.2. Exercise. *Repeat the above, but assume the enclosure is to be bordered on one side by a creek that runs straight across the farmer's property and thus only three sides of the fence will need to be built.*

The solution of maximum/minimum problems ordinarily requires the more high-powered techniques of the Calculus. But, as here, sometimes simple algebra is all that is needed. A nice exposition of such non-calculus optimisations is given by:

> Paul J. Nahin, *When Least is Best: How Mathematicians Discovered Many Clever Ways to Make Things as Small (or as Large) as Possible*, Princeton University Press, Princeton, 2004.

2. Extracting Square Roots

A somewhat more pedestrian application of the instance (1) of the Finite Binomial Theorem is its use in finding square roots. It is a simple iterative algorithm giving better and better approximations. As square roots are generally irrational, this is in theory the best one can hope for. In practice, of course, one can hope for ease of use and efficiency in the calculations.

A very general, but not all that efficient a method is the general *bisection method*. Given a number D, the square root of which is desired, one starts with two numbers a_1, a_2 such that

$$a_1^2 < D < a_2^2$$

and generates further approximations $a_3, a_4, a_5, \ldots$ as follows. Choose a_3 to be the midpoint between a_1 and a_2:

$$a_3 = \frac{a_1 + a_2}{2}.$$

If $a_3^2 = D$, one has the square root and one can stop the construction or declare $a_4 = a_3, a_5 = a_3, \ldots$ If $a_1^2 < D < a_3^2$, choose a_4 to be the midpoint between a_1 and a_3; otherwise $a_3^2 < D < a_2^2$ and one chooses a_4 to be the midpoint between a_3 and a_2. And so on.

The resulting sequence $a_1, a_2, a_3, \ldots$ converges to $\sqrt{D}$. In fact, a_{n+2} is within $(a_2 - a_1)/2^n$ of $\sqrt{D}$. However, the sequence can bounce around, a_{n+1} sometimes being farther from $\sqrt{D}$ than a_n.

As an example, let $D = 17$. Obvious choices of a_1, a_2 are $a_1 = 4, a_2 = 5$. Then $a_3 = (4+5)/2 = 9/2 = 4.5$ with a square greater than 17. Thus $a_4 = (4+9/2)/2 = 17/4 = 4.25$, again too large. Continuing in this fashion we generate the sequence $a_1, a_2, a_3, \ldots$:

$$4, 5, \frac{9}{2}, \frac{17}{4}, \frac{33}{8}, \frac{65}{16}, \frac{131}{32}, \frac{263}{64}, \ldots,$$

i.e.,

$$4,\ 5,\ 4.5,\ 4.25,\ 4.125,\ 4.0625,\ 4.09375,\ 4.109375,\ \ldots \qquad (10)$$

The convergence is a bit erratic and slow: a_1, a_3, a_4, a_6, a_7 are correct only in the integral digit; only a_5 and a_8 are correct to the first decimal; and only a_5 is correct to two decimals.

Using the information (1) of the given instance of the Finite Binomial Theorem affords a more efficient procedure than this. One begins with a single approximation a to $\sqrt{D}$ and sets $D = (a+b)^2$. By (1), this means $D = a^2 + 2ab + b^2$. If a^2 is close enough to D to begin with, this means that b^2 is negligible and D is approximately $a^2 + 2b$, whence

$$b \approx \frac{D - a^2}{2a}.$$

($\approx$ abbreviates approximate identity.) So one can take

$$a + \frac{D - a^2}{2a}$$

as one's next approximation and repeat, generating a sequence $a_1 = a, a_2, a_3, \ldots$ of successive approximations to $\sqrt{D}$.

For example, for $D = 17$, an obvious choice of a is 4. Then
$$b \approx \frac{17 - 4^2}{2 \cdot 4} = \frac{1}{8}$$
and
$$4 + \frac{1}{8} = \frac{33}{8}$$
is our next choice of a. The next b is
$$\frac{17 - \left(\frac{33}{8}\right)^2}{2 \cdot \frac{33}{8}} = -\frac{1}{528}$$
and the next choice of a is
$$\frac{33}{8} - \frac{1}{528} = \frac{2177}{528}.$$

Continuing in this manner, we get a sequence $a_1, a_2, a_3, \ldots$:
$$4, \frac{33}{8}, \frac{2177}{528}, \frac{9478657}{2298912}, \frac{179689877047297}{43581196642368}, \ldots \tag{11}$$

The fractions are hard to compare, but conversion to decimals yields
$$4, 4.125, 4.123\overline{106}, 4.1231056256176\ldots, 4.12310562561766055498214\ldots, \ldots \tag{12}$$
all the displayed digits of the last two numbers agreeing with those of $\sqrt{D}$.

The procedure based on identity (1) evidently yields a much more rapidly convergent sequence $a_1, a_2, a_3, \ldots$ than that obtained via the bisection method. However, each iteration is a painful computation in its own right. Consider, for example, the step from a_4 to a_5:
$$\frac{17 - \left(\frac{9478657}{2298912}\right)^2}{2 \cdot \frac{9478657}{2298912}} = \frac{17 \cdot 2298912^2 - 9478657^2}{2 \cdot 9478657 \cdot 2298912}.$$

This involves three hideous multiplications (I'm not counting the multiplications by 17 and 2.), and the next step from a_5 to a_6 will involve the same number of multiplicatons with integers with double the numbers of digits. And, of course, passage to the decimal representation requires a long division with multidigit numbers.

In this age of pocket calculators and computers, the disadvantage of such heavy calculation may not be immediately apparent. Those of us educated in the old days of hand calculation know, however, that multiplying large numbers is laborious—time-consuming and boring. Today we merely grab our calculators, enter the numbers to be multiplied, and press the ENTER button. The answers appear almost effortlessly. There is, however, one little problem with this: the accuracy of a result is only guaranteed to so many digits. One of the more popular calculators at the high school and general college level in the

United States is the *TI-84* from Texas Instruments. I have the obsolescent older version known as the *TI-83 Plus*. The above iteration can be performed on this calculator as follows: enter 4. The answer is, of course, 4. Now enter

ANS+(17−ANS^2)/(2ANS)

repeatedly. One gets the sequence

$$4.125, 4.123106061, 4.123105626, 4.123105626, \ldots$$

There is no improvement in the last step. This should not be too surprising as a_4 and a_5 agree to 13 places beyond the decimal point and the calculator only displays 9 of them in each case. The calculator actually stores a few more digits which one can recover by entering

ANS−4.123105,

immediately after the calculation of a_4 or a_5. Doing so reveals that the calculator stores a_4 and a_5 as 4.1231056256177, the last digit differing from the correct one of 6. One cannot even check directly on the *TI-83 Plus* that what I said is true: both 4.1231056256176^2 and 4.1231056256177^2 result in 17 and not 16.9... and 17.0..., the actual differences occurring well beyond even the stored digits.

The way to get around these limitations is to deal with integers and fractions, not decimals. On calculators like the *TI-83 Plus*, there is still the limitation of the restricted number of digits stored. One can get around this by treating numbers that are too long as lists of "digits" in bases larger than 10. For example, the numerator of a_5 in its fractional representation could be represented itself as a list,

{179 689877 47297},

of its digits in base 10^6. Programs can then handle the basic arithmetic operations. I have actually carried this out for base 100 once[1], and will do so for base 1000000 in section 2 of the Appendix, below, and can report that it is not difficult and that it can be quite handy. However, since doing this, I have acquired a *TI-89 Titanium* which has all this built-in. If one sets the Exact/Approx mode of the calculator to EXACT, enters 4 as initial value, and repeatedly enters[2]

ans(1)+(17−ans(1)^2)/(2ans(1))

as before, the sequence (11) will result. The sequence (12) of decimals is almost as easy. One multiplies the fraction by 10^n, i.e., one pads n 0's at the ends of the numerators, and applies the function iPart(). Thus, e.g., entering

iPart(9478657000000000000/2298912)

or

iPart(9478657/2298912∗10^13)

---

[1] Craig Smoryński, *Chapters in Mathematics; From π to Pell*, College Publications, London, 2012, chapter 8.

[2] ans(1) can be entered either by pressing the ANS key as on the *TI-83 Plus* or by typing it in one character at a time.

will yield 41231056256176, i.e., the estimate 4.1231056256176 after restoring the decimal point. One can check the correctness of these digits through the simple trick of squaring 41231056256176 and 41231056256177. The results are

$$16.\underbrace{999999999999}_{12}50069338142976 \text{ and } 17.\underbrace{000000000000}_{12}32531450655329,$$

respectively, yielding 16.9... and 17.0... when the decimals are reintroduced.

The rapidity with which the digits of the successive approximations settle down, i.e., the rapidity of the convergence of the sequence $a_1, a_2, a_3, \ldots$ to the square root of D is quite impressive. Ancient mathematicians may have seen this and accepted it as a fact. Modern mathematicians are a bit more demanding and want some justification. This isn't hard to provide in the present case: it amounts to our earlier observation that a sequence $a_1, a_2, a_3, \ldots$ for which $a_1^2 < D < a_2^2$ and a_{n+2} is always taken to be the mean of a_n, a_{n+1} converges to $\sqrt{D}$. For, if we look at

$$a + \frac{D - a^2}{2a} = \frac{2a^2 + D - a^2}{2a} = \frac{D + a^2}{2a} = \frac{1}{2}\left(\frac{D}{a} + a\right),$$

we see that the rule we've been following to obtain the next approximation is just such an average of a and D/a. And these numbers must straddle $\sqrt{D}$ as otherwise,

$$a, \frac{D}{a} < \sqrt{D} \Rightarrow D = a \cdot \frac{D}{a} < \sqrt{D}\sqrt{D} = D$$

and

$$\sqrt{D} < a, \frac{D}{a} \Rightarrow D = a \cdot \frac{D}{a} > \sqrt{D}\sqrt{D} = D,$$

a contradiction either way. If, say $a < D/a$, then $\sqrt{D}$ lies in one of the intervals

$$\left[a, a + \frac{D - a^2}{2a}\right] \text{ and } \left[a + \frac{D - a^2}{2a}, \frac{D}{a}\right],$$

thus it is within

$$\frac{1}{2}\left(\frac{D}{a} - a\right)$$

of the new approximation $a + (D - a^2)/(2a)$. The next estimate will be within half of that, the following within another half, etc.

Notice that this argument does not presuppose that a is close to $\sqrt{D}$ at the outset. It may take a while, but convergence is assured no matter where we start from. For example, for $D = 17$, I could make the initial guess $a = 100$ for $\sqrt{17}$. Dragging out the calculator, I get the series of estimates

100, 50.085, 25.21221149, 12.94324396, 7.128335245, 4.7565919,

4.165289701, 4.123319236, 4.123105631, 4.123105626, ...

the last number now repeating itself indefinitely. And $a = .1$ yields the sequence

.1, 85.05, 42.62494121, 21.51188437, 11.15107261, 6.337794815,

4.510057898, 4.139705419, 4.123138907, 4.123105626, ...

Bad guesses offer slow starts, but the approximations soon become reasonable and thereafter converge rapidly.

3. Finding Square Roots One Decimal at a Time

This ease, speed, and accuracy was not always available. Working by hand was laborious and involved multiplication of many-digit numbers. For over a millennium and a half, however, there has been another method for finding square roots based on (1) that cuts down on the sizes of the numbers to be multiplied and produces the square root directly in decimal form, one digit at a time. The disadvantage is the large number of iterations that must be made if one wants the root taken to a great many decimal places. The advantages are the lessened pain in not dealing with as many large numbers, the relative simplicity of the method, and knowledge without appeal to advanced mathematics of the relative accuracy of the result at any stage in the operation. [How did I know a_4 would be accurate to exactly 13 places above? Or that a_5 would be accurate to 22? Well, I know the Calculus and was able to make a rough estimate on the number of reliable figures in the ratios and proceed from there. Working on the *TI-89 Titanium* in EXACT mode it doesn't take long to determine the exact degree of accuracy acquired. Working by hand would have been less pleasant.]

The method of generating successive digits of square roots goes back in one form or another to the 3rd century A.D. in China, and continued to be widely used until recently. Indeed, in my youth it was taught in elementary school. I found it somewhat tedious and until calculators arrived on the scene in the 1960s I preferred to use tables and logarithms to find square roots. However, it is a good little algorithm and learning its use may be of some pædagogical use in building mental discipline. It also makes a nice little programming exercise.

In layout, the application of the algorithm resembles long division, but with these differences: there is nothing taking the place of a divisor, and the number D of which one wants to find the square root is written in base 100. I should continue using 17 as D, but it doesn't illustrate all the complications, so I choose instead to use $D = 189574$, a number for which I have previously typeset the necessary diagrams. We shall find its square root to four decimal places. The first step is to write the number in base 100, adding a decimal point, and four pairs of 0s after the decimal point:

$$18\ 95\ 74 \,.\, 00\ 00\ 00\ 00\,.$$

One then places the whole thing under a radical sign and looks at the first centesimal digit $d = 18$. The largest square not exceeding 18 is $16 = 4^2$. One places 4 above the 18 and 16 below the 18, and then subtracts to obtain 2, which is placed below the 16 as in long division. One brings down the 95 to obtain 295, again as in long division. The result should look like FIGURE 2 below. We now have 1895 as a new d, and, imagining 0 as temporarily lying above 95, 40 as a. We want the largest single digit b such that $(a+b)^2 \leq d$, i.e.,

$$(40+b)^2 = 1600 + 80b + b^2 \leq 1895,$$

3. FINDING SQUARE ROOTS ONE DECIMAL AT A TIME

$$\begin{array}{r} 4\\ \sqrt{18\ 95\ 74\ .\ 00\ 00\ 00\ 00}\\ 16\\ \hline 2\ 95 \end{array}$$

FIGURE 2. First Step

i.e.,
$$80b + b^2 \leq 295.$$
b can be shown to be at most
$$\left[\frac{295}{80}\right]$$
where $[\cdot]$ denotes the greatest integer function. Now $295/80 = 3.6875$, so b will be at most 3. We first try $b = 3$.
$$80 \cdot 3 + 3^2 = 249 < 295,$$
so we choose 3, place it above 95, and $249 = 1895 - 40^2$ below 295. Subtract 249 from 295, bring down the 74, and repeat. The full calculation is given in FIGURE 3, below.

$$\begin{array}{r}
4\quad 3\quad 5\ .\ 4\quad 0\quad 0\quad 9\\
\sqrt{18\ 95\ 74\ .\ 00\ 00\ 00\ 00}\\
16\\ \hline
2\ 95\\
2\ 49\\ \hline
46\ 74\\
43\ 25\\ \hline
3\ 49\quad 00\\
3\ 48\quad 10\\ \hline
84\ 00\ 00\\
0\\ \hline
84\ 00\ 00\ 00\\
0\\ \hline
84\ 00\ 00\ 00\ 00\\
78\ 37\ 20\ 81\\ \hline
5\ 62\ 79\ 19
\end{array}$$

FIGURE 3. Final Result

1.3.1. Exercise. *Use the algorithm to find the integral parts of $\sqrt{1848}$ and $\sqrt{1849}$.*

The point here is that choosing
$$b = \left[\frac{d - a^2}{2a}\right]$$

can result in having $(a+b)^2 = d$ or $(a+b)^2 > d$. In the latter case one must then change one's choice to
$$b = \left[\frac{d-a^2}{2a}\right] - 1,$$
and, if this fails we subtract another 1, and another 1, ...

1.3.2. Exercise. *Use the algorithm to find $\sqrt{17}$ to several decimal places.*

I assume one will tire of the business and break off the calculation well before matching our earlier calculations to 13 or 23 decimal places. Using the calculator will make this more palatable. Elsewhere[3] I have shown how to do this on the *TI-83 Plus*. However, although I don't believe my program was conceptually inefficient, the implementation was extremely slow, successive digits requiring increasing amounts of time. One reason for this is that the *TI-83 Plus* does not allow for the definition of functions or the use of local variables. The result was a lot of copying of one variable into another as procedures had to be applied multitudes of times to different variables. And, since that calculator does not deal with exact integers, I had to run programs designed to do this. The net result is that it took a quarter of an hour to calculate $\sqrt{2}$ to 50 decimals. Compiling that program (actually, suite of programs) into an assembly language program ought to speed things up a bit. I have not done that, preferring instead to convert the program into SCHEME and running it on the computer. The result was the generation in about 1 second of a table of the square roots of the first 100 positive integers each calculated to 50 decimals.[4] While I don't imagine the *TI-89 Titanium* to be as fast as a desktop computer, I can report that, with its exact calculations and ability to program functions, it is fairly easy to implement the current algorithm and get the desired degree of accuracy.

The *TI-89 Titanium* is not a calculator for everybody. It has a very steep learning curve and is recommended only for those at the level of the Calculus and beyond. Most high school students as well as those college students taking no more than general education mathematics courses are better served with a *TI-83 Plus* or *TI-84*, on which, as I said, the automation of the task at hand is far from ideal. Thus, I urge most readers to skip over the following to the beginning of the next section, unless they wish to see if they should covet the precious *Titanium*.

I assume anyone who possesses a *TI-89 Titanium* is sufficiently sophisticated computationally as not to require an explanation of the following program defining a function nextPair() which takes a list $\{a, d\}$ one is likely to encounter in the course of finding the square root of a number D and produces the next $\{a, d\}$ pair. It is a stripped down version of a more general such program and is only guaranteed to work for integers D possessing one or two digits (i.e., a single centesimal digit).[5]

---

[3] Smoryński, *op. cit.*

[4] *Ibid.*, chapter 9.

[5] The program does not take into account any but the leading centesimal digit of D.

```
:nextPair(list1)
:Func
:Local a
:Local b
:Local d
:Local top
:Local bottom
:list1[1]→a
:list1[2]→d
:100∗d→d
:10∗a→a
:If a^2=d
:Return {a,d}
:d−a^2→top
:2∗a→bottom
:iPart(top/bottom)→b
:Lbl lbl1
:If b∗bottom≥top Then
:b−1→b
:Goto lbl1
:EndIf
:a+b→a
:Return {a,d}
:EndFunc
```

If you install this program on your *TI-89 Titanium* and enter

nextPair({4,17})

and then enter

nextPair(ans(1))

successively, you will see the successive values of a, d paired at the bottom of the screen. The interesting values will be the successon of a's:

$$41, 412, 4123, 41231, \ldots,$$

decimal-point-free approximations to $\sqrt{17}$. Entering nextPair(ans(1)) 12 times will give us the value of a_4 as listed in (12) above, and 21 successive presses of the ENTER button will give us the corresponding value of a_5.

This is not as convenient as our earlier use of the calculator as described on page 8, above. It doesn't converge as rapidly and one has to press the ENTER button frequently.

1.3.3. Exercise. *Program a function on the TI-89 Titanium that will take as input a one or two digit integer D and a positive integer n, and will iterate the application of* nextPair() *n times, producing the square root of D to n decimal places. Apply the program to $D = 2$, $n = 50$. How does its execution compare to the 15 minutes reported for the TI-83 Plus?*

As I said, nextPair() and, thus, the function of this exercise are only guaranteed to work for one and two digit numbers D.

1.3.4. Exercise. *Generalise the preceding exercise to program a function taking any positive integers D and n and producing the square root of D to n decimal places.*

In writing this program you will have to first produce the list of centesimal digits of D, then append the n zero centesimal digits at the end of this list, and finally replace the command 100*d→d by one which adds the appropriate centesimal digit.

I should also note that the decimal point will not appear anywhere in the program or its execution unless you wish to give the outcome not as a number but as a string in which the character "." is inserted in the correct position. If you wish to add this refinement, you might as well also write the program to take an arbitrary decimal fraction inputed as a string, converting it to the integer one gets by multiplying the decimal by an appropriate power of 100.

4. Third, Fourth, and Fifth Roots

The next instance of the Finite Binomial Theorem after the quadratic, the cubic equation does not represent as great a success story for the Theorem. We can repeat everything we did in the preceding sections on the quadratic equation, but only if we use more powerful tools. The problem is that the analogue to completing the square is not as successful in the cubic case.

The method of completing the square can be thought of as converting the quadratic equation
$$Ax^2 + Bx + C = 0$$
into a simpler and more easily solvable quadratic equation
$$Ay^2 + C' = 0$$
via the substitution $y = x + B/(2A)$, i.e., $x = y - B/(2A)$. If we try to do the same with a cubic equation,
$$Ax^3 + Bx^2 + Cx + D = 0, \tag{13}$$
substituting $x = y - a$, we obtain
$$A(y-a)^3 + B(y-a)^2 + C(y-a) + D =$$
$$= Ay^3 - 3Aay^2 + 3Aa^2y - Aa^3 + By^2 - 2Bay + Ba^2 + Cy - Ca + D.$$

Collecting like terms, we see that the coefficient of y^2 is $-3Aa + B$, and to make this 0 we must take $a = B/(3A)$. But the coefficient of the collected linear terms then becomes
$$3A\left(\frac{B}{3A}\right)^2 - 2B \cdot \frac{B}{3A} + C = \frac{B^2}{3A} - \frac{2B^2}{3A} + C = -\frac{B^2}{3A} + C,$$
which will generally not be 0. (Once one has specified A, B, only one choice of C will make the coefficient 0.) No substitution will reduce (13) to the desired simple form,
$$Ay^3 + D' = 0,$$
except in special cases.

4. THIRD, FOURTH, AND FIFTH ROOTS

All is not lost however. We have not ruled out the existence of a *cubic formula*; we have merely shown the failure of the most obvious approach to obtaining one. There is, in fact, a cubic formula obtainable from (13) by applying an additional substitution, a clever (i.e., devious) one. Thus, with a bit of effort, the cubic equation can be solved algebraically. This solution is not as pleasing or as easily learned as the quadratic formula and is nowadays consigned to courses in the History of Mathematics. Virtually every textbook in the general history of the field will contain the derivation and criticism of it.[6] There is also a *biquadratic formula* for solving the general equation of degree 4, but it is even less popular and is usually only found in the more thorough histories of mathematics. There is no general algebraic solution to equations of degrees 5 or higher, as was shown in the early 19th century by a young Norwegian mathematician named Niels Henrik Abel. We will encounter Abel's contributions to the Binomial Theorem in a later chapter. His work on the quintic is a matter of advanced mathematics and I can say nothing more about it here.

Today, we use numerical methods to solve any polynomial equation beyond the quadratic. We will consider this and its relation to the Finite Binomial Theorem in section 7, below. For now, let us simply consider the iterative extraction of cube roots.

If we begin with an approximation a to $\sqrt[3]{D}$, we can set $\sqrt[3]{D} = a + b$ and apply the cubic instance of the Finite Binomial Theorem to obtain

$$D = (a+b)^3 = a^3 + 3a^2b + 3ab^2 + b^3 \approx a^3 + 3a^2b,$$

if b is small enough. Thus

$$b \approx \frac{D - a^3}{3a^2} \quad \text{and} \quad \sqrt[3]{D} = a + b \approx a + \frac{D - a^3}{3a^2}$$

and we can use our calculator as before.

For $D = 17$, choose $a = 2$ and enter

2+(17−2^3)/(3∗2²)

and then

ANS+(17−ANS^3)/(3∗ANS²)

successively on the *TI-83 Plus* to get

2.75, 2.582644628, 2.571331512, 2.571281592, 2.571281591, 2.571281591,...,

the last item repeating itself from here on. And on the *TI-89 Titanium* one enters

2+(17−2^3)/(3∗2^2)

and

ans(1)+(17−ans(1)^3)/(3∗ans(1)^2),

---

[6] I immodestly refer the reader to Craig Smoryński, *History of Mathematics; A Supplement*, Springer Science+Business Media, LLC, New York, 2008, chapter 6, for a relatively detailed discussion. (The pretty picture on page 153 of that work is reason enough for consulting it.)

respectively, to get
$$\frac{11}{4}, \frac{625}{242}, \frac{121535591}{47265625}, \frac{179515519423897730270858 9}{69815581462735664101562 5}, \ldots$$
in EXACT mode. I leave it to the reader with the requisite computing power to find the remaining two terms and convert all of these to decimals.

It should come as no surprise that this method works for higher powers. To find $\sqrt[4]{D}$ for example, one starts with an estimate a of $\sqrt[4]{D}$ and sets $\sqrt[4]{D} = a+b$:
$$D = (a+b)^4 = a^4 + 4a^3 b + 6a^2 b^2 + 4ab^3 + b^4$$
$$\approx a^4 + 4a^3 b,$$
if a is sufficiently close to $\sqrt[4]{D}$. This leads to $b \approx (D - a^4)/(4a^3)$, and thus the choice of
$$a + \frac{D - a^4}{4a^3}$$
as one's next estimate. For the fifth root, the transition is from a to
$$a + \frac{D - a^5}{5a^4},$$
and, in general, given an approximation a to $\sqrt[n]{D}$, one iterates the passage from a to
$$a + \frac{D - a^n}{na^{n-1}}. \tag{14}$$

The efficacy of the procedure is testified to by the example given above for $D = 17, n = 3, a = 2$, but an example is far from a proof and we would like the comfort of a guarantee that the method always works. Well, as before, the numbers a and D/a^{n-1} straddle $\sqrt[n]{D}$ because $a^{n-1} \cdot D/a^{n-1} = D$. But the new approximation (14) is a weighted average of a and D/a^{n-1},
$$a + \frac{D - a^n}{na^{n-1}} = \frac{na^n}{na^{n-1}} + \frac{D - a^n}{na^{n-1}} = \frac{(n-1)a^n + D}{na^{n-1}} = \frac{n-1}{n} \cdot a + \frac{1}{n} \cdot \frac{D}{a^{n-1}},$$
and thus lies between the numbers a and D/a^{n-1}. The question now is: how far is the new approximation (14) from $\sqrt[n]{D}$? Assume, for the sake of definiteness that $a \leq \sqrt[n]{D}$. [The other case merely flips the endpoints of the intervals and reverses some signs.] Because it is in the interval $[a, D/a^{n-1}]$, $\sqrt[n]{D}$ must lie in one of the subintervals
$$\left[a, a + \frac{D - a^n}{na^{n-1}} \right] \text{ of length } \frac{D - a^n}{na^{n-1}} = \frac{1}{n} \left(\frac{D}{a^{n-1}} - a \right)$$
or
$$\left[a + \frac{D - a^n}{na^{n-1}}, \frac{D}{a^{n-1}} \right] \text{ of length } \frac{D}{a^{n-1}} - \frac{D - a^n}{na^{n-1}} = \frac{n-1}{n} \left(\frac{D}{a^{n-1}} - a \right).$$
Thus, k iterations of the procedure produces an approximation a_{k+1} within
$$\left(\frac{n-1}{n} \right)^k \left(\frac{D}{a^{n-1}} - a \right)$$

4. THIRD, FOURTH, AND FIFTH ROOTS

of $\sqrt[n]{D}$. As n is fixed and $(n-1)/n < 1$, this can be made as small as desired by choosing k large enough. That is, the sequence $a_1, a_2, a_3, \ldots$ of generated approximations to $\sqrt[n]{D}$ converges to $\sqrt[n]{D}$.

For larger values of n, the convergence can start off rather slowly. Let $D = 17$ as usual and $n = 5$. $a = 1$ is the largest integer the fifth power of which does not exceed 17. Using the *TI-83 Plus* or the *TI-89 Titanium* in APPROXIMATE mode, enter 1 and then successively enter

ANS+(17−ANS^5)/(5∗ANS^4).

The resulting sequence,

$$1,\ 4.2,\ 3.370926517,\ 2.723073088,\ 2.240294518,\ 1.927212205,$$
$$1.788237893,\ 1.763079668,\ 1.762340968,\ 1.762340348,\ \ldots,\quad (15)$$

takes several steps before even being correct in the first decimal place. The problem is that as n gets larger, even though b may be small, making b^2, b^3, b^4, b^5 negligible, b^2, b^3, and b^4 are being multiplied by powers of a and binomial coefficients. As n increases, one must work harder to speed up the convergence or to produce a digit-by-digit algorithm analogous to the one illustrated by FIGURES 2 and 3.

One can speed things up a bit by using more terms in the binomial expansion to determine b:

$$D = a^5 + 5a^4b + 10a^3b^2 + 10a^2b^3 + 5ab^4 + b^5$$
$$\approx a^5 + 5a^4b + 10a^3b^2$$
$$\approx a^5 + \left(5a^4 + 10a^3b\right)b$$
$$\approx a^5 + \left(5a^4 + 10a^3 \frac{D-a^5}{5a^4}\right)b,$$

using the first approximation to b,

$$\approx a^5 + \left(5a^4 + \frac{2(D-a^5)}{a}\right)b$$
$$\approx a^5 + \left(\frac{5a^5 + 2D - 2a^5}{a}\right)b$$
$$\approx a^5 + \frac{3a^5 + 2D}{a} \cdot b,$$

whence

$$b \approx \frac{(D-a^5)a}{3a^5 + 2D} \qquad (16)$$

and

$$a + b \approx a + \frac{Da - a^6}{3a^5 + 2D} = \frac{3aD + 2a^6}{3a^5 + 2D}.$$

If we now enter 1 and follow it with the succession of hits of

(3∗ANS∗17+2∗ANS^6)/(3∗ANS^5+2∗17),

we get the sequence

$$1.432432432,\ 1.734065187,\ 1.762325443,\ 1.762340348,\ \ldots$$

which converges more rapidly.

The jump from degrees 2 and 3 to degree 5 is not a great leap conceptually. And with modern pocket calculators even the inefficient extraction of $\sqrt[5]{17}$ is no real pain. But in the days of computation by hand, it would seem to have required great determination to carry out such an extraction. Here I might take the liberty of quoting J.L. Berggren on the history. He begins by quoting a passage from 'Umar al-Khayyāmī's *Algebra*:

> From the Indians one has methods for obtaining square and cube roots, methods which are based on the knowledge of individual cases, namely the knowledge of the squares of the nine digits $1^2, 2^2, 3^2$ (etc.) and their respective products, i.e. $2 \cdot 3$, etc. We have written a treatise on the proof of the validity of these methods and that they satisfy the conditions. In addition we have increased their types, namely in the form of the determination of the fourth, fifth, sixth roots up to any desired degree. No one has preceded us in this and these proofs are purely arithmetic, founded on the arithmetic of *The Elements*.[7]

Berggren continues

> 'Umar was neither the first mathematician nor the last who believed falsely that he was the originator of a method. In this case we know that Abu l-Wafā', who flourished over 100 years before 'Umar, in the late tenth century, wrote a work entitled *On Obtaining Cube and Fourth Roots and Roots Composed of These Two*. Of course, 'Umar may not have known of Abu l-Wafā' 's treatise, or it may be that Abu l-Wafā' simply pointed out that $\sqrt[4]{N} = (\sqrt{\sqrt{N}})$ and, since $\sqrt[3]{N}$ was already known from the Indians, roots such as the twelfth, for example $\sqrt[12]{N} = \sqrt[4]{\sqrt[3]{N}}$ could be calculated by known methods. Thus Abu l-Wafā' 's work may have been less innovative than that of 'Umar.[8]

1.4.1. Exercise. *Find $\sqrt[4]{17}$ to 9 decimals by first entering 2 and then repeatedly entering*

ANS+(17−ANS^4)/(4ANS^3).

Keep track of the list of estimates. Now find $\sqrt{17}$ to 9 decimals by first entering 4 and then repeatedly entering

ANS+(17−ANS^2)/(2ANS).

As soon as the answer repeats, assume it is correct as far as it goes and enter

ANS→D.

Then repeat entering

---

[7]J.L. Berggren, *Episodes in the Mathematics of Medieval Islam*, Springer-Verlag New York Inc., New York, 1986, p. 53.

[8]*Ibid.*

4. THIRD, FOURTH, AND FIFTH ROOTS

ANS+(D−ANS²)/(2ANS)

until an answer repeats. Which takes fewer steps, finding $\sqrt[4]{17}$ directly or finding $\sqrt{17}$ and then the square root of that?

Berggren points out that these higher root extractions of Abu l-Wafā (940 – 997 or 998) and 'Umar al-Khayyāmī (1040(?) – 1131(?)) are lost to us, and presents instead in some detail an extant fifth root extraction performed by Ghiyāth al-Dīn Jamshīd al-Kāshī (†1429).[9]

I shall merely refer the reader to Berggren for al-Kāshī's calculation, preferring instead to discuss the simpler square and cube root extractions of Hindu mathematicians. The digit-by-digit extraction of the square root was described by Âryabhaṭa (476 – 550?) in the *Âryabhaṭiya* (c. 499) in the words

> Always divide the even place by twice the square-root (up to the preceding odd place); after having subtracted from the odd place the square (of the quotient), the quotient put down at the next place (in the line of the root) gives the root.[10]

This is barely intelligible. Fortunately, Datta and Singh work through an example. As they do not explicitly state that their example comes directly from Âryabhaṭa, I shall choose, for readier comparison with our earlier digit-by-digit square root extraction the number $D = 189574$.

First I should explain that "place" just means a decimal digit. A centesimal digit has two places, termed odd and even. The odd place is the tens digit and the even the ones digit. In $189574 = 18\ 95\ 74$, the odd places are 1, 9, 7, and the even ones are 8, 5, 4.

The root extraction proceeds left to right one place at a time, exactly what is done depending on whether one is at an odd or even place. The exception is the starting step in which one finds one's initial estimate a to the square root of the first centesimal digit of D. For $D = 18\ 95\ 74\ .\ 00\ 00$, one quickly sees one should take $a = 4$ and subtract $a^2 = 16$: $d = 18 − 16 = 2$. Thereafter one repeats two steps.

Bring down the odd place of the second centesimal digit, i.e., 9, transforming d into 29. Divide this new d by $2a = 2 \cdot 4 = 8$. The quotient is $b = 3$. Subtract 3 times 8 ($2ab$) from 29 ($d = D − a^2$) to get 5. Now bring down the second place of the current centesimal digit, i.e., the 5 from 95 to get 55 and subtract $b^2 = 9$ to get 46. Set $d = 46$ and append b to a to yield a new $a = 43$.

Repeat the procedure: Bring down the odd place of the second centesimal digit appending it to the end of d: 467. Divide by $2a = 2 \cdot 43 = 86$, yielding a quotient $b = 5$. Subtract $2ab = 86 \cdot 5 = 430$ from 467 to get 37. Bring down

---

[9] *Ibid.*, pp. 53 – 63.

[10] Cited in Bibhutibhushan Datta and Avadesh Narayan Singh, *History of Hindu Mathematics*, Bharatiya Kala Prakashan, Delhi, 2004; here: vol. I, p. 170. The first edition was originally published in the 1930s. The short biographical entries on Datta (1888 – 1958) and Singh (1901 – 1954), pp. 404 – 405 and p. 523, respectively, in Joseph W. Dauben and Christoph J. Scriba, eds., *Writing the History of Mathematics: Its Historical Development*, Birkhäuser Verlag, Basel, 2002, contain more information on their mathematical backgrounds and work on the history of Indian science.

the even place 4 yielding 374 and subtract $b^2 = 25$ to get a new $d = 349$. As before, append b to the end of a to get a new root $a = 435$.

One continues in this manner.

Obviously, one would not work such a root extraction out in so wordy a fashion in actual computation. On paper one might again create something like a long division as in FIGURE 4 below.

```
              4    3    5 .   4    0
       √   1 8  9 5   7 4 . 0 0   0 0
           1 6
       8)   2   9  (3    Divide by 2a
                2 4       Subtract 3 · 2a
                5 5
                  9       Subtract $3^2$
            86) 4 6   7  (5   Repeat
                4 3   0
                  3   7 4
                      2 5
              870) 3   4 9   0  (4   Repeat
                   3   4 8   0
                             1   0 0
                                 1 6
                      8708) 8 4   0  (0   Repeat
                                  0
                            8 4   0
                                etc.
```

FIGURE 4. Final Result

It is not immediately evident what the advantage is of splitting the treatment of successive centesimal digits into two steps. Presumably, since the division by $2a$ does not take into account the even place of the centesimal digit, it is less likely the quotient b will be too large. Yet it can happen. For $D = \sqrt{2}$, the computation goes as in FIGURE 5, below.

This is not incorrect; it merely stands for

$$1.5\,{}^-8 = 1 + \frac{5}{10} + \frac{-8}{100} = 1 + \frac{4}{10} + \frac{10-8}{100} = 1 + \frac{4}{10} + \frac{2}{100} = 1.42.$$

It did, however, take this conversion and an extra step to secure $\sqrt{2}$ as $1.4\ldots$ Thus, one replaces b by $b - 1$ as in FIGURE 6, below.

1.4.2. Exercise. Find $\sqrt{2}$ to several additional places using Âryabhaṭa's procedure. Find $\sqrt{17}$ to several places.

The reader with a *TI-89 Titanium* and some programming experience might wish at this point to repeat Exercises 1.3.3 and 1.3.4 of the preceding section using the present procedure in place of the one used there.

4. THIRD, FOURTH, AND FIFTH ROOTS

$$
\begin{array}{r}
\phantom{\sqrt{}}\ 1\ .\ \ \ 5\ -8 \\
\sqrt{}\ 2\ .\ 0\ 0\ \ 0\ 0 \\
\phantom{\sqrt{}}\ 1 \\ \hline
\phantom{\sqrt{}}\ 1\ \ \ 0\ \ (5\ \text{Divide by } 2a = 2) \\
\phantom{\sqrt{}}\ 1\ \ \ 0 \\ \hline
\phantom{\sqrt{}}\ \ \ \ 0\ 0 \\
\phantom{\sqrt{}}\ -\ 2\ 5 \\ \hline
\phantom{\sqrt{}}\ -\ 2\ 5\ \ \ 0\ \ (-8\ \text{Divide by } 2a = 30) \\
\phantom{\sqrt{}}\ -\ 2\ 4\ \ \ 0 \\ \hline
\phantom{\sqrt{}}\ -\ 1\ \ \ 0 \\
\text{etc.}
\end{array}
$$

FIGURE 5. Too Large a Quotient

$$
\begin{array}{r}
\phantom{\sqrt{}}\ 1\ .\ \ \ 4\ \ \ 1 \\
\sqrt{}\ 2\ .\ 0\ 0\ \ 0\ 0 \\
\phantom{\sqrt{}}\ 1 \\ \hline
\phantom{\sqrt{}}\ 1\ \ \ 0\ \ (4 \\
\phantom{\sqrt{}}\ \ \ \ 8 \\ \hline
\phantom{\sqrt{}}\ \ \ 2\ 0 \\
\phantom{\sqrt{}}\ \ \ 1\ 6 \\ \hline
\phantom{\sqrt{}}\ \ \ \ 4\ \ \ 0\ (1\ \text{Divide by } 2a = 28) \\
\phantom{\sqrt{}}\ \ \ \ 2\ 8 \\ \hline
\text{etc.}
\end{array}
$$

FIGURE 6. Corrected Calculation

The cube root extraction algorithm considers D to be written in base 1000, each millesimal digit of which consists of three decimal digits which are now given a three-fold characterisation analogous to the odd/even one: second aghana, first aghana, and ghana. Datta and Singh describe the method of cube root extraction as follows:

The Operation. The first description of the operation of the cube-root is found in the *Āryabhaṭīya*. It is rather too concise:

"Divide the second *aghana* place by thrice the square of the cube-root; subtract from the first *aghana* place the square of the quotient multiplied by thrice the preceding (cube-root); and (subtract) the cube (of the quotient) from the *ghana* place; (the quotient put down at the next place (in the line of the root) gives the root."

As has been explained by all the commentators, the units place is *ghana*, the tens place is first *aghana*, the hundreds place is second *aghana*, the thousands place is *ghana*, the ten-thousands place is first *aghana* and so on. After making the places as *ghana*, first *aghana* and second *aghana*, the process begins with the subtraction of the greatest cube number from the figures as far as the last *ghana* place. Though this has not been explicitly mentioned in the rule, the commentators say that it is implied in the expression "ghanasya mûla vargena" etc. ("by the square of the cube-root" etc.)[11]

The procedure itself is analogous to that for extracting a square root. The cube root of the first millesimal digit is estimated by appeal to a table of cubes of the digits $0, 1, 2, \ldots, 9$. It is called the root a and is subtracted from the millesimal digit. The second aghana of the next millesimal digit is brought down and appended to the difference d resulting in a new quantity d to be divided by $3a^2$. The quotient is called b and $3a^2b$ is subtracted from the new d. The first aghana is brought down, appended to this difference, and $3ab^2$ is subtracted therefrom. The ghana is then brought down and appended to the difference from which b^3 is subtracted. The new difference is the new d and the result of appending b to the old root a is the new root a. All the steps involving the aghanas and the ghana are repeated as often as desired. If at any stage the difference is negative, one backs up to the most recent division and replaces b by $b - 1$.

I suppose I should illustrate this with an example. I find though that spreading it out like FIGURES 4 – 6 is not only a painful typesetting exercise, but is an easily confusing process as I tend to forget which step I am on. Instead, let me offer a sort of spreadsheet layout. First, one sets out a line

$$D: 189\,574.000\,000 \qquad a: 5 \qquad d: 189 - 125 = 64$$

Then one has the table:

(a)ghana	subtractor (d)	a	divisor	b	subtrahend	difference
2nd: 5	645	5	$3a^2$: 75	8	$3a^2b$: 600	45
1st: 7	457	5	–	8	$3ab^2$: 960	-503
Error: Go back.						
2nd: 5	645	5	–	7	$3a^2b$: 525	120
1st: 7	1207	5	–	7	$3ab^2$: 735	472
g: 4	4724	5	–	7	b^3: 343	4381
2nd: 0	43810	57	$3a^2$: 9347	4	$3a^2b$: 38988	4822
1st: 0	48220	57	–	4	$3ab^2$: 2736	45484
g: 0	454840	57	–	4	b^3: 64	454776
2nd: 0	4547760	574	$3a^2$: 988428	4	$3a^2b$: 3953712	594048
etc.						

This gives us $\sqrt[3]{189574} \approx 57.44$. One readily checks on one's calculator that $57.44^3 \approx 189514.8708$ and $57.45^3 \approx 189613.8686$ and we indeed have the first four digits of the cube root.

This procedure makes an interesting programming exercise for the calculator, particularly on the *TI-89 Titanium* which can handle the ever growing numbers with ease. Of course, if one doesn't want or need greater than 14 digits in one's cube roots, one is better off just using the exponent button on the calculator.

---

[11] *Ibid.*, pp. 175 – 176.

An alternative exercise would be to set up a spreadsheet like the above: Just type in the formulæ, and watch the numbers generate. When a minus sign appears in the last column, go back to the appropriate second aghana row and simply replace the formula defining b by the value $b - 1$.

At this stage the rôle of the binomial coefficients should be obvious and the procedure for extracting 4th and 5th roots should be clear enough.

1.4.3. Exercise. *Find $\sqrt[5]{17}$ to several decimal places by generalising the procedure just given.*

1.4.4. Exercise. *Repeat Exercise 1.4.1 using the one digit at a time algorithm.*

5. The Full Finite Binomial Theorem

It is time we stated the Finite Binomial Theorem in full.

1.5.1. Theorem (Finite Binomial Theorem). *Let n be a nonnegative integer. For any real numbers a and b,*

$$(a+b)^n = \sum_{k=0}^{n} \binom{n}{k} a^k b^{n-k} = \sum_{k=0}^{n} \binom{n}{k} a^{n-k} b^k, \qquad (17)$$

where the binomial coefficients $\binom{n}{k}$ *are defined by any of the following three methods:*

i. (*Arithmetical Triangle*)

$$\binom{0}{0} = 1, \quad \binom{1}{0} = \binom{1}{1} = 1$$

$$\binom{n+1}{0} = 1, \quad \binom{n+1}{k+1} = \binom{n}{k} + \binom{n}{k+1} \text{ for } k < n+1, \quad \binom{n+1}{n+1} = 1.$$

ii. (*Combinatorics*)

$$\binom{n}{k} = \begin{cases} 1, & k = 0 \\ \dfrac{n(n-1)\cdots(n-k+1)}{k(k-1)\cdots 1}, & k > 0. \end{cases}$$

iii. (*Horizontal Recursion*)

$$\binom{n}{0} = 1$$

$$\binom{n}{k+1} = \frac{n-k}{k+1}\binom{n}{k}.$$

The term "Arithmetical Triangle" was coined by Blaise Pascal and refers to any of several arrays displaying the binomial coefficients. FIGURE 7, below, presents two such arrays, a triangular one on the left and a rectangular one with all the entries not occurring in the triangle being 0s. In these representations the coefficient $\binom{n}{k}$ is the $(k+1)$-th entry of the $(n+1)$-th row. Thus, for $\binom{4}{2}$, the 5th row is

1 4 6 4 1

1. THE FINITE BINOMIAL THEOREM

					1						1	0	0	0	0	0	0
				1		1					1	1	0	0	0	0	0
			1		2		1				1	2	1	0	0	0	0
		1		3		3		1			1	3	3	1	0	0	0
	1		4		6		4		1		1	4	6	4	1	0	0
1		5		10		10		5		1	1	5	10	10	5	1	0
1	6		15		20		15		6	1	1	6	15	20	15	6	1

FIGURE 7. The Arithmetical Triangle

and its third entry is 6: $\binom{4}{2} = 6$.

When the coefficients are laid out this way, the pattern jumps out at the viewer. Each entry beyond the second row other than the first and last elements of the row, which are both 1, is the sum of the two nearest entries in the row above it. If one presents the triangle in rectangular form as in the right half of FIGURE 7, in effect defining $\binom{n}{k} = 0$ for $k > n$, every element outside the first row and column is the sum of the numbers directly above it and directly to the left of that.

The recursion cited in Theorem 1.5.1.i is thus that implicit in the construction of the Arithmetical Triangle. The Arithmetical Triangle, incidentally, was rechristened "Pascal's triangle" in honour of his study of its properties. The rise of multiculturalism has seen a return to Pascal's original nomenclature, which is less Eurocentric if not any more descriptive.

Our designation "Combinatorics" will be clarified when we come to the proof of Theorem 1.5.1.ii, and "Horizontal Recursion" simply refers to the fact that it defines the recursion that generates successive entries in the $(n+1)$-th row of the Arithmetical Triangle.

As stated here, the Finite Binomial Theorem is more a conglomeration of theorems than a theorem. The three main theorems are the assertions that

$$(a+b)^n = \sum_{k=0}^{n} \binom{n}{k} a^k b^{n-k}$$

for each of the three distinct definitions of the binomial coefficients. There is also the extra bit asserting

$$\sum_{k=0}^{n} \binom{n}{k} a^k b^{n-k} = \sum_{k=0}^{n} \binom{n}{k} a^{n-k} b^k. \tag{18}$$

This last identity is a simple matter of observation. Since $a + b = b + a$, it follows that $(a+b)^n = (b+a)^n$. But if we have

$$(a+b)^n = \sum_{k=0}^{n} \binom{n}{k} a^k b^{n-k},$$

5. THE FULL FINITE BINOMIAL THEOREM

for all values of a, b, it also holds when a and b are interchanged:

$$(b+a)^n = \sum_{k=0}^{n} \binom{n}{k} b^k a^{n-k} = \sum_{k=0}^{n} \binom{n}{k} a^{n-k} b^k.$$

Thus,

$$\sum_{k=0}^{n} \binom{n}{k} a^k b^{n-k} = (a+b)^n = (b+a)^n = \sum_{k=0}^{n} \binom{n}{k} a^{n-k} b^k.$$

There are two points to mentioning this. One is that it is natural to expand $(a+b)^n$ in two ways: as

$$\binom{n}{0} a^n b^0 + \binom{n}{1} a^{n-1} b^1 + \ldots + \binom{n}{n-1} a^1 b^{n-1} + \binom{n}{n} a^0 b^n,$$

where the k's are listed in their natural order and the powers of a are listed from highest to lowest degree, and as

$$\binom{n}{0} a^0 b^n + \binom{n}{1} a^1 b^{n-1} + \ldots + \binom{n}{n-1} a^{n-1} b^1 + \binom{n}{n} a^n b^0,$$

in which the index k of the binomial coefficient matches the degree of a. The identity (18) tells us the two expressions are the same.

The second point is the symmetry of the coefficients

$$\binom{n}{k} = \binom{n}{n-k} \qquad (19)$$

for $0 \leq k \leq n$. To conclude this from (18) requires us to think of a, b as variables and not as unspecified real numbers. It is a subtle point. For some values of a and b one can write

$$(a+b)^n = \sum_{k=0}^{n} c_k a^k b^{n-k} \qquad (20)$$

in more than one way. For example,

$$(2+3)^2 = 1 \cdot 2^2 + 2 \cdot 2 \cdot 3 + 1 \cdot 3^2 \text{ with } c_0 = 1, c_1 = 2, c_2 = 1$$
$$(2+3)^2 = 4 \cdot 2^2 + 1 \cdot 3^2 \text{ with } c_0 = 4, c_1 = 0, c_2 = 1.$$

However, the second set of coefficients will not yield $(a+b)^n$ for all a, b. We can, in fact, determine those values of a, b for which these coefficients work by simple algebra: Set $(a+b)^2 = 4a^2 + b^2$ and solve:

$$a^2 + 2ab + b^2 = 4a^2 + b^2$$
$$2ab = 3a^2,$$

whence $a = 0$ or $2b = 3a$, i.e., $b = \frac{3}{2}a$. In particular,

$$(0+5)^2 = 4 \cdot 0^2 + 1 \cdot 5^2$$
$$(4+6)^2 = 4 \cdot 4^2 + 1 \cdot 6^2,$$

but

$$(1+5)^2 = 36 \neq 29 = 4 \cdot 1^2 + 1 \cdot 5^2.$$

The point here is that the binomial coefficients are coefficients of a polynomial in two variables in a polynomial *identity* and thus work for all real values of the variables a and b and are thus unique. It is (18) viewed as a polynomial identity that allows us to conclude (19).

This last remark is not our central issue here. We will return to it later, however, in Chapter 3, section 5. For now suffice it to say that the binomial coefficients satisfy (17) for all real numbers a, b and they are the only coefficients that do, and we will only prove the first of these two assertions.

When I refer to proving the first assertion, namely (17), I mean of course that we will prove three versions of it, one for each of the three definitions given of the binomial coefficients. Since we are not assuming the uniqueness of the coefficients satisfying (20), we will also explicitly prove the equivalence of the three definitions.

Proof of 1.5.1.i. This is a straightforward induction on the exponent n.
For the basis, note that
$$(a+b)^0 = 1 = \binom{0}{0} a^0 b^0$$
and
$$(a+b)^1 = a + b = 1 \cdot a + 1 \cdot b = \binom{1}{0} a + \binom{1}{1} b.$$
For the induction step, assume
$$(a+b)^n = \sum_{k=0}^{n} \binom{n}{k} a^k b^{n-k}$$
and observe that
$$(a+b)^{n+1} = (a+b)(a+b)^n = (a+b)\left(\binom{n}{0} a^0 b^n + \ldots + \binom{n}{n} a^n b^0 \right).$$
In collecting like terms, that for $a^0 b^{n+1}$ is
$$b \binom{n}{0} a^0 b^n = 1 \cdot a^0 b^{n+1} = \binom{n+1}{0} a^0 b^{n+1},$$
and that for $a^{n+1} b^0$ is
$$a \binom{n}{n} a^n b^0 = 1 \cdot a^{n+1} b^0 = \binom{n+1}{n+1} a^{n+1} b^0,$$
while those for $a^{k+1} b^{(n+1)-(k+1)}$ for $k < n$ are
$$a \binom{n}{k} a^k b^{n-k} + b \binom{n}{k+1} a^{k+1} b^{n-k-1} = \left(\binom{n}{k} + \binom{n}{k+1} \right) a^{k+1} b^{n-k}$$
$$= \binom{n+1}{k+1} a^{k+1} b^{(n+1)-(k+1)},$$
using the recursion of Theorem 1.5.1.i defining the binomial coefficients. □

An alternative presentation of the proof of the induction step, one that is more painful to typeset, and possibly less readable, but which doesn't require one to think about why what I said about collecting the terms is true, proceeds as follows:

$$(a+b)^{n+1} = (a+b)(a+b)^n = (a+b)\sum_{k=0}^{n}\binom{n}{k}a^k b^{n-k}$$

$$= \sum_{k=0}^{n}\binom{n}{k}a^{k+1}b^{n-k} + \sum_{k=0}^{n}\binom{n}{k}a^k b^{n-k+1}$$

$$= \sum_{j=1}^{n+1}\binom{n}{j-1}a^j b^{n-j+1} + \sum_{k=0}^{n}\binom{n}{k}a^k b^{n+1-k}, \text{ for } j=k+1$$

$$= \sum_{k=1}^{n+1}\binom{n}{k-1}a^k b^{n+1-k} + \sum_{k=0}^{n}\binom{n}{k}a^k b^{n+1-k}, \text{ relabelling}$$

$$= \binom{n}{n}a^{n+1}b^0 + \sum_{k=1}^{n}\left(\binom{n}{k-1}+\binom{n}{k}\right)a^k b^{n+1-k} + \binom{n}{0}a^0 b^{n+1}$$

$$= a^0 b^{n+1} + \sum_{k=1}^{n}\binom{n+1}{k+1}a^k b^{n+1-k} + a^{n+1}b^0$$

using the recursion again,

$$= \binom{n+1}{0}a^0 b^{n+1} + \sum_{k=1}^{n}\binom{n+1}{k+1}a^k b^{n+1-k} + \binom{n+1}{n+1}a^{n+1}b^0$$

$$= \sum_{k=0}^{n+1}\binom{n+1}{k}a^k b^{n+1-k},$$

as was to be shown. □

Proof of 1.5.1.*ii*. $(a+b)^n$ can be described directly, without referring to the recursion, as the result of a two-step process: First, choose one term from each of the n factors,

$$\underbrace{(a+b)(a+b)\cdots(a+b)}_{n},$$

and take their product. If one lists all the a's first, each of these products will be of the form $a^k b^{n-k}$ for some $0 \le k \le n$. The second step is to collect like terms, i.e., to count how many ways one can choose k factors from which the a's came out of a total of n factors. Thus, if we define $_nC_k$ to be the number of ways of choosing k elements from a set of n elements, usually referred to as "the number of *combinations* of n things taken k at a time", we see that

$$(a+b)^n = \sum_{k=0}^{n} {_nC_k}\, a^k b^{n-k}.$$

This great insight is of little use unless we can compute $_nC_k$. To do this, we first compute $_nP_k$, the number of *permutations* of n things taken k at a time. A permutation is just an ordering, or arrangement, of objects in a set, and counting them is easy. How many ways can we order the numbers $1, 2, 3, 4, 5$?

Well, any of the numbers can be chosen to be the first number in our ordering. There are 5 such choices. For each of these possible choices there are 4 numbers remaining, hence 4 choices apiece, or $5 \cdot 4 = 20$ choices in all. Now two elements have been used up, leaving only 3 for the third choice, i.e., 3 possible choices for each of the 20 choices made so far. Thus, $5 \cdot 4 \cdot 3 = 20 \cdot 3 = 60$ possible ways of choosing 3 elements in succession. For the fourth element, there are 2 unchosen elements left to choose from, and after that only 1 for the last element of the ordering. The total number of ways of ordering the numbers $1, 2, 3, 4, 5$ is thus $5 \cdot 4 \cdot 3 \cdot 2 \cdot 1 = 120$. In general, the number of orderings of a set of n objects will be $n(n-1)\cdots 1$, the product of all the integers from n to 1.

If we only wish to choose k elements from a set of n elements and order them we simply stop the above procedure after k steps, as when we concluded above that there were $5 \cdot 4 \cdot 3 = 60$ ways of choosing 3 elements in succession from a set of 5 elements. Notice the total was a product of 3 factors, starting at 5, each successive factor being 1 less than the previous. In general, the number of permutations of n things taken k at a time is a product of k such decreasing factors:

$$_nP_k = \underbrace{n(n-1)\cdots(n-k+1)}_{k}.$$

When $k = n$, the number $_nP_n$ is abbreviated to $n!$ and called *n-factorial*.

There is another way of counting $_nP_k$. We can think of choosing a permutation of k things from a set of n objects as a two-step process: First, choose a set of k elements from the n elements, and then choose an ordering of the k elements. The first task can be done in $_nC_k$ ways and the second in $_kP_k = k!$ ways. Thus $_nP_k = {_nC_k} \cdot k!$. Given that we already know how to calculate $_nP_k$ and don't know how to calculate $_nC_k$, this is of little use in counting permutations. But it tells us how to calculate combinations. Division by $k!$ yields

$$_nC_k = \frac{_nP_k}{k!} = \frac{n(n-1)\cdots(n-k+1)}{k(k-1)\cdots 1},$$

as was to be proved.

There is still the task of showing that $_nC_k = \binom{n}{k}$. The simplest way of doing this is to show that $_nC_k$ obeys the recursion of 1.5.1.i.

Well, $_0C_0$ is the number of empty subsets of the empty set. Clearly there is only one, namely, the empty set:

$$_0C_0 = 1 = \binom{0}{0}.$$

Likewise, a single element set has only two subsets, the empty set and the singleton itself:

$$_1C_0 = 1 = \binom{1}{0}, \qquad _1C_1 = 1 = \binom{1}{1}.$$

For the actual recursion, let A be an n-element set for some $n \geq 1$, and let a not be in A. Clearly again there is only one empty subset of $A \cup \{a\}$ and only

5. THE FULL FINITE BINOMIAL THEOREM

one $(n+1)$-element subset of $A \cup \{a\}$:

$$_{n+1}C_0 = 1 = \binom{n+1}{0}, \quad _{n+1}C_{n+1} = 1 = \binom{n+1}{n+1}.$$

A $(k+1)$-element subset of $A \cup \{a\}$ is either a $(k+1)$-element subset of A or the union of a k-element subset of A with $\{a\}$:

$$_{n+1}C_{k+1} = {}_nC_{k+1} + {}_nC_k. \tag{21}$$

Assuming as induction hypothesis that $_nC_j = \binom{n}{j}$ for $j = 0, 1, \ldots, n$, (21) yields

$$_{n+1}C_{k+1} = \binom{n}{k+1} + \binom{n}{k} = \binom{n+1}{k+1},$$

and induction allows us to conclude $_nC_k = \binom{n}{k}$ for all n and all $k = 0, 1, \ldots, n$.
□

Proof of 1.5.1.*iii*. It is hard to imagine recognising the Horizontal Recursion from staring at the Arithmetical Triangle. Combinatorially, however, if one looks at permutations, it falls out. After having chosen k elements in order from an n-element set, there are $n - k$ elements left for choosing a $(k+1)$-th element:

$$_nP_{k+1} = (n-k) \cdot {}_nP_k.$$

But

$$_nC_{k+1} = \frac{_nP_{k+1}}{(k+1)!} = \frac{_nP_{k+1}}{(k+1)k!} = \frac{(n-k) \cdot {}_nP_k}{(k+1)k!} = \frac{n-k}{k+1} {}_nC_k. \quad \square$$

The Horizontal Recursion is not as celebrated as the generating recursion of the Arithmetical Triangle or the combinatorial definition of $_nC_k$ as $_nP_k/k!$ or the definition purely in terms of factorials derivative to the latter,

$$\binom{n}{k} = \frac{n(n-1)\cdots(n-k+1)}{k(k-1)\cdots 1}$$
$$= \frac{n(n-1)\cdots(n-k+1)(n-k)\cdots 1}{k(k-1)\cdots 1 \cdot (n-k)\cdots 1} = \frac{n!}{k!(n-k)!},$$

but it is quite useful. Simple as the basic recursion is, the calculation of binomial coefficients by using it or the combinatorial formulæ could be labour intensive. The importance of binomial probability, a subject that sadly will not be discussed in detail in the present book, led to the calculation of massive tables involving them in the 20th century. Those produced in the United States during the Great Depression by unskilled labour as work relief would have relied as much as possible on addition, thus on the basic arithmetic recursion of Theorem 1.5.1.i. With the advent of computers, however, arithmetic skill was not the issue. Memory was extremely limited in the early decades of computers and the machines were programmed to generate and print tables on paper. The Horizontal Recursion reduced the memory requirements in generating the coefficients line-by-line.[12]

---

[12]Sol Weintraub, *Tables of the Cumulative Binomial Probability Distribution for Small Values of p*, Macmillan Company, New York, 1963, pp. xv – xvi.

6. Combinatorics in Early India

There is a world of difference between a knowledge of the first instances of the Finite Binomial Theorem and knowledge of the full result, namely that the full result recognises a *pattern* that is common to all instances. The priority question one would like the history books to answer is: whom do we credit the Finite Binomial Theorem to, i.e., who first recognised that there was this common pattern and generality? The instance,

$$(a+b)^2 = a^2 + 2ab + b^2,$$

is already in Euclid's *Elements* (c. 330 B.C.), albeit expressed in geometric terms. No one would credit Euclid with knowledge of the Finite Binomial Theorem, though he would perhaps be cited as part of the pre-history of the Theorem as a precursor.

After Euclid the discussion becomes more complicated. Today it is standard to present the binomial coefficients in triangular form, as in FIGURE 7 of the immediately preceding section. As the recursion of Theorem 1.5.1.i is so immediate, should one find the Arithmetical Triangle in a discussion of binomial exponentiation, one can be fairly certain that the author of the discussion was well aware of the general Finite Binomial Theorem and not just the instances actually cited. This, of course, is not yet a proof of the theorem.

Also, to complicate the history somewhat, the Arithmetical Triangle pops up in other contexts in early mathematics, whence knowledge of this array of numbers without a discussion of binomial exponentiation cannot be taken as knowledge of the full Finite Binomial Theorem. In his now standard history of the Arithmetical Triangle[13], A.W.F. Edwards discusses "figurate numbers", "combinatorial numbers", and "binomial numbers". Figurate numbers date back to the Pythagoreans in the 6th century B.C., although their connexion with the Arithmetical Triangle took some centuries to be established. The combinatorial numbers and the Arithmetical Triangle were introduced in India before the beginning of the so-called Christian era. He credits the connexion of the Arithmetical Triangle with binomial expansions to Chinese and Persian (Islamic) scholars in or near the end of the first century of the second millennium.

There seems to be at present no definitive history of Hindu mathematics, at least not in English. The first major European work on the subject was given by Henry Thomas Colebrooke (1765 – 1837) with his translations[14] from Bhâskara II and Brahmagupta. Colebrooke was mathematically and linguistically capable, but the works he translated are too recent to interest us here. In the 1930s B. Datta and A.V. Singh projected a three-volume source book

---

[13] A.W.F. Edwards, *Pascal's Arithmetical Triangle: The Story of a Mathematical Idea*, Charles Griffin & Company Limited, London, 1987; 2nd edition: Johns Hopkins University Press, Baltimore, 2002.

[14] Henry Thomas Colebrooke, *Algebra with Arithmetic and Mensuration from the Sanscrit of Brahmegupta and Bháscara*, London, 1817. The book is available online at Google Books. There is a short biographical entry on Colebrooke in Dauben and Scriba, *op. cit.*, pp. 398 – 399.

on Hindu mathematics, two volumes of which appeared in 1935 and 1938. I quoted them in the last section and will cite them again shortly. The works of Colebrooke and Datta and Singh are generally reliable and, despite their age, useful. Radha Charan Gupta warns about unreliable Indian historiography:

> The monographs of BENOY KUMAR SARKAR [SARKAR 1918], like [KAYE 1925], are not considered reliable because the first makes exaggerated claims and the other denies even genuine ones.
>
> When India became independent from British rule in 1947, the need to boost national sentiment by making such exaggerated historical claims about Indian accomplishments was no longer necessary...
>
> However, some serious problems, including questions of accurate chronology, continue to hamper the historiography of mathematics in India, especially for the ancient period... Deliberate attempts to make exaggerated claims for India on the one hand, and to deprive Indians of even their original achievements on the other, still continue.[15]

George Gheverghese Joseph wrote a most interesting book on the development of mathematics outside of Europe.[16] He too addresses the issue of distortion in the history of Indian mathematics:

> Ancient Indian history raises many problems. The period before the Christian era takes on a haziness which seems to have prompted opposing reactions. There are those who make excessive claims for the antiquity of Indian mathematics, and others who go to the opposite extreme and deny the existence of any 'real' Indian mathematics before about 500 AD. The principal motive of the former is to emphasize the uniqueness of Indian mathematical achievements where, if there was any influence, it was always a one-way traffic from India to the rest of the world. The motives of the latter are more mixed. For some their Eurocentrism (or Graeco-centrism) is so deeply entrenched that they cannot bring themselves to face the idea of independent developments in early Indian mathematics, even as a remote possibility.
>
> A good illustration of this blinkered vision is provided by a widely respected historian of mathematics at the turn of this century, Paul Tannery. Confronted with the evidence from Arab sources that the Indians were the first to use the sine function as we know it today, Tannery devoted himself to seeking ways in which the Indians could have acquired the concept from the Greeks. For Tannery, the very fact that the Indians knew and used sines in their astronomical calculations was sufficient evidence that they must have had it from

---

[15]Radha Charan Gupta, "India", in Dauben and Scriba, *op. cit.*, p. 315.

[16]George Gheverghese Joseph, *The Crest of the Peacock: Non-European Roots of Mathematics*, I.B. Tauris, London, 1991; paperback edition: Penguin Books, London, 1992.

the Greeks. But why this tunnel vision? The following quotation from G.R. Kaye (1915) is illuminating:

> The achievements of the Greeks in mathematics and art form the most wonderful chapters in the history of civilisation, and these achievements are the admiration of western scholars. It is therefore natural that western investigators in the history of knowledge should seek for traces of Greek influence in later manifestations of art and mathematics in particular.

It is particularly unfortunate that Kaye is still quoted as an authority on Indian mathematics. Not only did he devote much attention to showing the derivative nature of Indian mathematics, usually on dubious linguistic grounds (his knowledge of Sanskrit was such that he depended largely on indigenous 'pandits' for translations of primary sources), but he was prepared to neglect the weight of contemporary evidence and scholarship to promote his own viewpoint. So, while everyone else claimed that the Bakshali Manuscript... was written or copied from an earlier text dating back to the first few centuries of the Christian era, Kaye insisted it was no older than the twelfth century AD. Again, while the Arab sources unanimously attributed the origin of our present-day numerals to the Indians, Kaye was of a different opinion. And the distortions that resulted from Kaye's work have to be taken seriously because of his influence on Western historians of mathematics, many of whom remained immune to findings which refuted Kaye's inferences and which established the strength of the alternative position much more effectively than is generally recognized.

This tunnel vision is not confined to mathematics alone. Surprised at the accuracy of information on the preparation of alkalis contained in an early Indian textbook on medicine (*Susruta Samhita*) dating back to a few centuries BC, the eminent chemist and historian of the subject, M. Berthelot (1827 – 1909)[17], suggested that this was a later insertion, after the Indians had come into contact with European chemistry!

While non-European chauvinism (on the part of, for example, the Arabs, Chinese and Indians) does persist, the 'arrogant ignorance' (as J.D. Bernal (1969) described the character of Eurocentric scholarship in the history of science) is the other side of the same coin. But the latter tendency has done more harm than the former because it rode upon the political domination imposed by the West, which imprinted its own version of knowledge on the rest of the world.[18]

---

[17] A typo: Berthelot died in 1907.
[18] Joseph, *op.cit.*, pp. 215 – 216.

The words "blinkered vision", "tunnel vision" and "distortion" may be a bit negative, but I fancy them more in the nature of euphemisms than arguments *ad hominem* in intention. For, as far as I can tell, the facts back him up. And, for the most part his book is free of overstatement.

Scattered throughout Joseph's book are tables offering timelines for historical and mathematical developments. The short entries in a table must inevitably oversimplify and perhaps overstate. His Table 8.1 for India has the entry for the period 800 – 200 B.C.:

> Vedic mathematics continues during the earlier years but declines with ending of ritual sacrifices; emergence of Jaina mathematics: number theory, permutations and combinations, the binomial theorem; astronomy.[19]

Crediting Vedic mathematics with the Finite Binomial Theorem is a matter of some controversy. I quote Paul Luckey:

> The Indians after all, as we yet know—and this agrees with the statement of 'Omar Ḥaiyāmī[20]—only extracted square and cubic roots, and based on translated sources and descriptions by those researching these sources their procedures for the formation of squares and cubes of multidigit numbers and correspondingly too their procedures for the extraction of square and cubic roots based itself on the finished formula of the binomial theorem for $n = 2$ and $n = 3$, as with the mathematicians of the Renaissance. They possessed though the binomial coefficients and their means of generation in the form of the so-called Pascal triangle in a special expression.
>
> Already very early, in the beginnings of the *Chandaḥ-sūtra* of Piṅgala (according to DATTA and SINGH before 200 B.C.), they treated in their metrics the determination of the number of possible combinations of long and short syllables to an n-syllable metric foot. ν short and $n - \nu$ long syllables yield through their various possible distributions in n places $\binom{n}{\nu}$ n-syllable metric feet (exactly as n-place dyadic numbers can be formed from ν zeros and $n - \nu$ ones). E.g., one has for $n = 3, \nu = 2$
>
> $$\text{the metrical feet } \smile\smile-,\ \smile-\smile,\ -\smile\smile$$
>
> (or the dyadic numbers *001, 010, 100*).
>
> The possibilities of realising these with meaningful words and the melodious sound of the corresponding verses is obviously ignored by this setting out of the combinatorial reasoning. The successive formation and numerical determination of these possible metrical feet for $n = 1, 2, \ldots$ led to the Pascal triangle on the basis of its formation law
>
> $$\binom{n}{\nu} = \binom{n-1}{\nu-1} + \binom{n-1}{\nu},$$

---

[19] *Ibid.*, p. 218.
[20] I.e., 'Umar al-Khayyāmī.

further to the recognition
$$\binom{n}{0} + \binom{n}{1} + \ldots + \binom{n}{n} = 2^n$$
in the expression that the number of n-syllable metric feet comes to 2^n (just as the number of the n-place dyadic numbers from $00\ldots 0$ to $11\ldots 1$).

A.N. SINGH says in connexion with this, "the development of $(a+b)^n$ for whole values of n" would have been known in India since very early times. He produces no instances for $n > 3$ at the place cited. From such prosodic teachings it appears to me certain that the formation and combinatorial applications of the numbers $\binom{n}{\nu}$ is proven, but not the knowledge of the binomial theorem. In the opinion of L. ALSDORF this teaching was never given a connexion to metrics. G. CHAKRAVARTI however established that such combinatorial contemplations would also be employed for other things from various domains of Indian life and in Indian science.[21]

Coming to the defence of ancient Indian mathematics, A.K. Bag cites the Arithmetical Triangle and its appearance in Europe and China before he comments on India:

> Singh pointed out the existence of an identical triangular array of binomial coefficients, known as *meru-prastāra*, in Piṅgala's *Chandaḥ-sūtra* (200 B.C.), which received only a passing notice in a footnote in Needham's admirable review[22] with the remark, based on Luckey, that it 'has nothing to do with binomial coefficients.'[23]

Bag then discusses the general combinatorial approach, citing examples and quoting a 10th century commentator Halāyudha:

> Here the method of pyramidal expansion (*meru-prastāra*) of the (number of) combinations of one, two, etc., syllables formed of short (and long sounds) are explained. After drawing a square on the top, two squares are drawn below (side by side) so that half of each is extended on either side. Below it three squares, below it (again) four squares are drawn and the process is repeated till the desired pyramid is attained. In the (topmost) first square the symbol for one is to be marked. Then in each of the two squares of the second

---

[21] Paul Luckey, "Die Ausziehung der n-ten Wurzel und der binomische Lehrsatz in der islamischen Mathematik", *Mathematische Annalen* 120 (1948), pp. 217 – 274; here: pp. 218 – 220. The names written in SMALL CAPS refer to specific works by the authors whose names are cited. I omit the bibliographic details. Luckey (1884 – 1949) was a mathematics teacher who studied Arabic upon retirement and wrote several important papers on Islamic mathematics and astronomy. *Cf.* Dauben and Scriba, *op. cit.*, pp. 474 – 475 for more.

[22] Joseph Needham, *Science and Civilization in China*, vol. 3 (1959), p. 137. Needham's massive history of Chinese science also covered some Indian and Islamic developments. More information on Needham's work in the history of Chinese science can be gleaned from the biographical entry in Dauben and Scriba, *op. cit.*, pp. 494 ? 495.

[23] Amulya Kumar Bag, "Binomial theorem in ancient India", *Indian Journal of the History of Science* 1, no. 1, (1966), pp. 68 – 74; here: p. 68.

line figure one is to be placed. Then in the third line figure one is to be placed on each of the two extreme squares. In the middle square (of the third line) the sum of the figures of the two squares immediately above is to be placed; this is the meaning of the term *pūrṇa*. In the fourth line one is to be placed in each of the two extreme squares. In each of the two middle squares, the sum of the figures in the two squares immediately above, that is, three, is placed. Subsequent squares are filled in this way. Thus the second line gives the expansion of combinations of (short and long sounds forming) one syllable; the third line the same for two syllables, the fourth line for three syllables, and so on.[24]

This clearly indicates a knowledge of the Arithmetical Triangle. But whether or not Piṅgala or any of his contemporaries applied it to the arithmetical problem of calculating $(a+b)^n$ is nowhere indicated. Certainly, if one is aware of the combinatorial significance of the entries in the Arithmetical Triangle and one thinks about the terms in the expansion of $(a+b)^n$, the Finite Binomial Theorem is immediate. But one has to consider the expansion problem first to make the application. Thus we see Luckey taking an agnostic stand (or: not taking a stand), Needham denying the Indians credit for the Finite Binomial Theorem, and Bag claiming credit on their behalf. A.W.F. Edwards, whose book on the Arithmetical Triangle is currently the definitive account, distinguished carefully between the numbers of this array as figurate numbers, combinatorial numbers, and binomial numbers, and sided with Needham against Bag on this:

> ...a writer on prosody by the name of Pingala (*ca* 200 B.C.) gave a rule, by all accounts very cryptically, for finding the number of combinations of n syllables taken one at a time, two at a time, three at a time, ..., all at a time...
>
> Pingala's rule, known as the *Meru Prastara*, is most succinctly given by his commentator Varahamihara who wrote (in A.D. 505) "It is said that the numbers are obtained by adding each with the one which is past the one in front of it, except the one in the last place"...
>
> In the tenth century another commentator Halayudha even gave the familiar "triangular" form of the Triangle...
>
> No comment is needed from us on this invention of the Combinatorial Triangle six centuries, ten centuries, or even seventeen centuries (reckon it as you will[25]) before it was used for the same purpose in the West, though we cannot overlook the reference to it by Needham—"it concerns prosodic combinations only and has nothing to do with binomial coefficients", for of course the whole import of the Arithmetical Triangle is that it is to do with combinations *and* figurate numbers *and* binomial coefficients.

---

[24]*Ibid.*, p. 72.
[25]The initial European appearance of the triangle came in the 16th century.

... All that the Hindus lacked was the appreciation that the coefficients were also those that occur in the binomial expansion. Brahmegupta got as far as $(a+b)^3$ in A.D. 628, which was one power further than Euclid, but that was that. The Arithmetical Triangle still awaited its Pascal, and he was not to be found in India.[26]

Brahmagupta (598 – after 665) was a leading mathematician of his age. He is of special interest here for having discovered the Horizontal Recursion for the rows of the Arithmetical Triangle and for cubing the binomial. Âryabhaṭa had done the latter earlier and had even, as we have already seen, inverted the process to find cube roots. But it is Brahmagupta who is believed to have influenced binomial development in the Islamic world:

> It is to Persia, however, that the European thread can be traced back. The *Al-bahir* of Al-Samawal[27] (died about 1180) is reported as containing a calculation of the coefficients resulting in the Binomial Triangle, which had been discovered by Al-Karaji some time soon after 1007. It is possible that al-Karaji was inspired to make his discovery by hearing of Brahmegupta's results for the cube of a binomial, for it is believed that Brahmegupta's work had been brought to Baghdad in the eighth century, and Al-Karaji, who worked in Baghdad, drew much else from Hindu sources.[28]

We have already taken note of later work on the Binomial Theorem by Islamic scholars in our earlier mention of al-Khayyāmī and al-Kāshī.

There were also developments in China, to which we next turn our attention.

7. The Chinese Application: Horner's Method

The crediting by European scholars of many results to Europeans can be annoying to non-Europeans who feel they are being robbed of their heritage. The justification can be practical (inaccessibility of primary sources) or theoretical (History of Science is the story of a development, not a mere chronology or list of credits.). Given the influence of Hindu mathematics on Islamic mathematics and of the latter on European mathematics, the developments in India cannot be ignored in the present exposition. It is not known that Chinese binomial developments had any influence outside China, but the story is known and interesting enough that it must be discussed here, even if only as a digression—a side story beginning with the words, "Meanwhile, in the Far East...".

Let me begin this digression with another digression, this one linguistic. It will not have escaped the reader's notice that the spelling of peoples' names has not been uniform among the various sources quoted. One of the major reasons for this is the existence of different transliteration schemes for conversion from

---

[26]Edwards, *op. cit.*, pp. 30 – 32.

[27]The *Al-Bāhir fi'l-Ḥisāb* [*The shining book of calculation*] of al-Samaw'al ben Yaḥyā ben Yahūda al-Maghribī is a commentary on the work of Abū Bakr ibn Muhammand ibn al Ḥusayn al-Karajī, whose own work on the subject is lost to us.

[28]Edwards, *op. cit.*, p. 52.

languages in alphabets other than the Latin. The differences can be small, due to different choices of representatives for specific sounds,

Brahmegupta = Brahmagupta, Bhaskara = Bhascara,

or the suppression of accents,

Âryabhaṭa = Aryabhata, Bhâskara = Bhaskara.

A transliteration scheme can be refined over time,

'Umar al-Khayyāmī = Omar Khayyam.

The biggest difference, however, comes from transliterations into different European languages. At various times in the past few centuries, the dominant languages of European scholarship have been German, French, and English. Some names are more familiar to English readers in their earlier French or German transliterations,

Hammurabi = Hammurapi, Chaikovsky = Tchaikovsky.

The Russian mathematician Chebyshev published in French and the French version of his name, Tchebycheff, is probably as familiar to the English reader as the proper English transliteration.

In the mid-twentieth century, the People's Republic of China introduced *Pinyin*, a scheme for writing Chinese in the Latin alphabet. The results can be wildly different from those produced by the older *Wade-Giles* transliteration scheme. While modern books on Chinese mathematics give the names in Pinyin, older English language reference books, which are still valuable resources, used Wade-Giles and it is not always easy to match names from reference to reference. In TABLE 1 below, I present the names of those Chinese mathematicians we will be discussing here.

TABLE 1. Some Chinese Mathematicians

Pinyin	Wade-Giles	Date
Jiǎ Xiàn	Chia Hsien	*fl. c.* 1050
Yáng Huī	Yang Hui	*fl. c.* 1261 – 1275
Zhū Shìjié	Chu Shih-chieh	*fl.* 1280 – 1303
Qín Jiǔsháo	Ch'in Chiu-shao	*c.* 1202 – *c.* 1261

The earliest extant work in mathematics proper from ancient China is the *Jiǔzhāng suànshù* [*Nine Chapters on the Mathematical Art*]. It is of unknown date and authorship, but is held to be no later than the first century A.D. and is of interest here for its inclusion of methods of square and cube root extraction, methods based on the binomial expansions of $(a+b)^2$ and $(a+b)^3$, respectively. And in the first century of the next millennium Jiǎ Xiàn worked out an improved method, extended it to extracting higher roots, and, indeed, extended the procedure to apply it to finding roots of polynomials.

Jiǎ Xiàn's work, the *Sī shū suǒ suàn* [*The Key to Mathematics*] no longer exists, but Yáng Huī reported on it two centuries later in works whose titles translate to *Reclassification of the Mathematical Methods in the* "*Nine*

Chapters" and *A Detailed Analysis of the Mathematical Methods in the "Nine Chapters"*. In that I opened this section with a comment on robbing a people of their cultural heritage, I cannot resist quoting the following remark on the binomial coefficients:

> Jiǎ Xiàn not only gave the method for finding these coefficients. In Volume 16 344 of the *Great Encyclopaedia of the Yǒng Lè Reign Period*, which was looted during the Boxer Rebellion in 1900 and is now in Cambridge University Library, there is a diagram from Yáng Huī's book *A Detailed Analysis of the Mathematical Methods in the 'Nine Chapters'*... Yáng Huī said that it 'appeared in the mathematical manual *The Key to Mathematics*... Jiǎ Xiàn used this method'; that is to say, this diagram was invented by Jiǎ Xiàn in the eleventh century.[29]

In the caption to the reproduction of this diagram in the third volume of his massive history of Chinese science and technology, Joseph Needham describes it thus:

> The oldest extant representation of the 'Pascal' Triangle in China, from ch. 16,344 of the MS. *Yung-Lo Ta Tien* of +1407 (Cambridge University Library). As the seventh column of characters from the right shows, it was taken by Yang Hui in his *Hsiang Chieh (Chiu Chang Suan Fa Tsuan Lei)* of +1261 from an earlier book, *Shih So Suan Shu*. The text adds that the Triangle was made use of by Chia Hsien (*fl. c.* +1100).[30]

The diagram in question shows the rows of the triangle up to the sixth power, represented as circled numerals. A second famous Chinese graphical representation from about four decades later appeared in the *Precious Mirror of the Four Elements* (c. 1303) of Zhū Shìjié. This diagram presents the coefficients up to the 8th power as circled numerals placed in a triangular lattice. It is more attractive than Yáng Huī's cruder diagram and was even used on a postage stamp issued by Liberia in 2000 in celebration of Chinese achievements on the occasion of the coming of the new millenium. A picture of this stamp appears as FIGURE 8, below.

---

[29]Lǐ Yan and Dù Shíràn, (John N. Crossley and Anthony W.-C. Lun, translators), *Chinese Mathematics; A Concise History*, Oxford University Press, Oxford, 1987, pp. 121 – 122. The classic books on Chinese mathematics are Yoshio Mikami, *The Development of Mathematics in China and Japan*, 2nd. ed., Chelsea Publishing Company, New York, 1974 and Needham, *op. cit.* To this I might add Ulrich Libbrecht, *Chinese Mathematics in the Thirteenth Century*, MIT Press, Cambridge (Mass.), 1973, for its information on the period in question and the mathematician Qín in particular. Another reference, of course, is Joseph, *op. cit.* Mikami, Needham, Libbrecht, and Joseph use Wade-Giles transliteration or some variant. The biographical entries on Lǐ (1892 – 1963) and Mikami (1875 – 1950), pp. 464 – 465 and pp. 484 – 486, respectively, in Dauben and Scriba, *op. cit.*, give more information on the contributions to the history of oriental mathematics of these scholars.

[30]Needham, *op. cit.*, p. 136. The diagram is also reproduced in Lǐ and Dù, *op. cit.*, p. 122.

7. THE CHINESE APPLICATION: HORNER'S METHOD

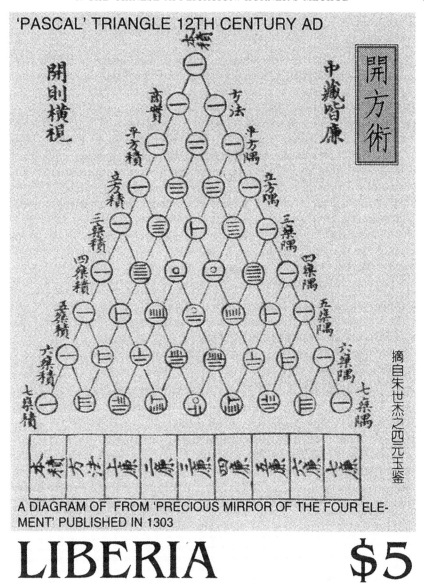

FIGURE 8. The Arithmetical Triangle, c. 1303

There is a curious typographical error in the diagram, as the figures for $_7C_3$ and $_7C_4$, which should both equal 35, read 34 and 35, respectively. The description given by Needham contains another amusing error:

> The 'Pascal' Triangle as depicted in +1303 at the front of Chu Shih-Chieh's *Ssu Yuan Yü Chien*. It is entitled 'The Old Method

Chart of the Seven Multiplying Squares' and tabulates the binomial coefficients up to the sixth power.[31]

Evidently, Needham had the older diagram in mind when he wrote that it tabulated up to the sixth power. A third error connected with this diagram occurs on the Liberian stamp where it places the year 1303 in the 12th century.[32]

By the mid-13th century, Jiǎ Xiàn's method had culminated in a procedure now incorrectly called *Horner's Method* or sometimes the *Ruffini-Horner Method* after William George Horner (1786 – 1837) and Paolo Ruffini (1765 – 1822). The eponymous names initially signify European ignorance of earlier Chinese work and later, perhaps, Eurocentric skepticism In introducing Qín Jiǔsháo's use of the procedure in 1247 in his great work *Shùshū jiǔzhāng* [*Mathematical Treatise in Nine Sections*], Libbrecht discusses the slow acceptance of Chinese priority for the method. According to him the first comparison between the Chinese and Horner's works was in a series of articles, "Jottings on the science of the Chinese: Arithmetic", published in the *North China Herald* by Alexander Wylie, a linguist working in China for the London Missionary Society,[33] in 1852. He quotes Wylie:

> It appears some have thought proper to dispute the right of Horner to the invention, and it will perhaps be an unexpected occurrence to our European friends to find a third competitor coming forward from the Celestial Empire, with a very fair chance of being able to establish his claim to priority.[34]

Libbrecht further explained that Wylie gave only one example, and Karl Leonhard Biernatzki, a protestant minister working in Germany for the Central Association for the China Mission,[35] in his 1856 translation published in the *Journal für die reine und angewandte Mathematik* (aka, *Crelle's Journal*) miscopied the leading coefficient of the example given, causing a change in the procedure. Biernatzki's translation being the only source then available to continental scholars, Moritz Cantor (1824 – 1920), the leading German historian of mathematics,[36] was unconvinced:

> An approximation method for equations of higher degree seems to have existed in which a similarity with the so-called Horner method is thought to be discernible, but which in the text before us has

---

[31]Needham, *op. cit.*, p. 135. Lǐ and Dù, *op. cit.*, p. 127, and Joseph, *op. cit.*, p. 188, also reproduced the diagram. The phrase "seven multiplying squares" meant the 8th power: one square is the second power, two multiplying squares the third, and so forth.

[32]If one were picky, one might also note that "IF FROM" and "ELEMENT" in the description at the bottom should read "FROM" and "ELEMENTS", respectively. The diagram is a jinx.

[33]*Cf.* also the short biographical entry in Dauben and Scriba, *op. cit.*, pp. 569 – 570.

[34]Libbrecht, *op. cit.*, p. 177.

[35]*Cf.* also the short biographical entry in Dauben and Scriba, *op. cit.*, p. 363.

[36]Cantor's work is discussed in the biographical entry in Dauben and Scriba, *op. cit.*, pp. 387 – 391.

7. THE CHINESE APPLICATION: HORNER'S METHOD

been treated too sketchily to permit us to venture either supporting or denying this opinion.³⁷

In favour of the Chinese, Libbrecht quotes Mikami:

> It was Mikami who in 1912³⁸ provided the first full explanation of the procedure followed by Ch'in Chiu-shao for solving the equation $-x^4 + 763,200x^2 - 40,642,560,000 = 0$. He says "...who can deny the fact of Horner's illustrious process being used in China at least nearly six long centuries earlier than in Europe...?³⁹

I quote Mikami at slightly greater length:

> In learning what the Chinese practised in extracting the root of an equation as we have described, we cannot but be reminded of the celebrated method of Horner for the approximate solution of numerical equations. See how identical the two are in their points of principle. If there is anything that differs in the two methods, it should be sought for in the different ways of arrangement, which we owe naturally to the difference in the systems of operation. The Chinese solved their equations on the calculating board, while Horner applied a written mode of calculation. At any rate, who can deny the fact of Horner's illustrious process being used in China at least nearly six long centuries earlier than in Europe, for Horner's paper was published in 1819? We of course don't intend in any way to ascribe Horner's invention to a Chinese origin, but the lapse of time sufficiently makes it not altogether impossible that the Europeans could have known of the Chinese method in a direct or indirect way. Moreover the Europeans had been for more than two centuries in active cooperation with the Chinese to transplant the Occidental sciences into the land of the Middle Empire. Among them there were not a few who were versed in mathematics. Why could they not have come into contact with the method of the Chinese algebra, that lay openly before their eyes?⁴⁰

There were still some doubters. One problem may stem from his admission that

> Although the successive steps are indicated in Ch'in's work in separate diagrams, we shall here explain the process in an abridged form, because it is too tedious to reproduce all the passages followed by the author.⁴¹

Another problem is that a few pages earlier Mikami's book contained another equation with 3 typographical errors. Gino Loria, the dean of Italian historians

---

³⁷Libbrecht, *op. cit.*, p. 177, cites p. 586 of Cantor's four volume *Vorlesungen über die Geschichte der Mathematik*, 1880 – 1908. I found it on p. 685 of the 3rd edition (1907) of volume 1.

³⁸Mikami's book was first published in 1913.

³⁹Libbrecht, *op. cit.*, pp. 177 – 178.

⁴⁰Mikami, *op. cit.*, p. 77.

⁴¹*Ibid.*, p. 75.

of mathematics,[42] apparently did not spot the errors and declared in 1921 that "the result stated by Ch'in, $x = 9$, does not satisfy his equation at all"[43] and he further expressed doubts about the method's coinciding with Horner's:

> But we still have too meager information as to the details of the work for us to be able to affirm confidently that Horner's method was known to the Chinese in the 13th century. We can only say that this method, or one practically identical with it, was known at that time, and we must await further evidence before affirming or denying the priority of the Chinese in its discovery.[44]

The American mathematical historian D.E. Smith, who co-authored a work on Japanese mathematics with Mikami in 1914, and accepted Chinese authorship of Horner's Method in 1912, was sufficiently influenced by Loria to express doubts in 1931.

When Libbrecht published his treatment of the work of Qín Jiǔsháo in 1973, he wrote, "Thus historians of mathematics, having no access to Chinese studies, cannot yet make a general and accurate judgment as to Horner's method in China,"[45] and presented in gory detail every step in the process used by Qín in solving $-x^4 + 763200x^2 - 40642560000 = 0$, leaving no doubt that, although the display of the results differs, the actual computational steps are identical to those of what is commonly called Horner's Method. As he says of the Chinese calculation,

> This method is exactly the same as that described by Horner. Of course, Horner provides a full analysis of his method, perhaps overly intricate and sophisticated; moreover, his field of application is much wider than that of the Chinese mathematicians.[46]

It is now customary, in discussing the Chinese application of Horner's Method, to use Qín's example, and, being loath to break such a fine tradition, I shall do the same. Libbrecht says "this equation is very well suited for a first explanation, because the root is a whole number".[47] It is the first equation of the *Mathematical Treatise in Nine Sections* having more than two terms and of degree higher than 2 to occur in the book, and it is worked out in detail, with no fewer than 21 diagrams included to depict the successive configurations of counting rods representing successive steps in the calculation. I suspect the popularity of this example is thus due to its having been given the most thorough explanation of the procedure in the old Chinese texts.

---

[42] Loria's work is reviewed in the biographical entry in Dauben and Scriba, *op. cit.*, pp. 464 – 474.

[43] Libbrecht, *op. cit.*, p. 178.

[44] *Ibid.*, p. 178.

[45] *Ibid.*, p. 179.

[46] *Ibid.*, p. 191. Libbrecht adds a footnote: "For that reason we do not follow Horner but give preference to the modern representation." I discuss Horner's "overly intricate" presentation in Smoryński, *History...*, *op. cit.*, Chapter 7, and return ever so briefly to it in the appendix.

[47] Libbrecht, *op. cit.*, p. 180.

7. THE CHINESE APPLICATION: HORNER'S METHOD

Today, of course, one always uses modern numerals in presenting the solution, but with that exception one can find the Chinese algorithm given in varying degrees of faithfulness to the original. Libbrecht offers the most faithful discussion, including all 21 diagrams illustrating the numbers on the calculating board and repeating the rules followed taking one from one configuration to the next. He also presents, for comparison, the modern version of Horner's method.[48] Lǐ and Dù combine steps, presenting only 8 diagrams before applying the modern procedure.[49] Joseph only gives diagrams for the first few steps, but offers a nice modern algebraic discussion.[50] Mikami offers a few diagrams and descriptions of the rules as they apply to the given diagrams, and remarks on Horner's Method but does not give the details for comparison.[51]

The central ingredient of the process is explained today as a substitution of variables. If we have an approximation a to a root $a + b$ of a polynomial $P(x)$, and we set $y = x - a$, then $x = y + a$ and b is a root of the polynomial $Q(y) = P(y+a)$. There are several techniques of finding the coefficients of Q. The Chinese used two, one based on the binomial expansion and one based on a form of polynomial division called *synthetic division*.

The binomial approach is fairly straightforward. Suppose we have a polynomial

$$P(x) = a_1 x^n + a_2 x^{n-1} + \ldots + a_n x + a_{n+1}$$

and a number a. Let $y = x - a$, so $x = y + a$ and $Q(y) = P(x)$. To find the coefficients $b_1, b_2, \ldots, b_{n+1}$ of $Q(y)$,

$$Q(y) = b_1 y^n + b_2 y^{n-1} + \ldots + b_n y + b_{n+1},$$

we would make the substitution

$$Q(y) = P(y+a) = a_1(y+a)^n + a_2(y+a)^{n-1} + \ldots + a_n(y+a) + a_{n+1},$$

and expand using the Finite Binomial Theorem:

$$Q(y) =$$

$$a_1 \binom{n}{0} y^n + a_1 \binom{n}{1} a y^{n-1} + \ldots + a_1 \binom{n}{n-1} a^{n-1} y + a_1 \binom{n}{n} a^n$$

$$+ a_2 \binom{n-1}{0} y^{n-1} + \ldots + a_2 \binom{n-1}{n-2} a^{n-2} y + a_2 \binom{n-1}{n-1} a^{n-1}$$

$$\vdots$$

Collecting like terms we see

$$b_1 = a_1 \binom{n}{0}$$

$$b_2 = a_1 \binom{n}{1} a + a_2 \binom{n-1}{0}$$

---

[48] Ibid., pp. 181 – 189.
[49] Lǐ and Dù, op. cit., pp. 131 – 135.
[50] Joseph, op. cit., pp. 200 – 204.
[51] Mikami, op. cit., pp. 74 – 77.

$$b_3 = a_1 \binom{n}{2} a^2 + a_2 \binom{n-1}{1} a + a_3 \binom{n-2}{0}$$

$$\vdots$$

$$b_k = a_1 \binom{n}{k-1} a^{k-1} + a_2 \binom{n-1}{k-2} a^{k-2} + \ldots + a_k \binom{n-k+1}{0}.$$

That is,

$$b_k = \sum_{i=1}^{k} a_i \binom{n-i+1}{k-i} a^{k-i}. \qquad (22)$$

For a more concrete example take Qín's

$$P(x) = -x^4 + 763200x^2 - 40642560000 \qquad (23)$$

for which we have somehow obtained 800 as an estimate of the root. Writing $x = y + 800$ and plugging this into P yields

$$Q(y) = P(y+800) = -(y+800)^4 + 763200(y+800)^2 - 40642560000.$$

We have but to expand the binomial and collect like terms:

$$-y^4 - 4 \cdot y^3 \cdot 800 - 6 \cdot y^2 \cdot 640000 - 4 \cdot y \cdot 512000000 - 409600000000$$
$$763200 \cdot y^2 + 763200 \cdot 2 \cdot y \cdot 800 + 763200 \cdot 640000$$
$$- 40642560000,$$

i.e.,

$$-y^4 - 3200y^3 - 3840000y^2 - 2048000000y - 409600000000$$
$$763200y^2 + 1221120000y + 488448000000$$
$$- 40642560000,$$

which on addition yields

$$Q(y) = -y^4 - 3200y^3 - 3076800y^2 - 826880000y + 38205440000. \qquad (24)$$

The second approach iterates the use of long division. Again assuming $y = x - 800$ we divide P by $x - 800$:

$$P(x) = (-x^3 - 800x^2 + 123200x + 98560000)(x - 800) + 38205440000$$
$$= (-x^3 - 800x^2 + 123200x + 98560000)y + 38205440000.$$

Obviously, one can do the same for the quotient

$$P(x) = ((-x^2 - 1600x - 1156800)(x - 800) - 826880000)y + 38205440000$$
$$= (-x^2 - 1600x - 1156800)y^2 - 826880000y + 38205440000,$$

and so on:

$$= ((-x - 2400)(x - 800) - 3076800)y^2 - 826880000y + 38205440000$$
$$= (-x - 2400)y^3 - 3076800y^2 - 826880000y + 38205440000$$
$$= ((-1)(x - 800) - 3200)y^3 - 3076800y^2 - 826880000y + 38205440000$$

$$= -y^4 - 3200y^3 - 3076800y^2 - 826880000y + 38205440000,$$

which agrees with (24).

The usual method of dividing a polynomial $P(x) = a_0 x^n + a_1 x^{n-1} + \ldots + a_{n-1}x + a_n$ by $x - a$ covers a lot of paper as in FIGURE 9, below, where

$$\begin{array}{r} A_0 x^{n-1} \quad + \quad A_1 x^{n-2} \quad + \ldots \\ \hline x-a \,\overline{)\, a_0 x^n \; + \; a_1 x^{n-1} \; + \; a_2 x^{n-2} \; + \ldots + a_{n-1}x + a_n} \\ A_0 x^n \; - \; aA_0 x^{n-1} \\ \hline (a_1 + aA_0)x^{n-1} \; + \; a_2 x^{n-2} \\ A_1 x^{n-1} \quad - \quad aA_1 x^{n-2} \quad + \\ \hline (a_2 + aA_1)x^{n-2} \; + \\ \vdots \end{array}$$

FIGURE 9.

$A_0, A_1, \ldots, A_n$ are defined by

$$\left. \begin{array}{l} A_0 = a_0 \\ A_1 = a_1 + aA_0 \\ A_2 = a_2 + aA_1 \\ \quad \vdots \\ A_{k+1} = a_{k+1} + aA_k \\ \quad \vdots \end{array} \right\}. \tag{25}$$

Now the simplicity of the recursion allows a simpler representation as in FIGURE 10, below. The first row is given and the rest is generated one column at a time.

$$\begin{array}{c|cccccc} a & a_0 & a_1 & a_2 & \ldots & a_{n-1} & a_n \\ & & aA_0 & aA_1 & \ldots & aA_{n-2} & aA_{n-1} \\ \hline & A_0 & A_1 & A_2 & \ldots & A_{n-1} & A_n \end{array}$$

FIGURE 10.

The bottom element of the first column is simply $A_0 = a_0$. One multiplies it by a and puts the product into the second row of the next column, adds the two elements of the column to get the bottom entry $A_1 = a_1 + aA_0$. One repeats the process.

I have added a vertical bar to the bottom row to separate the coefficients $A_0, A_1, \ldots, A_{n-1}$ of the quotient from the remainder A_n. This simple procedure for dividing $P(x)$ by $x - a$ has come to be called *synthetic division* and is quite convenient. Notice that, since

$$P(x) = (A_0 x^{n-1} + A_1 x^{n-2} + \ldots + A_{n-2}x + A_{n-1})(x - a) + A_n,$$

we have $P(a) = A_n$, and thus have a relatively simple procedure for calculating $P(a)$. For the moment what is of interest, however, is its iteration in FIGURE 11, below, from which we read off $Q(y) = N_0 y^n + \ldots + B_{n-1} y + A_n$. For the

$$
\begin{array}{c|ccccc}
a & a_0 & a_1 & a_2 & \ldots & a_{n-1} & a_n \\
  &   & aA_0 & aA_1 & \ldots & aA_{n-2} & aA_{n-1} \\
\hline
  & A_0 & A_1 & A_2 & \ldots & A_{n-1} & A_n \\
  &   & aB_0 & aB_1 & \ldots & aB_{n-2} & \\
\hline
  & B_0 & B_1 & B_2 & \ldots & B_{n-1} & \\
  & \vdots & & & & & \\
\hline
  & N_0 & & & & & \\
\end{array}
$$

FIGURE 11.

example at hand, the process goes as in FIGURE 12, below, and the remain-

$$
\begin{array}{c|cccccc}
800 & -1 & 0 & 763200 & 0 & -40642560000 \\
 & & -800 & -640000 & 98560000 & 78848000000 \\
\hline
 & -1 & -800 & 123200 & 98560000 & 38205440000 \\
 & & -800 & -1280000 & -925440000 & \\
\hline
 & -1 & -1600 & -1156800 & -826880000 & \\
 & & -800 & -1920000 & & \\
\hline
 & -1 & -2400 & -3076800 & & \\
 & & -800 & & & \\
\hline
 & -1 & -3200 & & & \\
\hline
 & -1 & & & & \\
\end{array}
$$

FIGURE 12.

ders $-1, -3200, -3076800, -826880000, 38205440000$ are just the coefficients of $Q(y)$.

If one consults Libbrecht, one will find the numbers of the first, third, fifth, seventh, and ninth rows represented in Qín's diagrams and the instructions outlining just these steps.

1.7.1. Exercise. *Apply this procedure to the polynomial $P(x) = x^5$ and a. Circle the entries directly below the horizontal lines. Is there a pattern to the coefficients?*

These are not the only methods of performing the substitution. Isaac Newton, whose own method of finding roots is now presented geometrically, originally appealed to substitution. About the method of substitution he wrote

7. THE CHINESE APPLICATION: HORNER'S METHOD

What hard work there is here will be found in substituting one group of quantities for another. That you might accomplish by various methods, but I think the following way exceedingly expedite, especially when the numerical coefficients consist of several figures. Suppose $p + 3$ has to be substituted for y in this equation

$$y^4 - 4y^3 + 5y^2 - 12y + 17 = 0:$$

since this may be resolved into the form

$$\overline{\overline{\overline{y - 4} \times y : +5 \times y} : -12 \times y} : +17, \tag{26}$$

the new equation may be generated thus:

$(p - 1)(p + 3) = p^2 + 2p - 3;\ (p^2 + 2p + 2)(p + 3) = p^3 + 5p^2 + 8p + 6;$
$(p^3 + 5p^2 + 8p - 6)(p + 3) = p^4 + 8p^3 + 23p^2 + 18p - 18;$ and
$p^4 + 8p^3 + 23p^2 + 18p - 1 = 0,$

which was required.[52]

The strange-looking formula (26), in more modern presentation, reads

$$(((y - 4)y + 5)y - 12)y + 17$$

and the next displayed batch of formulæ is the sequence of multiplications, from the inside out, performed when y is replaced by $p + 3$:

$$(y - 4)y = (p + 3 - 4)(p + 3) = (p - 1)(p + 3) = p^2 + 2p - 3$$
$$(p^2 + 2p - 3 + 5)(p + 3) = (p^2 + 2p + 2)(p + 3) = \text{etc.}$$

Newton's approach is comparable to that of iterating synthetic division in that each step involves multiplication by a ($= 3$ in Newton's example) and an addition. If one compares the procedures side-by-side, one sees that, with the exception of one extra multiplication of 1 by a in the synthetic division approach, the multiplications performed are identical and all the additions are the same. (*Cf.* TABLE 2, below.) Essentially, the only difference is the order in which the operations are performed.

Horner used a fourth method of expanding $P(x)$ into a Taylor series around 800:

$$Q(y) = P(x) = P(800) + P'(800)(x - 800) + \frac{P''(800)}{2!}(x - 800)^2$$
$$+ \frac{P'''(800)}{3!}(x - 800)^3 + \frac{P''''(800)}{4!}(x - 800)^4$$

---

[52] Jean-Luc Chabert (ed.), *A History of Algorithms; From the Pebble to the Microchip*, Springer-Verlag, Heidelberg, 1999, p. 174. I have taken the liberty of displaying and labelling one expression (26) for the purpose of referring to it in discussion. Chabert's book, incidentally, is a source book with many translated extracts from the literature dealing with algorithmic calculation. Of particular relevance here are Chapter 6 devoted to Newton's Method and Chapter 7 on other methods of finding roots, including sections on "Mediaeval Binomial Algorithms" and "The Ruffini-Budan Schema". (François Budan independently discovered a special case of the Ruffini-Horner Method.)

1. THE FINITE BINOMIAL THEOREM

TABLE 2. Comparing the Procedures

```
Synthetic Division                         Newton
3 | 1  -4   5  -12   17                    p  -  4
       3  -3    6  -18                        +  3
    1  -1   2   -6 | -1                    p  -  1
       3   6   24                       ×  p  +  3
    1   2   8 | 18                      p² -  p
       3  15                              3p -  3
    1   5 | 23                          p² + 2p -  3
       3                                         +  5
    1   8                                p² + 2p +  2
                                      ×  p  +  3
    1                              p³ + 2p² + 2p
                                     3p² + 6p + 6
                                  p³ + 5p² + 8p + 6
                                              -  12
                                              ⋮
```

$$= P(800) + P'(800)y + \frac{P''(800)}{2!}y^2 + \frac{P'''(800)}{3!}y^3 + \frac{P''''(800)}{4!}y^4.$$

The reader already familiar with the Calculus is invited to verify that this will again yield (24). For the rest, I will do so at the end of Chapter 2, section 9, Example 2.9.4 on page 128, after discussing Taylor's Theorem.

Today, all the hard work has been done by various programmers and we can let machines do the work for us. On the *TI-89 Titanium*, one enters successively

    `⁻x^4+763200x^2−40642560000→p(x)`

and

    `expand(p(y+800))`,

following which

    `⁻y^4−3200·y^3−3076800·y^2−826880000·y+38205440000`

will magically appear on the screen.

All this is very nice, but one may be wondering what this has done for us regarding the problem of finding a root of the polynomial (23). We know that 800 is our first approximation to a root and our next approximation will be obtained by adding to 800 the approximation to a root of (24). This is the second mysterious part of the procedure,[53] and most authors do not explain it. According to Libbrecht[54] one divides the constant term 38205440000 by the coefficient −826880000 of the linear term and changes the sign of the quotient,

---

[53]The first is how Qín arrived at 800 in the first place.
[54]Libbrecht, *op. cit.*, p. 184, instructions to Diagram 40.

7. THE CHINESE APPLICATION: HORNER'S METHOD

obtaining 46.2... With the Chinese, the procedure was a one-digit-at-a-time affair, so Qín chose 40, ignoring the excess. He then applied synthetic division using Q and 40, as in FIGURE 13, below. We thus see that 40 is a root y of Q,

```
40│−1  −3200   −3076800    −826880000    38205440000
  │     −40    −129600    −1282560000   −38205440000
  ├────────────────────────────────────────────────
   −1  −3240  −3206400    −955136000│         0
```

FIGURE 13.

whence $x = a + y = 800 + 40 = 840$ is a root of the original polynomial.

Had we taken 46 as an approximate root to Q, the ratio of the constant term to the linear coefficient of the new polynomial would have been 5.9... and we could have rounded this to 6 and changed the sign to get -6, which would have turned out to be the root of the new version of Q, yielding $800 + 46 - 6 = 840$ as the root of P.

Why is the negative of the ratio of the constant to the linear term a good choice for an approximation to the root of the next polynomial? This was answered first in 1829 by Augustin Louis Cauchy who gave conditions under which a sequence of such successive approximations converged to a root.[55]

There is a *caveat* to this: Unless the estimate a is very close to the root, the negative ratio may still be too large. In that case, one tries the next lower integer with one significant digit in the right place. For example, to solve $x^2 - 224 = 0$ the calculation would proceed as in TABLE 3, below. The greatest

TABLE 3.

a	resulting polynomial	ratio	b	$a+b$
10	$y^2 + 20y - 124$	6.2...	6	too large
			5	too large
			4	14
4	$z^2 + 28z - 28$	1.0	1	too large
			.9	14.9
.9	$w^2 + 29.8w - 1.99$	.066...	.06	14.96
.06	$v^2 + 29.92v - .1984$	.0066	.006	14.966

integer in the negative ratio may thus not equal the digit sought, but it can narrow down the search so that one needn't test all 10 digits, as we will do in TABLE 4, below.

---

[55]The procedure in question is the specialisation to polynomials of a more general procedure due to Newton. If $Q(x) = a_0 x^n + a_1 x^{n-1} + \ldots + a_{n-1} x + a_n$, and a is an approximation to the root that is close enough, the next Newtonian approximation to the root turns out to be $b = a - a_n/a_{n-1}$. Since $a = 800$ is an approximation to the root of $P(x)$ and $y = x - 800$, the root of Q is being approximated by $b = a - 800 = 800 - 800 = 0$. Thus the next approximation to a root of $Q(y)$ is $b = a - a_n/a_{n-1} = 0 - a_n/a_{n-1} = -a_n/a_{n-1}$.

This does not yet explain the choice of 800 for Qín's initial approximation. One might note that P is essentially a quadratic polynomial in the variable x^2, making 763200 the coefficient of what is essentially a linear term and yielding

$$-\frac{-40642560000}{763200} \approx 53232.83019$$

as an approximation to the root of

$$R(z) = -z^2 + 763200z - 40642560000,$$

and $\sqrt{53252.83019} \approx 230.7657474$ should thus be an approximation to a root of $P(x)$, which indeed it is: P has four roots: ± 240 and ± 840. The Chinese used negative numbers in their calculations but only looked for positive roots, which would be 240 and 840 in this case. Qín found the larger root starting with 800 as his initial estimate. The 230^+ suggests 200 as an initial approximation and the procedure will quickly yield 240 as a root.

But how did Qín come up with 800 as an estimate? To me the most obvious answer[56] to this question is given by counting the digits. The integer part of any solution to (23) must have three digits. If x had only two digits, x^2 could have at most 4 digits ($99^2 = 9801$) and $763200x^2$ at most 10 ($763200 \cdot 99^2 = 7480123200$). And x^4 can have at most 8 digits, whence the difference between $763200x^2$ and x^4 is at most a 10 digit number, while -40642560000 has 11 digits. And the smallest four digit number is 1000 and $1000^2 = 1000000 > 763200$, so any number x with four or more digits will result in $-x^4 + 763200x^2$ being negative. So a zero should lie somewhere between 100 and 1000. To find the first digit of a root of $P(x)$ one would make a simple table as in TABLE 4, below. And the sign changes between 200 and 300 and between 800 and 900

TABLE 4.

x	100	200	300	400
$P(x)$	$-3.3\ldots \times 10^{10}$	$-1.1\ldots \times 10^{10}$	$1.9\ldots \times 10^{10}$	$5.5\ldots \times 10^{10}$
x	500	600	700	800
$P(x)$	$8.7\ldots \times 10^{10}$	$1.0\ldots \times 10^{11}$	$9.3\ldots \times 10^{10}$	$3.8\ldots \times 10^{10}$
x	900	1000		
$P(x)$	$-7.8\ldots \times 10^{10}$	$-2.7\ldots \times 10^{11}$		

tell us there are two roots between 100 and 1000, one with leading digit 2 and one with leading digit 8. Thus, our initial estimate of the root would be 200 or 800. If one started filling in this table going from right to left, after calculating $P(1000), P(900), P(800)$, one would stop and take 800 as one's initial estimate.

---

[56] Well, the most obvious answer really is that he derived his equation from the roots and, knowing 840 in advance started with 800 and proceeded from there. So I guess the proper question should be: How *would* Qín have come up with 800 as an estimate had he not done so?

One could do the same, of course, with $Q(y)$. We know that $y < 100$ and could tabulate the values $Q(0), Q(10), \ldots, Q(100)$, looking for a sign change or, less likely, but what happens to be the case in this example, the root 40 itself.

In an algebra course in which synthetic division is taught one might be given the initial digit and the number of digits ("Tom has a polynomial $P(x) = \ldots$ for which he knows the root lies between 800 and 900. Find, to the nearest whole number, the root of the polynomial.") and be asked only to continue from there by making such tables and looking for changes of sign.

At first sight this procedure seems little or no better than the bisection method discussed earlier in this chapter. With synthetic division, however, it is not too painful for hand computation: each multiplication in each evaluation is by essentially a single digit number. And on the calculator the use of lists makes calculating the table a snap. I calculated TABLE 4 by entering $P(x)$ in Y_1 on the *TI-83 Plus* (or y1 on the *TI-89 Titanium*) and applied it to the list $\{100, 200, 300, 400, 500, 600, 700, 800, 900, 1000\}$. Somehow Qín knew how to bypass the tables in successive steps by zeroing in on the location of the root more quickly. Ignoring all but the first digit gave him the next digit of the root. Later mathematicians, armed with analyses of the rapidity of convergence, improved on the method by estimating how many digits their zeroing in rules secured. Their procedures, however, involved multiplication by multidigit numbers and no one today would carry them out by hand.

8. Pascal at Long Last

I suppose not too many years ago a treatise on the Binomial Theorem would have begun with Blaise Pascal (1623 – 1662) in the mistaken belief that he was its and the Arithmetical Triangle's discoverer. He was not: We have seen that the Chinese and Moslem scholars knew both and that the Arithmetical Triangle was known in India as well centuries earlier. Both were also known and published in Europe before Pascal made his contribution. A.W.F. Edwards documents this nicely in chapters 4 ("The combinatorial numbers in the West") and 5 ("The binomial numbers") of his history of the Arithmetical Triangle. With respect to the Arithmetical Triangle, he cites Abraham ben Meir ibn Ezra (*c.* 1090 – *c.* 1164 – 1167) who around 1140 studied the numbers of combinations of 7 things taken r at a time, adding that

> The methodical way in which Rabbi ben Ezra arrived at the combinatorial numbers for $n = 7$ leaves little doubt that he could have handled other numbers with equal facility, but did not give a general formulation.[57]

Edwards continues with Levi ben Gerson (1288 – 1344):

> Then in 1321 another Jew, Levi ben Gerson, who lived in France, wrote on permutations and combinations, and gave three of the principal Hindu rules: $n!$ as the number of arrangements of n things; $n(n-1)(n-2)\ldots(n-r+1)$ for the number of arrangements of n

---

[57]Edwards, *op. cit.*, p. 34.

things taken r at a time; and, for the number of combinations of n things taken r at a time,

$$^nC_r = \frac{n(n-1)(n-2)\ldots(n-r+1)}{1.2.3\ldots r}.$$

He commented on the fact that $^nC_r = {^nC_{n-r}}$. His proofs constitute the first known examples of explicit mathematical induction.[58]

Edwards points out that it is not unreasonable to assume "that Hebrew writers had acquired some knowledge of the Hindu combinatorial rules"[59]. Indeed, Jewish scholars having thrived in the Moslem world from Persia to Spain (where Rabbi ben Ezra lived), it would be surprising if they hadn't acquired some of the Hindu knowledge that had been transmitted to the Arab world.

However, in the West the principal spur to combinatorial enumeration in the Middle Ages was gaming with dice.[60]

Two of the most influential mathematicians of 16th century Europe, Niccolò Tartaglia (1499 or 1500 – 1557) and Girolamo Cardano (1501 – 1576) had a keen interest in gambling and wrote about combinatorics. In 1556 Tartaglia published his *General trattato di numeri et misure* in which one finds the Arithmetical Triangle presented in rectangular form in connexion with the discussion of the number of ways dice can fall.[61] This was in Book 1 of the *General trattato*; in Book 2 he presents the Arithmetical Triangle in triangular form in connexion with the Binomial Theorem.[62] Edwards informs us

> We need not be surprised that Tartaglia does not comment on the fact that the first four rows of Fig. 14[63] give the figurate numbers...; he must have known it because his rule of formation is exactly the same as that of Nicomachus, and he had described the triangular numbers at the beginnings of Book 1 of the second part of the *General trattato*. His habit of not giving cross-references even extends to omitting any comment on the fact that the numbers contained in the Binomial Triangle he gives in Book 2 (see Fig. 20, page 53)[64] are exactly those given in Book 1 in connexion with enumerating the throws of dice. It is impossible to believe he was not aware of the identity.[65]

---

[58]*Ibid.*, pp. 34 –35.

[59]*Ibid.*, p. 36.

[60]*Ibid.*

[61]Edwards, *op. cit.*, p. 39, reproduces the page in facsimile. The diagram is similar to Pascal's (FIGURE 16, below). Tartaglia fills in 8 rows and 7 columns, including all the cells of the table and not just those lying in complete diagonals.

[62]Edwards, *op. cit.*, p. 53, for a facsimile reproduction. But for his leaving out the outer 1's (i.e., the coefficients $\binom{n}{0}$ and $\binom{n}{n}$), this is the modern layout.

[63]The facsimile reproduction cited in footnote 61, above. The rows referred to correspond to the diagonals from left to right of the Arithmetical Triangle as given in FIGURE 7 on page 24, above.

[64]The facsimile cited in footnote 62.

[65]Edwards, *op. cit.*, p. 43.

8. PASCAL AT LONG LAST

Cardano's *Opus novum* of 1570 again discusses the numbers of combinations of n things taken r at a time, giving the Horizontal Recursion, and presenting the Arithmetical Triangle in triangular form.[66] As Edwards notes, Cardano pointed out that the figurate numbers are the same as the combinatorial numbers,

> that is, the same relation described for the figurate numbers, which we noticed in the *Trigonometria Britannica* of Briggs... Quite possibly Briggs obtained it from this account of Cardano's, even though he was not interested in using the figurate numbers for combinatorial purposes. But whereas Briggs explicitly connected the figurate and binomial numbers, Cardano did not, so we can only *infer* that the latter knew the binomial theorem for positive integral exponents.[67]

We are getting nearer to Pascal, but first I must digress to discuss *figurate numbers*, something I have studiously avoided doing thus far in this book, though the phrase has come up. The reason is that I find them singularly uninteresting. The figurate numbers are the numbers of dots that can be arranged in specific geometric shapes. One starts with the triangular numbers as in FIGURE 14 and then the square numbers as in FIGURE 15. There are also

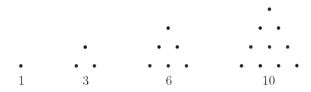

FIGURE 14. Triangular Numbers

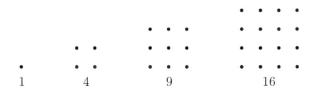

FIGURE 15. Square Numbers

pentagonal numbers, hexagonal numbers, etc.

Figurate numbers date back to the Greeks, most notably to Nicomachus of Gerasa (*fl. c.* 100) whose *Introduction to Arithmetic* was widely used from

---

[66] A facsimile reproduction appears on p. 44 of Edwards, *op. cit.* Cardano's version differs from FIGURE 16 below only in omitting the first column and including two additional diagonals.

[67] Edwards, *op. cit.*, p. 45.

his day until its replacement by a Latin version of Boethius during the Middle Ages. Such was its fame that it was included with the more substantial works of Euclid, Archimedes, and Apollonius of Perga in the volume on Greek mathematics in the series of Great Books of the Western World. In the 16th century they were being taken up again in Europe. Indeed, in 1523 Tartaglia had already derived his table of figurate numbers and developed some of their properties although he didn't publish this work until 1556. By then, however, Peter Apian (1495 – 1552) had published the Arithmetical Triangle on the title page of his arithmetic book of 1527, and Michael Stifel (*c.* 1487 – 1567) had published two works, *Arithmetica integra* (1544) and *Deutsche Arithmetica* (1545) in which he gave the first moderately extensive European publication of the Arithmetical Triangle. This was a table of figurate numbers but he used it to determine the binomial numbers in the extraction of roots, prompting Edwards to write, "If the identity of the binomial numbers and the figurate numbers... may be said to have had a discoverer in the West, he was surely Stifel".[68]

And on it goes. Between Stifel and Pascal, Edwards's account interpolates Henry Briggs, Thomas Harriot[69], and others, most significant among whom as relates to Pascal being Marin Mersenne (1588 – 1648). Mersenne's *Harmonicorum libri XII* published in Latin in 1636 and in French as *Harmonie universelle* in two volumes in 1636 and 1637 contained much information on the theory of combinations and a variant of the Arithmetical Triangle.[70] Of these books Edwards reports

> With these works the theory of combinations assumes its modern form. Not only did Mersenne bring together all the results then known, but through these volumes..., and no doubt through his numerous correspondents as well, he broadcast them throughout Europe.[71]

Of Mersenne's influence on Pascal, Edwards begins with Pascal's father Etienne Pascal:

> In 1635 Etienne was one of the founders of Marin Mersenne's "Academy", the finest exchange of mathematical information in Europe at the time. To this informal Academy he introduced his son at the age of fourteen...

> Thus the young and impressionable Pascal was introduced to the circle of mathematicians just as Mersenne's books containing "Pascal's Triangle" and its combinatorial applications were coming from the presses..., and he continued to visit Mersenne even when living out of Paris. Moreover, Etienne Pascal was particularly noted for

---

[68]*Ibid.*, p. 7.

[69]Harriot serves as an example of legitimate variations in the spelling of proper names due, not as on page 37 to varying transliteration schemes, but to the lack of a conventional orthography. Spelling was not as fixed in the 16th century as it is today. Thus one finds Hariot, Harriot, Harriott, and even Harryot.

[70]Facsimile reproduction on p. 46 of Edwards, *op. cit.*

[71]Edwards, *op. cit.*, p. 47.

his knowledge of music, to which Mersenne paid tribute in both the *Harmonicorum libri XII* and the *Harmonie universelle*, having already dedicated his treatise on the organ to him. It is unthinkable that these works were not in Etienne's house from the moment they were published, and we need look no further for the young Pascal's principal source. Yet there is a tendency amongst historians of science to view Pascal's *Traité* as a work of inspiration rather than consolidation.[72]

I believe it is safe to agree that Pascal's contribution was not original discovery, independent rediscovery, or even initial consolidation. One can argue that, even restricting one's attention to specifically European developments, Pascal's importance for the Binomial Theorem has been greatly overstated in the literature. Edwards cites two reasons for this, one being the fact that "Pascal was... a little forgetful about his sources"[73] and the other a traditional "reliance that has been placed on Isaac Todhunter's influential *History of the Mathematical Theory of Probability* published in 1865"[74]. With regard to this latter, Edwards adds, "In defence of Todhunter's neglect of the early writers it may be said that the rest of the title is '...*from the time of Pascal to that of Laplace*', which not all of his followers have noticed".[75]

Failure to cite one's sources was a habit Todhunter's teacher Augustus de Morgan found particularly prominent among the French (and particularly unnecessary by one as original as Laplace). Pascal's work, the *Traité du triangle arithmétique, avec quelques autres petits traités sur la même matière* [*Treatise on the Arithmetical Triangle with other small tracts on the same matter*], written in 1654 and published in 1665, is, however, a more important and more original work than those cited by Stifel, Tartaglia, *et al.* in that it is one of the founding works on the Theory of Probability and will have been read by more mathematicians and historians of mathematics and science than these other works. And the *Traité* has been more accessible.[76] English language readers can for example find a translation of sizable excerpts in the well-known source books of Smith[77] and Struik[78], and the contents of the *Traité* are discussed in detail in chapters 6 and 7 of Edwards's book on which I lazily rely for most of the information of this discussion.

So, one may ask, why so much emphasis on Pascal? The historians answer. First, Smith:

> Although Pascal... was not the originator of the arithmetic triangle, such an arrangement of numbers having been anticipated, his name

---

[72]*Ibid.*, pp. 57 – 58.

[73]*Ibid.*, p. 58.

[74]*Ibid.*, p. 47.

[75]*Ibid.*, p. 48.

[76]Thanks to digitisation and the internet, more and more works are easily accessible.

[77]David Eugene Smith (ed.), *A Source Book in Mathematics*, 1929; reprinted: Dover Publications, Inc., New York, 1959.

[78]Dirk J. Struik (ed.), *A Source Book in Mathematics, 1200 – 1800*, Harvard University Press, Cambridge (Mass.), 1969.

56   1. THE FINITE BINOMIAL THEOREM

has been linked with the triangle by his development of its properties, and by the applications which he made of these properties. The historical interest of the work is to be found, perhaps, in its bearing on probability discussions and on the early development of the binomial theorem.[79]

And the famous French historian René Taton gives a more detailed explanation in the *Dictionary of Scientific Biography*:

> Although the principle of the arithmetical triangle was already known,[80] Pascal was the first to make a comprehensive study of it. He derived from it the greatest number of applications, the most important and original of which are related to combinatorial analysis and especially to the study of the problems of stakes[81]. Yet it is impossible to appreciate Pascal's contribution if it is considered solely from the perspective of combinatorial analysis and the calculus of probability. Several modern authors have shown that Pascal's letters to Fermat and the *Traité du triangle arithmétique* can be fully understood only when they are seen as preliminary steps toward a theory of decision.[82]

If Pascal did not originate either the Arithmetical Triangle or the Binomial Theorem, he at least provided a milestone in their history. In giving a systematic survey of the Triangle's properties, he marks the end of an era, and his discussion of the combinatorial coefficients in connexion with probability, together with his more purely mathematical approach, ushers in a new era of the development of binomial matters.

For comparison, I started looking up some of those authors cited by Taton in his footnote referred to in footnote 80, below. The books by Jacques Peletier and Jean Trenchant can be found online. These are commercial arithmetic books teaching the basic arithmetic operations—addition, subtraction, multiplication, and division—and eventually techniques of root extraction, printing versions of the Arithmetical Triangle for computational purposes. Oughtred's

---

[79]Smith, *op. cit.*, p. 67.

[80]Taton adds a footnote: "This figure, in more or less elaborated forms that were equivalent to lists of coefficients of the binomial theorem, appeared as early as the Middle Ages in the works of Naṣir al-Din al Tūsī (1265) and Chu Shih-chieh (1303). The arithmetical triangle reappeared in the sixteenth and seventeenth centuries in the writings of Apian (1527), Stifel, Scheubel, Tartaglia, Bombelli, Peletier, Trenchant, and Oughtred. But Pascal was the first to devote to it a systematic study linked to many questions of arithmetic and combinatorial analysis." (vol. 10, p. 338)

[81]Popularly called the "problem of points" it had been investigated without success by a number of mathematicians, most famously Luca Pacioli, Tartaglia, and Cardano, before Pascal's successful solution and his correspondence with Pierre de Fermat on the subject. This correspondence is taken, with some justification, as the starting point of the mathematical Theory of Probability.

[82]Charles Coulston Gillispie (ed.), *Dictionary of Scientific Biography*, Charles Scribner's Sons, New York, 1970 – 1980, vol. 10, p. 335.

works, *Arithmeticæ in numeris et speciebus institutio* (1631) and the later edition *Clavis mathematicæ* (1667), are more algebraic but introduce the Arithmetical Triangle in discussing the arithmetic operations. Pascal's approach is completely different. He does not introduce the Arithmetical Triangle by reference to a particular application, but introduces a minor generalisation of the binomial coefficients by their defining recursion, then proceeds to derive various identities involving them with again no application in sight. Only after he has finished his exposition of the triangle itself does he consider applications in additional tracts appended to the main one.

Pascal begins his treatment by giving rules for filling in the entries of a triangular table. A number he calls the *generator* is placed in the upper-leftmost cell of the table and the number put into every other cell is the sum of the numbers in the cells immediately above and to the left of the given cell. When there is no cell above a given cell or to the left of it, 0 is assumed for the missing value. Starting with the generator 1, he presents the table, which is reproduced minus some extra lines and labels in FIGURE 16, below. Rotating

1	1	1	1	1	1	1	1	1	1
1	2	3	4	5	6	7	8	9	
1	3	6	10	15	21	28	36		
1	4	10	20	35	56	84			
1	5	15	35	70	126				
1	6	21	56	126					
1	7	28	84						
1	8	36							
1	9								
1									

FIGURE 16. Pascal's Arithmetical Triangle

the diagram 45° in a clockwise direction we see the Arithmetical Triangle in its now traditional representation.

Letting P^i_j denote the element of the i-th row and j-th column for integers $i, j \geq 1$, we see quickly that

$$P^i_j = \binom{i+j-2}{j-1},$$

[or, if we number the rows and columns starting with 0 instead of 1,
$$P^i_j = \binom{i+j}{j}.$$
]

Pascal does not use this notation, but introduces some no longer used terminology. The (horizontal) rows are called *parallel ranks*, their respective cells being *cells of the same parallel rank*. The (vertical) columns are called *perpendicular ranks* and their respective cells are *cells of the same perpendicular rank*. And the diagonals travelling upward from an entry of the leftmost column are called *bases*, their respective entries *cells of the same base*. The bases are just the rows of the rotated Arithmetical Triangle, I.e., the numbers

$$\binom{n}{0}, \binom{n}{1}, \ldots, \binom{n}{n}$$

for $n = 0, 1, \ldots$

Having defined his triangle and these terms, Pascal next sets out to derive a number of identities involving the numbers P^i_j, all stated in his archaic terminology. These are cited as "corollaries".

Pascal's first corollary is stated thus:

> *Corollary* 1.—In every arithmetic triangle, all the cells of the first parallel rank and of the first perpendicular rank are equal to the generator.[83]

Less prosaically, for all $n \geq 1$,
$$P^1_n = P^n_1 = 1, \tag{27}$$
if we take the generator to be 1. In terms of binomial coefficients this reads
$$\binom{n}{0} = \binom{n}{n} = 1. \tag{28}$$

Of the instances of (28), Pascal's construction only assumes the first, $\binom{0}{0} = 1$, i.e., the upper-leftmost cell contains the generator.

To understand Pascal's proof of this corollary, we must note that the labels not reproduced in FIGURE 16 for the cells of the first parallel rank in his diagram are G (doing double duty as the generator and the name of the cell it is initially placed into), $\sigma, \pi, \lambda, \mu$, etc., and those of the cells of the first perpendicular rank are G, ϕ, A, D, etc. Given this, his explanation of the validity of his corollary reads as follows:

> For, by the construction of the triangle, each cell is equal to that of the cell which precedes it in its perpendicular rank, added to that which precedes it in its parallel rank. Now the cells of the fist parallel rank have no cells which precede them in their perpendicular ranks, nor those of the first perpendicular rank in their parallel ranks; consequently they are all equal to each other and thus equal to the generating first number.
>
> Thus ϕ equals G + zero, that is, ϕ equals G.

---

[83]Smith, *op. cit.*, p. 69.

Likewise A equals ϕ + zero, that is, ϕ.

Likewise σ equals G + zero, and π equals σ + zero.

And likewise for the others.[84]

Pascal's rationale seems to be that, if there is no cell above or to the left of a given cell, one is adding *nothing*, i.e., 0. Today we would not accept this and either conclude that the value sought is *undefined* or that there has been a hidden assumption of the initial values

$$P_n^0 = P_0^n = 0, \text{ i.e., } \binom{n}{-1} = \binom{-1}{n} = 0. \tag{29}$$

This is less elegant than simply assuming Corollary 1 outright as the proper basis for the recursion—as we did in the statement of the Finite Binomial Theorem (Theorem 1.5.1 on page 23, above). On the other hand, today we might think in terms of a spreadsheet in which we start our table somewhere off the first row and first column by placing a generator in the upper left-hand corner of the table and pasting the additive instruction in the other cells, and we leave the row above and the column to the left untouched. They will by default contain 0's and Pascal's reasoning correctly tells us that the first row and column of the table are filled with copies of the generator.

All that said, his informal proof does not meet modern standards of exposition. Today we would be more formal, giving notation for the numbers P_j^i, stating formally that one was going to prove (27), perhaps stating loosely the idea given in the first paragraph of Pascal's proof, and then, the induction being trivial, probably not giving the full ritual proof by induction, but stating more explicitly than in the last 4 lines of Pascal's proof that it is an induction. For, as becomes clear when he gets to Corollary 12, Pascal, though he does not use the phrase "mathematical induction", was fully aware of the principle and knew he was using it here.

Correction: Today we would only proceed as above if we were discussing the entries in a spread sheet. Mathematically, we wouldn't prove Corollary 1, but would assume it as part of the recursive definition.

With one or two exceptions, the rest of Pascal's initial tract does not merit such detailed treatment. It concerns the derivation of a number of identities involving the binomial coefficients and are best presented here, in modern notation, as an enumeration of simple exercises:

Corollary 2. $P_{j+1}^{i+1} = P_1^i + P_2^i + \ldots + P_{j+1}^i$, i.e.,

$$\binom{i+j}{j} = \binom{i-1}{0} + \binom{i}{1} + \ldots + \binom{i+j-1}{j}.$$

Corollary 3. $P_{j+1}^{i+1} = P_j^1 + P_j^2 + \ldots + P_j^{i+1}$, i.e.,

$$\binom{i+j}{j} = \binom{j}{j-1} + \binom{j+1}{j-1} + \ldots + \binom{i+j}{j-1}.$$

Corollary 4. $P_{j+1}^{i+1} - 1 = \sum_{k=1}^{i} \sum_{l=1}^{j} P_l^k$.

---

[84] *Ibid.*

Corollary 5. $P^{i+1}_{j+1} = P^{j+1}_{i+1}$, i.e.,

Corollary 6. $\binom{i+j}{i} = \binom{i+j}{j}$, or $\binom{n}{r} = \binom{n}{n-r}$.

Corollary 7. $\displaystyle\sum_{i=0}^{n+1} \binom{n+1}{i} = 2\sum_{i=0}^{n} \binom{n}{i}$.

Corollary 8. $\displaystyle\sum_{k=0}^{n} \binom{n}{i} = 2^n$.

Corollary 9. $\displaystyle\sum_{i=0}^{n+1} \binom{n}{i} - 1 = \sum_{k=0}^{n}\sum_{i=0}^{k} \binom{k}{i}$.

Corollary 10. $\displaystyle\sum_{i=0}^{k+1} \binom{n+1}{i} = \sum_{i=0}^{k+1}\binom{n}{i} + \sum_{i=0}^{k}\binom{n}{i}$.

Corollary 11. $\binom{2n+1}{n+1} = 2\binom{2n}{n}$ for $n \geq 1$.

The corollaries thus far given are not particularly interesting in themselves. But they do make nice little exercises which the more energetic reader might enjoy playing with. At this point Pascal changes gears. First he adds a note:

> NOTE.—All these corollaries are on the subject of the equalities which are encountered in the arithmetic triangle. Now we shall consider those relating to proportions; and for these, the following proposition is fundamental.[85]

The "following proposition" is the recursive portion of the Horizontal Recursion:

Corollary 12. $\binom{n}{k+1} \Big/ \binom{n}{k} = \dfrac{n-k}{k+1}$.

Corollary 12 is an important but far from novel result. The proof is what is of interest here. It is an induction, again not novel in itself: the proof of Corollary 1 was an induction. What is novel here and of historical interest is his statement of the Principle of Mathematical Induction and his not yet fully ritualised inductive proof.

> Although this proposition has an infinite number of cases, I will give a rather short demonstration, assuming two lemmas.
>
> *Lemma 1:* which is self-evident, that this proportion is met with in the second base; for it is apparent that ϕ is to σ as 1 is to 1.
>
> *Lemma 2:* that if this proportion is found in any base, it will necessarily be found in the following base.
>
> From which it will be seen that this proportion is necessarily in all the bases: for it is in the second base by the first lemma; hence by

---

[85] *Ibid.*, p. 72.

8. PASCAL AT LONG LAST

the second, it is in the third base, hence in the fourth, and so on to infinity.

It is then necessary only to prove the second lemma in this way. If this proportion is met with in any base, as in the fourth...[86]

Pascal then proceeds to show that if the proportion of Corollary 12 holds for $n = 4$ it must also hold for $n = 5$, and follows this with the remark:

The same may be demonstrated in all the rest, since this proof is based only on the assumption that the proportion occurs in the preceding base, and that each cell is equal to its preceding plus the one above it, which is true in all cases.[87]

In Struik's source book this is accompanied by the footnote,

This seems to be the first satisfactory statement of the principle of complete induction. See H. Freudenthal, "Zur Geschichte der vollständigen Induktion", *Archives Internationales de Sciences 22* (1953), 17 – 37.[88]

There are many candidates for the first use of mathematical induction, but only two I am aware of for the first explicit statement of the principle—the present one by Pascal and a later statement by Jakob Bernoulli in the *Acta Eruditorum* in 1686 where the passage from a number a to its successor $a + 1$ is stated even more explicitly and, though it is used in a specific proof, the generality of the method is stated.[89]

Returning to Pascal, I note that in his source book Struik refers at this point to Smith's source book for the remaining corollaries. I shall depart from following this most excellent example only in referring also to Chapter 6 of Edwards's history of the Arithmetical Triangle for the list of all the corollaries stated in terms of figurate numbers. Edwards cites earlier European occurrences of some of the corollaries; Corollary 12, for example, was given earlier by Cardano in 1570, Briggs in 1600, and Fermat in 1636. It will again be rediscovered by Newton in 1664, and was, of course, known centuries earlier by Brahmagupta.

Pascal finished this first tract with a "Problem" which amounts to establishing the explicit formula

$$\binom{n}{r} = \frac{n(n-1)\cdots(n-r+1)}{r(r-1)\cdots 1}.$$

This is, of course, not the end of the Pascalian story. For, the *Traité* contains a few additional tracts on applications of the Arithmetical Triangle. The applications of the binomial coefficients as combinatorial coefficients to counting problems and probability are the most significant of these, although Pascal also finds the sums of powers and uses this to find the areas under the curves $y = x^n$ for positive integers n. Both sets of applications are relevant to the Binomial

---

[86] *Ibid.*, p 73
[87] *Ibid.*, pp. 73 – 74.
[88] Struik, *op. cit.*, p. 25.
[89] John Crossley, *The Emergence of Number*, World Scientific Publishing Co. Pte. Ltd., Singapore, 1987, pp. 45 – 46.

Theorem. Probability, however, will not be discussed in the present book as such a discussion would consume a great deal of space, and the area problem, which was the point of departure for Newton's discovery of his generalisation of the Binomial Theorem, will be touched on in the next chapter.

Of most direct relevance to our story is a short 3-page section Pascal included on the Finite Binomial Theorem: "Pour trouver les puissances des Binomes & Apotomes" ["To find the powers of binomials and apotomes"[90]]. He adds nothing here worth mentioning on its own merit, the earlier commercial arithmetics having done just as well. He simply explains how to use the binomial coefficients $\binom{4}{0}, \binom{4}{1}, \ldots, \binom{4}{4}$ to find $(A+1)^4, (A+2)^4, (A+3)^4, (A-1)^4, (A-1)^4$, and evaluates a couple of these for specific values of A. For example, having found $(A+1)^4$ by using the cited coefficients, he finds $(A+2)^4$ by listing the powers of 2 below the power of $(A+1)^4$, and multiplies the corresponding entries as in FIGURE 17, below. He evaluates this for $A = 1$ by evaluating the individual

$$\begin{array}{ccccccccc}
1A^4 & + & 4A^3 & + & 6A^2 & + & 4A & + & 1 \\
& & 2 & & 4 & & 8 & & 16 \\
\hline
1A^4 & + & 8A^3 & + & 24A^2 & + & 32A & + & 16
\end{array}$$

FIGURE 17.

terms as follows:

1 times the square-square of the unit A	1
8 times the cube of the unit	8
24, 1^2	24
32, 1	32
Plus	16
Thus the sum	$\overline{81}$

And he repeats this for $A = 2$.

He finishes with the remark:

> I am not giving a demonstration of all this, because others have already treated it, like Hérigogne; besides, the matter is self-evident.[91]

---

[90]Smith (*op. cit.*, p. 76) explains, "By 'apotome', Pascal means a binomial which is the difference between two terms."

[91]Smith, *op. cit.*, p. 79. Smith translates this section of Pascal's treatise in its entirety.

CHAPTER 2

Newton's Binomial Theorem

1. Newton's Discovery

Shortly after his reading in 1664 of the *Arithmetica infinitorum* (1656) of John Wallis, Isaac Newton discovered the generalisation of the Finite Binomial Theorem to fractional exponents, allowing the expression of the result of raising a binomial to a fractional power as an infinite series.

The meaning of fractional exponents goes back to the 14th century and Nicole Oresme, who reasoned from the exponential law
$$(x^m)^n = x^{mn},$$
that one must have
$$(x^{1/n})^n = x^{(1/n)n} = x^1 = x,$$
whence $x^{1/n}$ would be the n-th root of x. Likewise $x^{m/n}$ would be the m-th power of the n-th root of x, or, equivalently, the n-th root of the m-th power of x. Later, the exponential law,
$$x^m \cdot x^n = x^{m+n},$$
would lead to the definition of x^{-n} as $1/x^n$. In short, for positive real numbers x, the expression x^p was given a meaning for any rational number p serving as an exponent.

In discussing the Binomial Theorem in the case of a rational exponent p, one usually takes the binomial to be $1 + x$. This is no real restriction as, using Newton's notation, one has
$$(P + PQ)^{m/n} = P^{m/n}(1 + Q)^{m/n}$$
and one can consider Q to be x. Bearing this in mind, let us formally state Newton's result:

2.1.1. Theorem (Newton's Binomial Theorem). *Let p be any rational number.*
$$(1+x)^p = 1 + px + \frac{p(p-1)}{2 \cdot 1} x^2 + \frac{p(p-1)(p-2)}{3 \cdot 2 \cdot 1} x^3 + \ldots, \quad (30)$$

i.e.,
$$(1+x)^p = \sum_{k=0}^{\infty} \binom{p}{k} x^k,$$
where we define
$$\binom{p}{k} = \frac{p(p-1)\cdots(p-k+1)}{k(k-1)\cdots 1} \text{ for } k > 0 \text{ and } \binom{p}{0} = 1.$$

Today, we would be careful to specify the domain of convergence of (30), but in the days of Newton convergence was an afterthought. One carried out the calculations formally and then expected an equation like (30) to hold for a given value of x should the series converge. For p not a nonnegative integer, the series in question converges for $|x| < 1$, does not converge for $|x| > 1$, and, depending on p, may or may not converge for one of $x = \pm 1$. When p is a nonnegative integer, the identity

$$\binom{p}{k} = 0 \text{ for } k > p$$

guarantees convergence for all x:

$$\sum_{k=0}^{\infty} \binom{p}{k} x^k = \sum_{k=0}^{p} \binom{p}{k} x^k + \sum_{k=p+1}^{\infty} \binom{p}{k} x^k = \sum_{k=0}^{p} \binom{p}{k} x^k + \sum_{k=p+1}^{\infty} 0 x^k = \sum_{k=0}^{p} \binom{p}{k} x^k.$$

Thus (30) yields the ordinary Finite Binomial Theorem:

$$(1+x)^p = \sum_{k=0}^{\infty} \binom{p}{k} x^k = \sum_{k=0}^{p} \binom{p}{k} x^k. \tag{31}$$

Hence, not mentioning convergence, Newton's result incorporates the Finite Binomial Theorem without a complex disjunction ("for all $|x| < 1$ or for p a nonnegative integer") as stated condition.

One might imagine (31) to be the inspiration behind Newton's generalisation. It represents the finite binomial power $(1+x)^p$ as an infinite series whose coefficients $\binom{p}{k}$ are each rational expressions of p, k, thus meaningful for all values of p. If one suspects the existence of such a uniformly represented series for rational values of p as well, the only choice is (30). This is not how Newton arrived at his version of the Binomial Theorem. The route was more circuitous than this. One reason is that in Europe power series had barely been considered before Newton. Another is that he was working on a different problem and things just sort of fell into place.

As I said, Newton discovered the result around 1664/1665. He didn't publicly announce it however until 13 June 1676 when he wrote about it in a letter to Henry Oldenburg, Secretary of the Royal Society, to be forwarded to Gottfried Wilhelm Leibniz. In this letter he stated the result and cited some sample applications. In a later letter[1] of 24 October 1676, also intended for Leibniz, Newton described his method of discovery:

> In the beginning of my mathematical studies, when I was perusing the works of the celebrated Dr. Wallis (see his Arith. of Infinites, prop. 118, and 121, also his Algebra, chap. 82), and considering the series by the interpolation of which he exhibits the area of the circle and hyperbola; for instance in this series of curves, whose common base or axis is x, and the ordinates respectively $(1-xx)^{\frac{0}{2}}, (1-$

---

[1]These letters are of such importance in the history of mathematics that they have been given names. The June letter is the *Epistola prior* and the October one the *Epistola posterior*.

1. NEWTON'S DISCOVERY

$xx)^{\frac{1}{2}}, (1-xx)^{\frac{2}{2}}, (1-xx)^{\frac{3}{2}}, (1-xx)^{\frac{4}{2}}$, &c^2; I perceived that if the area of the alternate curves, which are[3]

$$x - \frac{1}{3}x^3,$$

$$x - \frac{2}{3}x^3 + \frac{1}{5}x^5,$$

$$x - \frac{3}{3}x^3 + \frac{3}{5}x^5 - \frac{1}{7}x^7,$$

&c;

could be interpolated, we should be able to obtain the areas of the intermediate ones; the first of which, $(1-xx)^{\frac{1}{2}}$, is the area of the circle: now in order to [do] this, it appeared that in all the series the first term was x; that the 2nd terms $\frac{0}{3}x^3, \frac{1}{3}x^3, \frac{2}{3}x^3, \frac{3}{3}x^3$, &c, were in arithmetical progression; and consequently that the first two terms of all the series to be interpolated would be

$$x - \frac{\frac{1}{2}x^3}{3}, \quad x - \frac{\frac{3}{2}x^3}{3}, \quad x - \frac{\frac{5}{2}x^3}{3}, \quad \&c.$$

Now for the interpolation of the rest, I considered that the denominators $1, 3, 5, 7$, &c, were in arithmetical progression; and that therefore only the numeral coefficients of the numerators were to be investigated. But these in the alternate areas, which are given, were the same with the figures of which the several powers of 11 consist: viz. $11^0, 11^1, 11^2, 11^3$, &c; that is

> the first   1,
> the second 1, 1
> the third   1, 2, 1
> the fourth 1, 3, 3, 1
> the fifth   1, 4, 6, 4, 1
> &c.

I enquired therefore how, in these series, the rest of the terms may be derived from the first two being given; and I found that by putting m for the 2d figure or term, the rest would be produced by the continued multiplication of the terms of this series,

$$\frac{m-0}{1} \times \frac{m-1}{2} \times \frac{m-2}{3} \times \frac{m-3}{4} \times \frac{m-4}{5} \ \&c.$$

---

[2] It is curious how, although Newton, among others, used exponential notation, he wrote "xx" here and not x^2. I've read that it is because the latter, being no shorter than the former, is not really an abbreviation and thus serves no purpose in this context. Being lazy, I would justify it as being marginally easier to typeset. However, today the exponent is not regarded as an abbreviation but as a function.

[3] Missing here is the first area, which should be x.

For instance, if the 2d term $m = 4$; then shall $4 \times \dfrac{m-1}{2}$, or 6, be the 3rd term; and $6 \times \dfrac{m-2}{3}$, or 4 the 4th term; and $4 \times \dfrac{m-3}{4}$ or 1, the 5th term; and $1 \times \dfrac{m-4}{5}$, or 0, the 6th; which shows that in this case the series terminates.

This rule therefore I applied to the series to be interpolated. And since, in the series for the circle, the 2d term was $\dfrac{\frac{1}{2}x^3}{3}$ I put $m = \frac{1}{2}$, which produced the terms $\dfrac{1}{2} \times \dfrac{\frac{1}{2}-1}{2}$ or $-\dfrac{1}{8}$; $-\dfrac{1}{8} \times \dfrac{\frac{1}{2}-2}{3}$ or $+\dfrac{1}{16}$; $\dfrac{1}{16} \times \dfrac{\frac{1}{2}-3}{4}$ or $-\dfrac{5}{128}$; and so on ad infinitum. And hence I found that the required area of the circular segment is

$$x - \dfrac{\frac{1}{2}x^3}{3} - \dfrac{\frac{1}{8}x^5}{5} - \dfrac{\frac{1}{16}x^7}{7} - \dfrac{\frac{5}{128}x^9}{9} - \&c.$$

And in the same manner might be produced the interpolated areas of the other curves: as also the area of the hyperbola and the other alternates in this series

$$(1-xx)^{\frac{0}{2}}, (1-xx)^{\frac{1}{2}}, (1-xx)^{\frac{2}{2}}, (1-xx)^{\frac{3}{2}}, \&c.$$

And in the same way also may other series be interpolated and that too if they should be taken at the distance of two or more terms.

This was the way then in which I first entered upon these speculations; which I should not have remembered, but that in turning over my papers a few weeks since, I chanced to turn my eyes on those relating to this matter.

Having proceeded so far, I considered that the terms

$$(1-xx)^{\frac{0}{2}}, (1-xx)^{\frac{2}{2}}, (1-xx)^{\frac{4}{2}}, (1-xx)^{\frac{6}{2}}, \&c,$$

that is,

$$\begin{array}{l} 1 \\ 1 - x^2 \\ 1 - 2x^2 + x^4 \\ 1 - 3x^2 + 3x^4 - x^6, \\ \&c, \end{array}$$

might be interpolated in the same manner as the areas generated by them: and for this, nothing more was required but to omit the denominators $1, 3, 5, 7, \&c$, in the terms expressing the areas; that is, the coefficients of the terms of the quantity to be interpolated $(1-xx)^{\frac{1}{2}}$, or $(1-xx)^{\frac{3}{2}}$, or generally $(1-xx)^m$, will be produced

1. NEWTON'S DISCOVERY

by the continued multiplication of the terms of this series

$$m \times \frac{m-1}{2} \times \frac{m-2}{3} \times \frac{m-3}{4} \text{ \&c.}$$

Thus, for example, there would be found

$$(1-xx)^{\frac{1}{2}} = 1 - \frac{1}{2}x^2 - \frac{1}{8}x^4 - \frac{1}{16}x^6 \text{\&c.}$$

$$(1-xx)^{\frac{3}{2}} = 1 - \frac{3}{2}x^2 + \frac{3}{8}x^4 + \frac{1}{16}x^6 \text{\&c.}$$

$$(1-xx)^{\frac{1}{3}} = 1 - \frac{1}{3}x^2 - \frac{1}{9}x^4 - \frac{5}{81}x^6 \text{\&c.}$$

Thus then I discovered a general method of reducing radical quantities into infinite series, by the theorem which I sent in the beginning of the former letter, before I knew the same by the extraction of roots.

But having discovered that way, this other could not long remain unknown: for to prove the truth of those operations, I multiplied $1 - \frac{1}{2}x^2 - \frac{1}{8}x^4 - \frac{1}{16}x^64$ &c, by itself, and the product was $1 - x^2$. But as this was a certain proof of those conclusions, so I was naturally led to try conversely whether these series, which were thus known to be the roots of the quantity $1 - x^2$, could not be extracted out of it after the manner of arithmetic; and upon trial I found it to succeed. The process for the square root is here set down.

$$\begin{array}{r} 1 \; - \; x^2 \; \big(\; 1 - \tfrac{1}{2}x^2 - \tfrac{1}{8}x^4 - \tfrac{1}{16}x^6 \text{ \&c} \\ \underline{1 } \\ 0 \; - \; x^2 \\ \underline{- \; x^2 \; + \; \tfrac{1}{4}x^4 } \\ -\tfrac{1}{4}x^4 \\ \underline{-\tfrac{1}{4}x^4 \; + \; \tfrac{1}{8}x^6 \; + \; \tfrac{1}{64}x^8} \\ -\tfrac{1}{8}x^6 \; - \; \tfrac{1}{64}x^8 \end{array}$$

These methods being found, I laid aside the other way by interpolation of series, and used these operations only as a more genuine foundation. Neither was I ignorant of the reduction by division, which was so much easier.[4]

---

[4] I quote from the English translation by Charles Hutton in his: *A Philosophical and Mathematical Dictionary*, vol. 1, London, 1815, pp. 230 – 231. I have taken some liberties with the layout. Another translation can be found in: David Eugene Smith (ed.), *A Source Book in Mathematics*, 1929; reprinted: Dover Publications, Inc., New York, 1959.

2. Discussion of Newton's Remarks

Not all of Newton's remarks will be thoroughly intelligible to the reader who may be unfamiliar with the Calculus. Indeed, any discussion of Newton's Binomial Theorem quickly involves one in topics of the subject—areas, differentiation, infinite sums. Fortunately, one doesn't need too deep a knowledge of the Calculus to understand Newton's contribution to the Binomial Theorem and I shall endeavour to provide the necessary background in the next few sections. A few things, however, can be said first.

Newton considers the sequence of curves

$$y = (1 - x^2)^0 = 1, \quad y = (1 - x^2)^1 = 1 - x^2, \quad y = (1 - x^2)^2 = 1 - 2x^2 + x^4,$$

the areas beneath which between 0 and x were known to be

$$x, \quad x - \frac{1}{3}x^3, \quad x - \frac{2}{3}x^3 + \frac{1}{5}x^5, \quad \ldots$$

How these results had been obtained is not relevant to understanding Newton's reasoning concerning the Binomial Theorem.[5] He knew these results and sought to interpolate an expression for the area under $y = (1 - x^2)^{1/2}$ between 0 and x, i.e., the area of that portion of the semi-circle of radius 1 centered at the origin as in FIGURE 1, below. He did this by considering the infinite array of

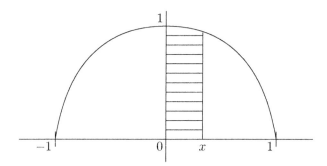

FIGURE 1. Area of a Circular Segment

coefficients of x^{2k} for $k = 0, 1, \ldots$ in $y = (1 - x^2)^n$ as in the first part of TABLE 1, below. The first thing he noticed was that the first column was constant; the second that the denominators, going from left to right, form the progression $1, 3, 5, 7, \ldots$ of odd integers; and the third thing was that, after suppressing the minus signs, the numerators formed Pascal's triangle, as in the second half of TABLE 1.

It appears, as he does not cite any reference and describes these entries in terms of powers of 11 instead of as coefficents of binomial powers, that Newton has, in fact, rediscovered the Finite Binomial Theorem on his own. The result has been the subject of multiple discovery and a Newtonian rediscovery is entirely plausible.

---

[5]But *cf.* sections 4 and 6, below, for the explanations.

2. DISCUSSION OF NEWTON'S REMARKS

TABLE 1. Newton's Tables

n	$k=0$	$k=1$	$k=2$	$k=3$	...	n	$k=0$	$k=1$	$k=2$	$k=3$	...
0	1	0	0	0	...	0	1	0	0	0	...
1	1	−1/3	0	0	...	1	1	1	0	0	...
2	1	−2/3	1/5	0	...	2	1	2	1	0	...
3	1	−3/3	3/5	−1/7	...	3	1	3	3	1	...
4	1	−4/3	6/5	−4/7	...	4	1	4	6	4	...
5	1	−5/3	10/5	−10/7	...	5	1	5	10	10	...

The next step was to determine how to generate successive entries of individual rows of the denominator-free half of TABLE 1. Obviously the first entry was 1 and the second was n. Finding successive elements amounts to expressing $\binom{n}{k+1}$ in terms of $\binom{n}{k}$:

$$\binom{n}{k+1} = \frac{n(n-1)\cdots(n-k+1)(n-(k+1)+1)}{(k+1)k(k-1)\cdots 1}$$
$$= \frac{n(n-1)\cdots(n-k+1)(n-k)}{(k+1)k(k-1)\cdots 1}$$
$$= \frac{n-k}{k+1} \cdot \frac{n(n-1)\cdots(n-k+1)}{k(k-1)\cdots 1} = \frac{n-k}{k+1}\binom{n}{k}.$$

This observation was not new even in Europe: it can be found in the *Trigonometrica Britannica* (1633) of Henry Briggs (1561 – 1630). Newton, however, applied it to fractional exponents. In particular, choosing $n = \frac{1}{2}$, he would interpolate a row in the second half of TABLE 1, starting with $1, \frac{1}{2}$ and then successively inserting

$$\frac{\frac{1}{2}-1}{1+1} \cdot \frac{1}{2} = \frac{-\frac{1}{2}}{2} \cdot \frac{1}{2} = -\frac{1}{8}$$

$$\frac{\frac{1}{2}-2}{2+1} \cdot \left(-\frac{1}{8}\right) = \frac{\frac{-3}{2} \cdot \frac{-1}{2}}{3} = \frac{3}{16\cdot 3} = \frac{1}{16},$$

etc. Reintroducing the denominators and the signs results in

$$1, \ -\frac{1/2}{3}, \ \frac{-1/8}{5}, \ -\frac{1/16}{7}, \ \frac{-5/128}{9}, \ -\frac{15/512}{11}, \ \ldots$$

And the area of the circular segment becomes

$$x - \frac{1}{6}x^3 - \frac{1}{40}x^5 - \frac{1}{112}x^7 - \frac{5}{1152}x^9 - \frac{15}{5632}x^{11} - \ldots, \qquad (32)$$

which for $x = 1$ gives .7924011319 as an estimate of the area of the quarter circle, i.e., $4 \times .7924011319 = 3.1696045276$ as a slight overestimate of the area of a circle.

At this point, Newton realised that the areas were no more than a distraction and redid his interpolations for the binomial powers themselves, which amounted in practice to considering only the second part of TABLE 1. The interpolated row is just the sequence

$$1, \frac{1}{2}, -\frac{1}{8}, \frac{1}{16}, -\frac{5}{128}, \frac{15}{512}, \ldots$$

and his expansion of $y = (1 - x^2)^{1/2}$ becomes

$$1 - \frac{1}{2}x^2 - \frac{1}{8}x^4 - \frac{1}{16}x^6 - \frac{5}{128}x^8 - \frac{15}{512}x^{10} - \cdots$$

Multiplying this by itself gives the array of TABLE 2, below. Thus Newton

TABLE 2. Squaring the Series

series × 1	1	$-\frac{1}{2}x^2$	$-\frac{1}{8}x^4$	$-\frac{1}{16}x^6$	$-\frac{5}{128}x^8 - \cdots$
series × $-\frac{1}{2}x^2$		$-\frac{1}{2}x^2$	$+\frac{1}{4}x^4$	$+\frac{1}{16}x^6$	$+\frac{1}{32}x^8 + \cdots$
series × $-\frac{1}{8}x^4$			$-\frac{1}{8}x^4$	$+\frac{1}{16}x^6$	$+\frac{1}{64}x^8 + \cdots$
series × $-\frac{1}{16}x^6$				$-\frac{1}{16}x^6$	$+\frac{1}{32}x^8 + \cdots$
series × $-\frac{5}{128}x^8$					$-\frac{5}{128}x^8 + \cdots$
sum	1	$-x^2$	$+ 0x^4$	$+ 0x^6$	$+ 0x^8 + \cdots$

demonstrated that his series for $(1 - x^2)^{1/2}$ did yield $1 - x^2$ upon squaring.

For good measure, Newton finished up applying the familiar technique of root extraction. Without any limitation on the coefficients, which are allowed to be as large as desired and positive or negative, the process is actually easier than it is for numbers, where digits are restricted to the collection $0, 1, 2, \ldots, 9$ and one must make sure the choice of b in taking one from one estimate a of the square root of D to $a + b$ as the next does not result in $(a + b)^2$ being larger than D. Thus, in finding the square root of $1 - x^2$ one first chooses 1 as one's estimate a. Next one takes half of $1 - x^2 - 1^2 = -x^2$ as $2b$ and subtracts

$$(1 - x^2) - \left(1 - \frac{x^2}{2}\right)^2 = (1 - x^2) - \left(1 - x^2 + \frac{x^4}{4}\right) = -\frac{x^4}{4}.$$

Half of this is $-x^4/8$, so one subtracts

$$(1 - x^2) - \left(\left(1 - \frac{x^2}{2}\right) - \frac{x^4}{8}\right)^2 =$$

$$= (1 - x^2) - \left(\left(1 - \frac{x^2}{2}\right)^2 - 2\left(1 - \frac{x^2}{2}\right)\frac{x^4}{8} + \frac{x^8}{64}\right)$$

$$= (1-x^2) - \left(1 - \frac{x^2}{2}\right)^2 + 2\left(1 - \frac{x^2}{2}\right)\frac{x^4}{8} - \frac{x^8}{64}$$

$$= -\frac{x^4}{4} + \frac{x^4}{4} - \frac{2x^6}{2 \cdot 8} - \frac{x^8}{64}$$

$$= -\frac{x^6}{8} - \frac{x^8}{64},$$

thus securing $1 - \frac{1}{2}x^2 - \frac{1}{8}x^4$ as the first three terms of the square root and suggesting $b = -\frac{1}{16}x^6$ as the fourth.

What one has is not a proof but a *control*—a check on the result of his reasoning from his interpolative assumptions. A valid proof would not be forthcoming until the first quarter of the 19th century, although there would be numerous attempts to establish the validity of the result in the 18th century. We will discuss this matter later in Chapter 3. Our primary concerns for the rest of the current chapter are a bit of background material on the Calculus and a description of the place of Newton's Binomial Theorem in the Calculus. But first let us digress to use Newton's Binomial Theorem for some actual root extractions.

3. Application of Newton's Theorem

Leibniz followed Newton by about a decade, but being less reclusive he spread his results around more publicly. He had begun to make inroads in infinite series when Newton sent the letter of 13 June 1676 to him via Oldenburg. In it Newton described his Binomial Theorem as follows:

The Extractions of Roots are much shortened by the Theorem

$$\overline{P + PQ}|\frac{m}{n} = P\frac{m}{n} + \frac{m}{n}AQ + \frac{m-n}{2n}BQ + \frac{m-2n}{4n}CQ$$
$$+ \frac{m-3n}{4n}DQ + \&c. \qquad (33)$$

where $P + PQ$ stands for a Quantity whose Root or Power or whose Root of a Power is to be found, P being the first Term of that quantity, Q being the remaining terms divided by the first term, and $\frac{m}{n}$ the numerical Index of the powers of $P + PQ$... Finally, in place of the terms that occur in the course of the work in the Quotient, I shall use A, B, C, D &c. Thus A stands for the first term $P^{\frac{m}{n}}$; B for the second term $\frac{m}{n}AQ$; and so on.[6]

The notation is a bit strange to the modern eye and the first occurrence of $4n$ as a denominator is a typographical error and should be replaced by $3n$.

At first sight the expression given for $(P + PQ)^{m/n}$ is a bit odd, a good exercise in decipherment for the budding maths historian. Closer examination

---

[6]David Eugene Smith, *op. cit.*, p. 225 of the first volume of the Dover reprint.

reveals, however, that his strange expression for the expected

$$(P+PQ)^{m/n} = P^{m/n} + \frac{m}{n}P^{m/n}Q + \frac{\frac{m}{n}\left(\frac{m}{n}-1\right)}{2\cdot 1}P^{m/n}Q^2 + \ldots \qquad (34)$$

is written with computation in mind. Writing $q = m/n$,

$$(P+PQ)^{m/n} = P^q(1+Q)^q = P^q\left(1 + qQ + \frac{q(q-1)}{2}Q^2 + \ldots\right),$$

we see that any term after the first is

$$P^q\binom{q}{k+1}Q^{k+1} = \frac{q-k}{k+1}P^q\binom{q}{k}Q^k Q,$$

where $P^q\binom{q}{k}Q^k$ is the term $A, B, C, \ldots$ preceding it. That is, P^q is the first term in the series and the term in which Q occurs to degree $k+1$ is obtained from the degree k term by multiplying by

$$\frac{q-k}{k+1}Q = \frac{\frac{m}{n}-k}{k+1}Q = \frac{m-nk}{(k+1)n}Q.$$

Thus,

$$B = A \cdot \frac{m - 0n}{(0+1)n}Q = \frac{m}{n}AQ$$

$$C = B \cdot \frac{m-n}{2n}Q = \frac{m-n}{2n}BQ,$$

etc.

Thus (33) offers the following instructions for calculating $D^{m/n}$:

(1) Write $D = P + PQ$, where $P^{m/n}$ is known, and $0 \leq Q < 1$.

(2) Calculate successive terms in the series:
$$t_0 = P^{m/n}$$
$$t_{k+1} = \frac{m-n}{2n}Qt_k.$$

(3) Sum the given terms $t_0, t_1, \ldots, t_k$ for as many terms, $k+1$, as one wishes to use.

Formula (34) instructs one to

(1) Write $D = P + PQ$, where $P^{m/n}$ is known, and $0 \leq Q < 1$.

(2) Calculate successive binomial coefficients using the horizontal recursion
$$b_0 = 1$$
$$b_{k+1} = \frac{m-n}{2n}b_k.$$

(3) Calculate successive powers of Q
$$Q^0 = 1$$
$$Q^{k+1} = Q^k Q.$$

3. APPLICATION OF NEWTON'S THEOREM

(4) Multiply the two lists together: $b_0 Q^0, b_1 Q^1, \ldots$

(5) Add as many terms as desired: $b_0 Q^0 + b_1 Q^1 + \ldots + b_k Q^k$.

(6) Multiply the sum by $P^{m/n}$.

The multiplication by $P^{m/n}$ can be done earlier, at step 2 by setting $b_0 = P^{m/n}$, at step 3 by replacing Q^0 by $P^{m/n}$, or at step 4 by multiplying each product $b_k Q^k$ by $P^{m/n}$. For calculation by hand, it generally saves some effort by postponing the multiplication by $P^{m/n}$ until the final step.

The obvious question to ask is: how accurate is Newton's assessment of the efficiency of these procedures? This is a question for the numerical analyst and it requires a bit of Calculus to begin to approach the solution. What we can do, however, is work through a couple of examples and see how many steps it takes to achieve the same accuracy as our calculators' exponentiation buttons. Assuming $P, P^{m/n}$ are rational, with a *TI-89 Titanium* and its ability to deal exactly with rational numbers, we can even go beyond the accuracy of the built-in exponentiation function just as we did in the previous chapter.

The first procedure based on Newton's formula (33) can be implemented as follows. To estimate $\sqrt{17}$, for example, we choose

$$m = 1, \ n = 2, \ P = 16, \ P^{1/2} = 4, \ Q = \frac{1}{16}$$

and store the following

    1→M
    2→N
    1/16→Q.

We could store the 4 in a variable P or simply use 4 itself as we only need it in the last step. Next we set the graphics mode of the calculator to Seq on the *TI-83 Plus*. [Instructions for the *TI-89 Titanium* will be given later.] One then enters the equation editor and sets

    nMin=0
    u(n)=(M−(n−1)N)/(nN)*u(n−1)*Q
    u(nMin)={1}.

One now chooses some number, say 8, for the degree of the last term in the expansion to be used in estimating $\sqrt{17}$, and enters

    4*sum(u(0,8))

to get this estimate. Better yet, enter

    4*cumSum(u(0,8))

to generate the sequence of estimates involving an additional term in each successive step:

    4, 4.125, 4.123046875, 4.12310791, 4.123105526, 4.12310563,
                                      4.123105625, 4.123105626, 4.123105626.

If one checks beyond the displayed digits in the calculator, one sees that the last two numbers are not actually equal even though the displayed digits agree with each other and with the displayed digits of $\sqrt{17}$ as obtained on entering $\sqrt{(}17)$ or $17\wedge(1/2)$.

Comparing this with our calculations of $\sqrt{17}$ of Chapter 1, we see that it did better than the bisection method starting with 4 and 5 as straddling estimates which yielded the less impressive sequence (10) of estimates on page 6, but did not do as well as the sequence (12) on page 7 obtained by direct appeal to the Finite Binomial Theorem.

[The procedure is a bit more complicated on the *TI-89 Titanium*. The calculator will not distinguish between v and V as variables, and n is reserved in defining sequences, so we must use a variable other than n for the denominator. So store

    1→m
    2→k
    1/16→q.

Then set the graphics mode of the calculator to SEQUENCE and in the equation editor define

    u1=(m−(n−1)∗k)/(n∗k)∗u(n−1)∗q
    ui1=1,

being sure to include the ∗ between n and k (the calculator interprets nk as a variable name, not as a multiplication). There is no place in the editor to determine the initial value of n, so enter

    0→nMin.

I don't know if it is necessary to enter a value for nMax, but entering any value greater than or equal to the number of terms one wants to include in one's expansion will certainly suffice. To generate the sequence, enter

    seq(u1(i),i,0,8)→list1

and then evaluate

    4∗cumSum(list1).

As in Chapter 1, if one sets the calculator to EXACT mode, one will have some horrendous fractions from which the decimals can be calculated.]

The implementation of the second procedure goes as follows. Do not bother changing the graphics mode of the calculator, store m, n, q in the variables M, N, Q. This time, however, let us look for an estimate of $\sqrt[5]{17}$. So $m = 1, n = 5$, and choose $P = 1.7^5$, so that

$$Q = \frac{17 - 1.7^5}{1.7^5}.$$

Store this number in Q. Again assume we are are going to use 9 terms of the series (stopping at degree 8) to estimate $\sqrt[5]{17}$. We generate the list of binomial coefficients in two steps, first the factors and then the coefficients themselves as cumulative products:

    augment({1},seq((M−KN)/((K+1)N),K,0,7))→L_1
    e^(cumSum(ln(L_1)))→L_1.

3. APPLICATION OF NEWTON'S THEOREM

[A word or two of explanation: the first line generates the list
$$\left\{1, \frac{m}{n}, \frac{m-n}{2n}, \frac{m-2n}{3n}, \ldots, \frac{m-7n}{8n}\right\}.$$
The second line calculates the cumulative product, which, unlike the sum, is not a built-in function. Using e^() and ln() is, of course, cheating and we should actually replace this use by a program performing the multiplications. I will do this a bit later.]

Exponentiation to a positive integral power is unproblematic, so the second list can be generated by appeal to the exponentiation button without any feelings of guilt[7]:

seq(Q^I,I,0,8)→L$_2$.

and finally one enters

1.7∗cumSum(L$_1$∗L$_2$)

to get the sequence

1.7, 1.767083249, 1.76178894, 1.762415692, 1.76232913,

1.76234211 − 2.596017572 × $10^{-18}i$, 1.762340061 − 2.596017572 × $10^{-18}i$,

1.762340396 − 2.529029435 × $10^{-18}i$, 1.76234034 − 2.551498344 × $10^{-18}i$.

And the reader can see that I had accidentally left the calculator in complex mode. This was serendipitous as the first incarnation of L$_1$ had negative terms and taking logarithms of L$_1$ in the second step would have resulted in an ERROR message had I not done so. The imaginary parts of the last four elements of the list are negligible, due to some rounding errors in taking the complex logarithms or during the exponentiations. If we drop them, the real parts, at least for the digits displayed, agree with those obtained by applying the procedure used based on Newton's preferred (33).

Entering 17^(1/5) in the calculator results in 1.762340348 and we see that all the displayed digits in the last estimate obtained from 9 terms are correct, but more steps need to be taken to match the built-in function. Comparison with the calculations of $\sqrt[5]{17}$ given in Chapter 1, section 4, is none too flattering. We got slightly better results starting with 1 as our initial estimate and applying the Finite Binomial Theorem in a relatively crude manner, and we did much better when we used a little sophistication in our application of the Theorem.

We might try replacing 1.7 by a better first estimate for $\sqrt[5]{17}$. Choosing 1.76 for $P^{1/5}$ gives $(17 - 1.76^5)/1.76^5$, so store this latter number in Q:

(17−1.76^5)/1.76^5→Q

and recalculate. Using the first procedure we have but to enter

1.76∗cumSum(u(0,8)).

The result is

1.76, 1.76234658, 1.762340323, 1.762340348,

the last entry repeating itself thereafter.

---

[7] Unless it be for inefficiency: *Cf.* page 290.

Applying the second method requires more typing. First we had to define L_1, then modify it, then define L_2, and finally take the cumulative sum of the product of the two lists and multiply the result by 1.76. At this stage we realise we should have written a program to begin with:

```
PROGRAM:NEWTROOT
:ClrHome
:Disp "PLEASE ENTER AN","EXPONENT 1/N:"
:Input "N=",N
:ClrHome
:Disp "PLEASE ENTER"
:Input "D=",D
:ClrHome
:Disp "PLEASE ENTER AN","ESTIMATE FOR"
:Input "D^(1/N)=",P
:ClrHome
:P^N→A
:(D−A)/A→Q                           8
:Disp "HOW MANY TERMS","IN THE","EXPANSION?"
:Input K
:ClrHome
:augment({1},seq((1−IN)/((I+1)N),I,0,K−1))→L₁
:For(I,1,K−1)
:L₁(I+1)∗L₁(I)→L₁(I+1)
:End
:seq(Q^I,I,0,K−1)→L₂
:P∗cumSum(L₁∗L₂)→L₃
:DelVar A
:DelVar N
:DelVar D
:DelVar P
:DelVar K
:DelVar I
:DelVar Q
:ClrList L₁,L₂
:L₃ .                                9
```

The program is a bit long because, instead of assuming one has stored the necessary information, it asks interactively for the user to enter the information, cleans up the screen several times, and finishes clearing memory with the deletion of unnecessary variables. Without these housekeeping tasks, the program is shortened considerably.

If one runs the program, entering 5 for N, 17 for D, 1.76 for D^(1/N), and 9 for the number of terms (1 higher than the degree), the screen will display the

---

[8] This and the previous line eliminate the need to calculate P^N twice.

[9] This command does nothing. Ending the program without it merely results in **Done** appearing on the screen. **Disp** L_3 will display a bit of L_3 and then **Done** immediately below it. Ending the program with this line results in a horizontally scrollable display of L_3.

sequence

$$1.76, \ 1.76234658, \ 1.762340323, \ 1.762340348,$$

the last entry repeating as before. The numbers agree with those obtained by the first method but for the lack of imaginary numbers.

I leave it to the reader with a *TI-89 Titanium* to rewrite the program for that calculator. He might wish to run the program in EXACT mode and use multiplication by large powers of 10 and iPart() as in Chapter 1 to see how many digits accuracy successive entries in L_3 will yield.

Once one has a program in one's calculator one looks for an excuse to run it repeatedly:

2.3.1. Exercise. *Run the program a number of times, always entering 5 for N, 17 for D, and 40 for the number of terms to be used for the expansion. When asked for an estimate for $D^\wedge(1/N)$ use the values $1, 1.5, 1.6, 1.69, 1.7, 1.75, 1.76, 1.79, 1.8, 1.9$, and 2. Make a table with these values in one row and below them write down how many terms of the expansion were required for 1.762340348 to show up on the list.*

This exercise is certainly not as toilsome as multiplying several multidigit numbers, but constantly having to enter 5, 17, and 40 is a bit boring, and in scrolling a list of 40 numbers looking for 1.762340348 it is easy to lose track of where you are:

2.3.2. Exercise. *Automate the process: Store $5, 17$, and 40 in the variables N, D, and K, respectively. Store $\{1, 1.5, 1.6, 1.69, 1.7, 1.75, 1.76, 1.79, 1.8, 1.9, 2\}$ in the list L_4. Remove all the commands in NEWTROOT requesting values for N, D, K, the estimate for $D^\wedge(1/N)$ to be stored in P, and the number of terms in the expansion. Now write a new program which first stores a list of twelve 0's in L_5, and then, for each entry in L_4 places the entry in P, runs the modified version of NEWTROOT, and checks the first position in L_3 for which 1.762340348 will be displayed, and if this is the i-th position puts i into the appropriate position of L_5 (the first entry if P is 1, the second if P is 1.5, etc.). [Remarks: The test $L_3(I)=1.762340348$ will not work because $L_3(I)$ has additional digits stored in the calculator. How would you check that the displayed value will be 1.762340348? You can finish your program with the command List▶Matr(L_3, L_5)→[A] to display a vertically scrollable version of the table. A 0 in L_5 means that the desired value did not appear even after 40 terms were used in the expansion.]*

Numerical analysts can determine in advance how quickly a given level of accuracy can be achieved. But the impression to be got from these exercises should be that, if one starts closely enough to to the root, the convergence will be rapid, but if one is not close to begin with, the initial behaviour of the procedure can be abysmal.

78   2. NEWTON'S BINOMIAL THEOREM

4. A Brief Calculus Primer, I: Area

Newton's treatment of binomial powers began with the consideration of the areas under the curves
$$y = (1-x^2)^0, \ y = (1-x^2)^{1/2}, \ y = (1-x^2)^1, \ y = (1-x^2)^{3/2}, \ \ldots$$
Every other one of these expanded to a polynomial
$$y = (1-x^2)^0 : \ y = 1$$
$$y = (1-x^2)^1 : \ y = 1 - x^2$$
$$y = (1-x^2)^2 : \ y = 1 - 2x^2 + x^4$$
and so on. And calculating the areas under these curves reduced quickly enough to calculating those beneath the curves $y = x^n$ for positive integers n. These areas had been determined by a number of individuals before Newton got involved. One of these was Wallis.

In Wallis's day Europe, having awoken from the intellectual slumber of the Middle Ages, was rapidly developing. Its mathematicians had already surpassed the ancient Greeks and were constantly testing their muscles on problems the Greeks could not handle or even express. The area under the graph of a polynomial of degree greater than 2 was one such problem. There were two separate parts to the problem of determining areas. First there was that of discovery—finding a method for obtaining the results—and then that of proving one's results. Until Newton and Leibniz "invented the Calculus", i.e., until they produced a standard algorithm for this discovery process, every mathematician it seems had his own heuristic method. As to the matter of proof, once you knew the result it wasn't too hard to reason that it had to be correct. That part was not as exciting as the initial search and was progressively ignored.

The modern definition of the area under a curve $y = f(x)$ between two limits $x = a$ and $x = b$, better termed the area trapped between $y = f(x)$ and the x-axis, was given for *continuous*[10] functions f by Augustin Louis Cauchy at the end of the first quarter of the 19th century and stated in greater generality by Bernhard Riemann in the middle of that century. Before Newton had come on the scene, however, Pierre de Fermat had used something along these lines and other methods "anticipated" it. Riemann's definition goes something like this.

Let f be a function and assume that $f(x) \geq 0$ on an interval $[a,b]$. To estimate the area under the curve $y = f(x)$ between the limits $x = a$ and $x = b$, first partition the interval into n subintervals by choosing $x_1, x_2, \ldots, x_{n-1}$ with $a < x_1 < x_2 < \ldots < x_{n-1} < b$. Writing $x_0 = a, x_n = b$, choose an element

---

[10]The exact meaning needs not concern us here. Suffice it to say that those functions one writes down using the usual algebraic and trigonometrical operations are continuous wherever they are defined, and that they are well-behaved. The traditional classroom gloss is that a function is continuous if it can be graphed without lifting one's pencil off the paper; its graph contains no gaps or jumps. The official definition is that f is continuous at x_0 if, as x approaches x_0, $f(x)$ approaches $f(x_0)$, i.e., $f(x_0)$ is closely approximated by $f(x)$ whenever x is close to x_0, ever more closely as x gets closer and closer to x_0.

$x_k^* \in [x_k, x_{k+1}]$ for $k = 0, 1, \ldots, n-1$, and let the rectangle of height $f(x_k^*)$ on the base $[x_k, x_{k+1}]$ approximate that portion of the area under $y = f(x)$ over the subinterval $[x_k, x_{k+1}]$. The sum of these approximations,

$$A_n = \sum_{k=0}^{n-1} f(x_k^*)(x_{k+1} - x_k), \qquad (35)$$

approximates the area under[11] $y = f(x)$ between $x = a$ and $x = b$. FIGURE 2, below, illustrates this for the choice $x_k^* = x_{k+1}$.

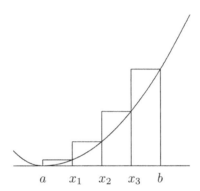

FIGURE 2. Approximating the Area

There will usually be an error, but as the number n of subintervals increases and the lengths of these subintervals decreases, the error decreases as well. With luck, a limiting value for such sums emerges. When this happens, the limiting value is called the *area* under the curve $y = f(x)$ between a and b, or the *definite integral* of f on $[a, b]$, and one writes[12]

$$\int_a^b f(x)\,dx$$

for this area. Moreover, the function f is said to be *integrable* if the definite integral always exists.

Fermat evaluated some simple integrals using clever choices of the partitions and the numbers x_k^*. In FIGURE 2 we have chosen x_k^* to be the right endpoint of the interval $[x_k, x_{k+1}]$. Insofar as f is strictly increasing on $[a, b]$, we have also chosen x_k^* where f assumes its maximum value on $[x_k, x_{k+1}]$ and the estimate A_n of (35) is thus an upper bound on the integral. In like manner, one can choose the x_k^*'s to be the left endpoints or where the minima occur. Riemann

---

[11] If $f(x_k^*)$ is allowed to be negative, $f(x_k^*)(x_{k+1} - x_k)$ is the negative of the area of the rectangle. If one assumes the region trapped between the curve and the x-axis lying below the x-axis to have negative area, the process again makes sense and one can drop the assumption that $f(x) \geq 0$ throughout the interval.

[12] The symbol "$\int$" is the italic form of the obsolete long-s, a variant of the letter "s", and stands for "sum".

proved that for well-behaved functions, the choice didn't matter: the same limit results.

In practice it is simplest in calculating areas directly, without using the Newton-Leibniz algorithms, to subdivide the interval $[a,b]$ into equal subintervals of length $(b-a)/n$, thus choosing

$$x_k = a + k\frac{b-a}{n}, \quad k = 0, 1, 2, \ldots, n,$$

and to choose x_k^* to be one of the endpoints.

We can illustrate the procedure using the parabola $y = x^2$ over an interval $[0, b]$ (i.e., taking $a = 0$ to simplify the algebra). Using right endpoints, the sum (2) is

$$A_n = \sum_{k=0}^{n-1} f\left((k+1)\frac{b-a}{n}\right) \cdot \frac{b-a}{n}, \text{ before taking } a = 0$$

$$= \sum_{i=1}^{n} \left(\frac{ib}{n}\right)^2 \cdot \frac{b}{n} = \frac{b^3}{n^3} \sum_{i=1}^{n} i^2, \text{ where } i = k+1 \text{ and } a = 0$$

$$= \frac{b^3}{n^3} \cdot \frac{n(n+1)(2n+1)}{6}$$

using the familiar formula for the sum of the first n squares,

$$= \frac{b^3}{6} \cdot \left(1 + \frac{1}{n}\right)\left(2 + \frac{1}{n}\right)$$

$$\approx \frac{b^3}{6} \cdot 1 \cdot 2 = \frac{b^3}{3},$$

since $1 + \frac{1}{n}$ and $2 + \frac{1}{n}$ are close to 1 and 2, respectively, for large n. The values of the approximations A_n thus have $b^3/3$ as their limiting value:

$$\int_0^b x^2\,dx = \frac{b^3}{3}.$$

2.4.1. Exercise. *Calculate the area trapped under the parabola $y = x^2$ between $x = a$ and $x = b$ by repeating the above for $a \neq 0$. For $0 < a < b$ note that the area in question is the difference of the areas trapped between 0 and b and between 0 and a, respectively. Use this to check the results of your calculation.*

The area under the parabola was as far as the ancient Greeks got. Wallis and his contemporaries[13] knew the formulæ for the sums

$$\sum_{k=1}^{n} k^m$$

for various positive integers m and knew that the sums were all expressed as $P(n)$ for polynomials of the form $P(x) = x^{m+1}/(m+1) + \text{lower degree terms}$.

---

[13]Including Pascal, who worked out these sums and areas in his *Traité du triangle arithmétique*, which we discussed earlier in the final section of Chapter 1.

By the above this meant one could derive

$$\int_0^b x^m dx = \frac{b^{m+1}}{m+1} \quad \text{and} \quad \int_a^b x^m dx = \frac{b^{m+1} - a^{m+1}}{m+1}$$

for any positive integer m. Wallis argued by interpolation that the same held for all rational $m \neq -1$.

An obvious question at this stage would be the areas of the segments of the circle $y = \sqrt{1 - x^2}$ for $0 \leq a < b \leq 1$. And this, of course, is where our discussion of areas started.

5. A Brief Calculus Primer, II: Slopes and Tangents

Tangents were discussed already by the classical Greeks. Whether or not he remembers it today, if the reader had a halfway decent Geometry course in high school, he learned that the tangent to a circle at a given point on the circle is perpendicular to the diagonal of the circle passing through the point. Thus through any point on the circle it is easy to draw the tangent line using only ruler and compass. It is also fairly easy to draw the tangents to parabolas, hyperbolas, and ellipses at points on these curves. I illustrate this with a parabola.

One way to define a parabola is as the locus of points equidistant from a given line called a *directrix* and a given point termed its *focus*. Suppose given a point P on the parabola, its focus F and directrix d as in FIGURE 3, below. Drop perpendiculars FO and PQ from F and P to d, label the vertex at the

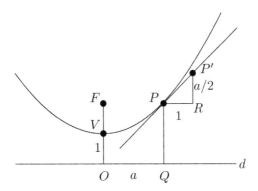

FIGURE 3. Tangent to a Parabola

intersection of FO with the parabola V, draw a line PR of length FV parallel to d, and at R draw a perpendicular to PR of length equalling half that of OQ. Call the endpoint of this line P'. The line PP' extended in both directions is the tangent to the parabola at P, as in the FIGURE.

This is, perhaps, not entirely clear, but with a little Algebra and Analytic Geometry it is readily verified. Choose d to be the x-axis and FO the y-axis.

Let OV be the unit 1, and P the point $\langle a, b \rangle$. Now, by definition, any point $\langle x, y \rangle$ on the parabola satisfies
$$(y - 2)^2 + (x - 0)^2 = (y - 0)^2 + (x - x)^2,$$
i.e., the squares of the distances from $\langle x, y \rangle$ to F and d are equal. Thus the points on the parabola satisfy
$$y^2 - 4y + 4 + x^2 = y^2,$$
whence $-4y + 4 + x^2 = 0$, i.e.,
$$y = \frac{x^2}{4} + 1. \tag{36}$$
But the equation of the line I claim to be the tangent is
$$\frac{y - b}{x - a} = \frac{b + \dfrac{a}{2} - b}{a + 1 - a} = \frac{a}{2},$$
i.e.,
$$y = \frac{a}{2}x - \frac{a^2}{2} + b. \tag{37}$$
Equating (36) and (37) to solve for points of intersection we have
$$\frac{x^2}{4} + 1 = \frac{a}{2}x - \frac{a^2}{2} + b,$$
i.e.,
$$\frac{x^2}{4} - \frac{a}{2}x + 1 + \frac{a^2}{2} - b = 0.$$
The quadratic formula yields
$$x = \frac{\dfrac{a}{2} \pm \sqrt{\dfrac{a^2}{4} - 4 \cdot \dfrac{1}{4}\left(1 + \dfrac{a^2}{2} - b\right)}}{\dfrac{2}{4}}$$
$$= a \pm 2\sqrt{\frac{a^2}{4} - 1 - \frac{a^2}{2} + b}$$
$$= a \pm 2\sqrt{b - \frac{a^2}{4} - 1}$$
$$= a$$
since $\langle a, b \rangle$, being on the parabola, satisfies (36). Thus $\langle a, b \rangle$ is the only point in common shared by the line PP' and the parabola. If we accept that the only lines crossing the parabola and intersecting it in only one point are vertical and that PP' is not vertical, this means the line must be a tangent line.

This is not exactly a *tour de force*, but if it is any indication, problems involving tangents can be quite tough—and I opted for the simple algebraic proof and not a geometric one. To be honest, I don't know where to begin a geometric verification of the correctness of the construction.

And other tangent problems are more difficult. For example, given a curve and a point not on the curve, to draw a line tangent to the curve and passing through the point. Ruler and compass constructions can be given whenever the curve is a conic section, but even for a circle it is not a trivial exercise. And for third or higher order polynomial curves the construction cannot in general be given by ruler and compass construction.[14]

2.5.1. Example. *Let P be a point outside a circle C centred at O. Let Q be the midpoint of OP and draw a circle C' of radius half the length of OP centred at Q. Let S be the point of intersection of the two circles. The triangle OSP is a right triangle as in* FIGURE 4, *below, because it is inscribed in the circle C' with a diameter as one of its sides. Thus $\angle OSP$ is a right angle and SP is*

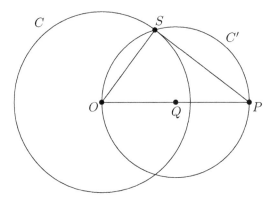

FIGURE 4.

perpendicular to a diameter of C, making it tangent to C.

As the exercise shows, if you know the seemingly unrelated fact that triangles inscribed in a circle with one side being a diameter are right triangles, it is not hard to draw a line tangent to a circle from a given point. But you have to know this fact first. One prefers a more direct algebraic approach, one in which one sits down and sets up some equations to be solved by familiar techniques.

Algebraically, the key to finding the equation of the tangent line to a curve $y = f(x)$ at a point $\langle a, f(a) \rangle$ on the curve is to know the slope $f'(a)$ of the line at $x = a$. One does this by finding the slopes of secant lines connecting the point $\langle a, f(a) \rangle$ with nearby points $\langle a + \Delta x, f(a + \Delta x) \rangle$, where Δx is a small increment[15]. The slope of this secant is just the ratio

$$\frac{\Delta y}{\Delta x} = \frac{f(x + \Delta x) - f(x)}{\Delta x},$$

which is called the *difference quotient*.

---

[14] *Cf.* Craig Smoryński, *History of Mathematics; A Supplement*, Springer Science + Business Media, LLC, New York, 2008, Chapter 4, for references and examples.

[15] "Δ" is the Greek letter "D" and stands for "difference". In my youth all textbooks used "Δx" for this increment. Today the fashion is to use the letter "h". I am inconsistent in my acceptance of the switch. I like "Δx" in explanations and "h" in calculations.

The Finite Binomial Theorem allows us to find these slopes easily for positive integral powers. Let $y = x^n$ for a positive integer n. Then

$$\frac{\Delta y}{\Delta x} = \frac{f(x+\Delta x) - f(x)}{\Delta x}$$

$$= \frac{x^n + nx^{n-1}\Delta x + \frac{n(n-1)}{2}x^{n-2}(\Delta x)^2 + \ldots + nx(\Delta x)^{n-1} + (\Delta x)^n - x^n}{\Delta x}$$

$$= nx^{n-1} + \left(\frac{n(n-1)}{2}x^{n-2} + \ldots + nx(\Delta x)^{n-3} + (\Delta x)^{n-2}\right)\Delta x.$$

For fixed x and n, the terms inside the large parentheses are readily bounded:

$$\binom{n}{k} < 2^n, \quad |x^k| < \max\{|x|^{n-1}, 1\}, \quad |\Delta x|^{n-k-1} < 1$$

so long as $|\Delta x| < 1$. There are $n-1$ such summands, whence the expression within the parentheses is less than $A = (n-1)2^n \max\{|x|^{n-1}, 1\}$ and

$$\left|\frac{\Delta y}{\Delta x} - nx^{n-1}\right| < A\Delta x,$$

for fixed A. Choosing Δx ever smaller will make $A\Delta x$ tend to 0. Thus, $\Delta y/\Delta x$ tends to nx^{n-1}.

This process of letting Δx tend to 0 and seeing how $\Delta y/\Delta x$ behaves is that of *taking the limit*. The operation of taking the limit of the difference quotient is called *differentiation* and we write

$$f'(x) = \lim_{\Delta x \to 0} \frac{\Delta y}{\Delta x} = \lim_{\Delta x \to 0} \frac{f(x+\Delta x) - f(x)}{\Delta x},$$

and the function $y' = f'(x)$ is called the *derivative* of $y = f(x)$. There are numerous notations for the derivative of f:

$$Df = f' = \frac{df}{dx} = \frac{d}{dx}f = \frac{dy}{dx} = y'$$

for the derivative as a function, and

$$Df(x) = f'(x) = \frac{df(x)}{dx} = \frac{df}{dx}(x) = \frac{dy}{dx}(x) = y'(x),$$

etc., for the value of the derivative at x.

The derivative of f is itself a function which may have its own derivative. When it does it is variously denoted:

$$D^2 f = f'' = \frac{d^2 f}{dx^2} = \frac{d^2}{dx^2}f = \frac{d^2 y}{dx^2} = y''.$$

The third derivative, f''', has similar notation. In the past lower case roman numerals replaced the primes for n-th derivatives for $n > 3$, but today arabic numerals enclosed in parentheses are used in their place:

$$f, f', f'' = f^{(2)}, f''' = f^{(3)}, f^{(4)}, f^{(5)}, \ldots$$

The Calculus derives its name from a series of rules for calculating derivatives and areas of functions set down independently by Newton and Leibniz. The rules for differentiation are easy to describe, some are easy to derive, and one can get quite proficient at applying them with enough practice. This is one of the main activities of the undergraduate Calculus course. The rules consist of a list of some initial derivatives followed by a list of rules for obtaining the derivatives of complex functions built up from simpler ones. TABLE 3 below collects a number of these.

TABLE 3. Some Differentiation Rules

Rule	$h(x)$	$h'(x)$	
I1	c	0	c a constant
I2	x^n	nx^{n-1}	n a positive integer
I3	$\sin x$	$\cos x$	
I4	$\cos x$	$-\sin x$	
I5	e^x	e^x	
I6	$\ln x$	$1/x$	$x > 0$
C1	$f(x) + g(x)$	$f'(x) + g'(x)$	
C2	$cf(x)$	$cf'(x)$	c a constant
C3	$f(x)g(x)$	$f'(x)g(x) + f(x)g'(x)$	
C4	$\dfrac{1}{f(x)}$	$-\dfrac{f'(x)}{f(x)^2}$	where $f(x) \neq 0$
C5	$\dfrac{f(x)}{g(x)}$	$\dfrac{f'(x)g(x) - f(x)g'(x)}{g(x)^2}$	where $g(x) \neq 0$
C6	$f(g(x))$	$f'(g(x))g'(x).$	

The derivations of these rules is not our main concern here. However, some of them are simple enough that their discussion should not take us too far afield. I1 is trivial: for $h(x) = c$,
$$\frac{h(x + \Delta x) - h(x)}{\Delta x} = \frac{c - c}{\Delta x} = \frac{0}{\Delta x} = 0$$
for all $\Delta x \neq 0$, whence the difference quotient is always 0 and tends[16] to 0 as Δx gets smaller and smaller:
$$h'(x) = \lim_{\Delta x \to 0} \frac{0}{\Delta x} = \lim_{\Delta x \to 0} 0 = 0.$$

---

[16]I am hoping that such vague terminology as "tends to", "gets smaller and smaller", "limit" is intuitive enough that a formal definition is not necessary. I believe this to be the case on historical grounds: a rigorous definition of the notion of limit only came about after a century and a half of working with limits and then not because people didn't know a limit when they saw one, but because such a definition was needed for rigorous proofs in more difficult cases than should appear in the present book, at least until Chapter 3.

We have already discussed I2. I3 and I4 are not needed here, but are worth mentioning because finding these derivatives involves proving

$$\lim_{x \to 0} \frac{\sin x}{x} = 1 \quad \text{and} \quad \lim_{x \to 0} \frac{1 - \cos x}{x} = 0,$$

the first two difficult limits that occur in the Calculus course. Most students are probably more convinced by a glance at the graphs of these functions or by tables of their values for ever smaller values of x than by the tricky geometric proofs given in class. In any event, students find they don't need to master the proof, but can get by knowing the results of TABLE 3 and applying them.

Likewise I5 is based on a difficult limit, one of the major hurdles in traditional Calculus courses:

$$\lim_{k \to \infty} \left(1 + \frac{1}{k}\right)^k = e,$$

or, more generally,

$$\lim_{k \to \infty} \left(1 + \frac{x}{k}\right)^k = e^x.$$

In an advanced book on Analysis, the author might define e^x by the infinite series,

$$e^x = \sum_{k=0}^{\infty} \frac{x^k}{k!},$$

and differentiate term-by-term to get

$$\frac{de^x}{dx} = \sum_{k=0}^{\infty} \frac{kx^{k-1}}{k!} = \sum_{k=1}^{\infty} \frac{kx^{k-1}}{k(k-1)!}$$

$$= \sum_{k=1}^{\infty} \frac{x^{k-1}}{(k-1)!} = \sum_{i=0}^{\infty} \frac{x^i}{i!}, \quad \text{for } i = k-1$$

$$= e^x.$$

The derivative of the logarithm will be discussed below.

As for the rules for dealing with functions h obtained by combining simpler functions, C1 and C2 are trivial. C3 is almost so:

$$f(x + \Delta x)g(x + \Delta x) - f(x)g(x) = f(x + \Delta x)g(x + \Delta x) - f(x)g(x + \Delta x)$$
$$+ f(x)g(x + \Delta x) - f(x)g(x)$$
$$= \big(f(x + \Delta x) - f(x)\big)g(x + \Delta x)$$
$$+ f(x)\big(g(x + \Delta x) - g(x)\big),$$

whence

$$\frac{f(x + \Delta x)g(x + \Delta x) - f(x)g(x)}{\Delta x} =$$

$$\frac{f(x + \Delta x) - f(x)}{\Delta x} g(x + \Delta x) + f(x) \frac{g(x + \Delta x) - g(x)}{\Delta x}.$$

5. A BRIEF CALCULUS PRIMER, II: SLOPES AND TANGENTS

Now as $\Delta x \to 0$, the difference quotients become derivatives and $g(x + \Delta x)$ approaches $g(x)$, so that in the limit this last equation becomes
$$(f \cdot g)'(x) = f'(x)g(x) + f(x)g'(x).$$
C4 is an easy corollary to C3: from $1 = f(x)(1/f(x))$, we have
$$0 = f'(x)\frac{1}{f(x)} + f(x)\frac{d}{dx}\left(\frac{1}{f(x)}\right),$$
whence
$$\frac{d}{dx}\left(\frac{1}{f(x)}\right) = -\frac{f'(x)}{f(x)} \cdot \frac{1}{f(x)},$$
i.e.,
$$\frac{d}{dx}\left(\frac{1}{f(x)}\right) = -\frac{f'(x)}{f(x)^2},$$
as desired.

C5 is a quick application of C3 and C4.

The last combination rule C6 is called the *Chain Rule* and is extremely powerful. As with C3 and C5, its form is not immediately obvious, but it is easily "derived" if we use the fractional notation for derivatives. Writing $y = g(x), z = f(y)$, we have
$$\frac{dz}{dy} \cdot \frac{dy}{dx} = \frac{dz}{\cancel{dy}} \cdot \frac{\cancel{dy}}{dx} = \frac{dz}{dx}.$$
This is hardly a valid proof, nor is the computation
$$\frac{dz}{dx} = \lim_{\Delta x \to 0} \frac{\Delta z}{\Delta x} = \lim_{\Delta x \to 0}\left(\frac{\Delta z}{\Delta y} \cdot \frac{\Delta y}{\Delta x}\right)$$
$$= \left(\lim_{\Delta x \to 0} \frac{\Delta z}{\Delta y}\right) \cdot \left(\lim_{\Delta x \to 0} \frac{\Delta y}{\Delta x}\right)$$
$$= \left(\lim_{\Delta y \to 0} \frac{\Delta z}{\Delta y}\right) \cdot \left(\lim_{\Delta x \to 0} \frac{\Delta y}{\Delta x}\right) = \frac{dz}{dy} \cdot \frac{dy}{dx}.$$

The problem is a subtle one: unless y is one-to-one, Δy can be 0 for some value of Δx. A correct general proof is a tiny bit deeper, but treating differential quotients as fractions has great heuristic value, and the cancellation of dy is a good mnemonic for remembering the Chain Rule.

One application of the Chain Rule is to inverse functions. For example e^x and $\ln x$ are inverse to one another:
$$x = e^{\ln x},$$
whence
$$1 = \frac{dx}{dx} = \frac{d}{dx}\left(e^{\ln x}\right) = e^{\ln x} \cdot \frac{d\ln x}{dx} = x \frac{d\ln x}{dx},$$
whence
$$\frac{d\ln x}{dx} = \frac{1}{x}$$
and we have derived I6.

Similarly, we can differentiate $x^{1/n}$, or more generally $x^{m/n}$:

$$\frac{d\left(x^{m/n}\right)^n}{dx} = n\left(x^{m/n}\right)^{n-1}\frac{dx^{m/n}}{dx} = nx^{m-\frac{m}{n}}\frac{dx^{m/n}}{dx}.$$

But $(x^{m/n})^n = x^m$, whence

$$\frac{d\left(x^{m/n}\right)^n}{dx} = mx^{m-1},$$

and thus

$$mx^{m-1} = n\left(x^{m-\frac{m}{n}}\right)\frac{dx^{m/n}}{dx}.$$

This yields

$$\frac{dx^{m/n}}{dx} = \frac{m}{n}x^{m-1-m+\frac{m}{n}} = \frac{m}{n}x^{\frac{m}{n}-1}.$$

Thus we have generalised I2 to all positive rational exponents.

For negative rational exponents a quick application of C5 (or C4 and C6) does the trick:

$$\frac{dx^{-q}}{dx} = \frac{d}{dx}\left(\frac{1}{x^q}\right) = \frac{0 \cdot x^q - 1 \cdot q \cdot x^{q-1}}{x^{2q}} = -qx^{q-1-2q} = -qx^{-q-1}.$$

I2 also holds for irrational exponents, but the proof lies a bit deeper. One must first define what is meant by x^r for general real numbers r. The simplest approach computationally, the one often used in advanced courses, is to define

$$x^r = e^{r\ln x},$$

where e^u is defined either as

$$\lim_{k\to\infty}\left(1+\frac{u}{k}\right)^k$$

or as the infinite series

$$e^u = \sum_{k=0}^{\infty}\frac{u^k}{k!}.$$

After establishing I5 and I6 using one of these definitions, one applies the Chain Rule again:

$$\frac{dx^r}{dx} = \frac{de^{r\ln x}}{dx} = e^{r\ln x}\frac{dr\ln x}{dx}, \quad \text{by I5, C5}$$

$$= e^{r\ln x}r\frac{d\ln x}{dx}, \quad \text{by C2}$$

$$= e^{r\ln x}r\frac{1}{x}, \quad \text{by C6}$$

$$= e^{r\ln x}re^{-\ln x} = re^{r\ln x - \ln x}$$

$$= re^{(r-1)\ln x} = rx^{r-1}.$$

This section has gone on for some pages, but there are still two things to be done. The first is to observe that with the rules given we can quickly find

the derivative of any polynomial function, a task that will have significance in later discussions in this book. It is very simple. Write
$$P(x) = a_n x^n + a_{n-1} x^{n-1} + \ldots + a_1 x + a_0.$$
Then
$$\begin{aligned}\frac{dP}{dx} &= \frac{d}{dx}(a_n x^n) + \frac{d}{dx}(a_{n-1} x^{n-1}) + \ldots + \frac{d}{dx}(a_1 x) + \frac{d}{dx} a_0, \text{ by C2}\\ &= a_n \frac{d}{dx} x^n + a_{n-1} \frac{d}{dx} x^{n-1} + \ldots + a_1 \frac{d}{dx} x + 0, \text{ by C3, I1}\\ &= a_n n x^{n-1} + a_{n-1}(n-1) x^{n-2} + \ldots + a_1, \text{ by I2}\\ &= n a_n x^{n-1} + (n-1) a_{n-1} x^{n-2} + \ldots + a_1.\end{aligned}$$

The second task is to use these rules to solve a tangent problem or two, something not relevant to the rest of the book, but which ties in with the opening discussion of this section.

A simple example is given in Example 2.5.1. Suppose the point O of FIGURE 4 is the origin, OP is the x-axis, and P is the point $\langle a, 0\rangle$. What are the coordinates of the point S? Let us say S is the point $\langle u, v\rangle$ on the circle $x^2 + y^2 = r^2$, i.e., $y = \sqrt{r^2 - x^2}$. The derivative of y is
$$\begin{aligned}y' &= \frac{d}{dx}(r^2 - x^2)^{1/2} = \frac{1}{2}(r^2 - x^2)^{-1/2} \frac{d(r^2 - x^2)}{dx}\\ &= \frac{1}{2} \cdot \frac{-2x}{\sqrt{r^2 - x^2}} = \frac{-x}{\sqrt{r^2 - x^2}} = \frac{-x}{y}.\end{aligned}$$
So the slope of SP is $-u/v$ and the equation of SP is
$$\frac{y - 0}{x - a} = -\frac{u}{v}.$$
Because $\langle u, v\rangle$ is on this line we have
$$\frac{v}{u - a} = -\frac{u}{v},$$
i.e., $v^2 = -u^2 + au$. Thus $au = u^2 + v^2 = r^2$ and we have
$$u = \frac{r^2}{a}.$$
Plugging this into the equation of the circle, we get
$$v = \sqrt{r^2 - u^2} = \sqrt{r^2 - \frac{r^4}{a^2}} = \frac{r}{a}\sqrt{a^2 - r^2}.$$
These are hardly the most exciting formulæ, but if we choose particular values of a, r, they will give definite numerical values. And, taking a different tack, in considering the angle $\theta = \angle SOP$, we know $u = r\cos\theta$, whence
$$\cos\theta = \frac{r}{a}, \quad \text{i.e., } \theta = \cos^{-1}\left(\frac{r}{a}\right),$$
a more memorable expression of the solution.

2.5.2. Exercise. *Find the equation of a line tangent to the parabola $y = x^2 - 2x + 3$ and passing through the point $\langle 2, 1\rangle$.*

2.5.3. Exercise. *How many lines passing through the point $\langle 4, -2 \rangle$ are tangent to the curve $y = x^3 - 6x^2 + 11x - 6$?*

6. A Brief Calculus Primer, III: Fundamental Theorem of Calculus

To better find the areas under certain curves Newton and Leibniz exploited a result from the geometric lectures of Isaac Barrow (1630 – 1677): the area and tangent problems are inverse to one another. This is termed the *Fundamental Theorem of Calculus* and is fundamental indeed.

If $f(x)$ is a differentiable function then f' is called the derivative of f. In the other direction, if F is a function such that $F' = f$, then F is called an *antiderivative* of f. In the Calculus it is shown that the antiderivative is unique up to an additive constant: if $F' = f = G'$, there is a constant C such that for all x
$$F(x) = G(x) + C.$$
Basically, the reason this is true is simple. If $F' = G' = f$, then $F'(x) - G'(x) = 0$ and the tangent lines to the curve $y = F(x) - G(x)$ are all horizontal, whence the curve itself must be horizontal, i.e., constant. A rigorous proof is a bit more involved than this, but the intuition is clear.

The Fundamental Theorem of Calculus is stated in two parts:

2.6.1. Theorem (Fundamental Theorem of Calculus). *Let f be continuous on the interval $[a, b]$.*
i. The function
$$F(x) = \int_a^x f(t) dt \qquad (38)$$
is an antiderivative of f on $[a, b]$.
ii. If G is any antiderivative of f on $[a, b]$, then, for $a \leq c \leq d \leq b$,
$$\int_c^d f(t) dt = G(d) - G(c).$$

A rigorous proof depends on some deep properties of continuous functions that I do not care to prove and which are usually stated without proof in textbooks on the Calculus. For example, every function that is continuous on a closed interval must assume a maximum and a minimum value somewhere in that interval. Accepting this I can explain the main idea of the proof quite simply. Let $y = f(x)$ and let F be the area function defined as in (38). The difference $F(x + \Delta x) - F(x)$ is the area between $y = f(x)$ and the x-axis between the bounds x and $x + \Delta x$. As in section 4, this is nicely approximated by the area $f(x) \Delta x$ of the rectangle of height $f(x)$ and base $\Delta x = x + \Delta x - x$. Thus
$$F(x + \Delta x) - F(x) \approx f(x) \Delta x.$$
In fact, if m is the minimum and M the maximum value of f in the subinterval $[x, x + \Delta x]$, then (assuming f nonnegative and $\Delta x > 0$ for simplicity)
$$m \Delta x \leq F(x + \Delta x) - F(x) \leq M \Delta x,$$

i.e.,
$$m \le \frac{F(x+\Delta x) - F(x)}{\Delta x} \le M.$$
Now, by continuity, m and M are very close to $f(x)$, and as Δx tends to 0, m and M tend to $f(x)$:
$$\lim_{\Delta x \to 0} \frac{F(x+\Delta x) - F(x)}{\Delta x} = f(x).$$
Thus we have $F'(x) = f(x)$, i.e., part i of the Theorem.

As I said, this is only the idea behind the proof, which requires quite a bit more detail to be rigorously established. But once one has proven part i, part ii follows quickly. If F is given by (38) and G is another antiderivative of f, there is a constant C such that $F(x) = G(x) + C$, and we see
$$\int_c^d f(t)dt = \int_a^d f(t)dt - \int_a^c f(t)dt = F(d) - F(c)$$
$$= (G(d) + C) - (G(c) + C) = G(d) - G(c).$$

Thus, the Fundamental Theorem of Calculus reduces the problem of finding an area to that of finding an antiderivative, evaluating it at a couple of points, and subtracting the values. For this reason one writes
$$\int f(x)dx$$
to denote an arbitrary antiderivative of f and calls it the *indefinite integral* of f. If F is a particular antiderivative of f, one writes
$$\int f(x)dx = F(x) + C$$
to remind us that F is unique only up to a constant.

We can illustrate this by reference to our example of section 4, namely the parabola $y = x^2$. We have
$$\int x^2 dx = \frac{x^3}{3} + C,$$
as one can verify by differentiation, whence the area below the parabola between a and b is
$$\left.\frac{x^3}{3}\right|_a^b = \frac{b^3}{3} - \frac{a^3}{3}.$$

This was much easier than calculating the sums as we did back in section 4, but it is not always this easy. The problem, of course, is finding the antiderivative. This is an inverse problem and is not the simple deterministic algorithm that differentiation is. It is, in fact, a partial, nondeterministic algorithm, with all the difficulties that entails.

One starts as with differentiation with a small table of functions and their antiderivatives. TABLE 4 below can be derived from TABLE 3.

When it comes to rules for integrating combinations of functions there is one main rule:

TABLE 4. A Table of Integrals

$f(x)$	$\int f(x)dx$	
0	C	
a	$ax + C$	a a constant
x^r	$\dfrac{x^{r+1}}{r+1} + C$	$r \neq -1$
$\dfrac{1}{x}$	$\ln x + C$	
e^x	$e^x + C$	
$\sin x$	$-\cos x + C$	
$\cos x$	$\sin x + C$	

2.6.2. Theorem (Linearity of the Integral). *For integrable functions f, g and constants a, b,*

$$\int (af(x) + bg(x))dx = a\int f(x)dx + b\int g(x)dx + C. \qquad (39)$$

For, by C1, C2, the derivative of the right-hand side of (39) is

$$a\frac{d}{dx}\int f(x)dx + b\frac{d}{dx}\int g(x)dx = af(x) + bg(x) = \frac{d}{dx}\int (af(x) + bg(x))dx.$$

Together with TABLE 4 this allows us to integrate any polynomial in standard form. On page 65, for example, we listed some integrals Newton cited from Wallis:

$$\int (1 - x^2)^1 dx = x - \frac{x^3}{3} + C$$

$$\int (1 - x^2)^2 dx = \int (1 - 2x^2 + x^4) dx = x - \frac{2x^3}{3} + \frac{x^5}{5} + C$$

$$\int (1 - x^2)^3 dx = \int (1 - 3x^2 + 3x^4 - x^6) dx = x - \frac{3x^3}{3} + \frac{3x^5}{5} - \frac{x^7}{7} + C.$$

Unfortunately, not all integrals are this easy. Multiplicative combinations that do not expand like binomials into additive combinations of functions from the TABLE can be quite tricky. Consider $h(x) = x\cos x^2$. It is not too hard to see that it is almost of the form $g'(f(x))f'(x)$ where $f(x) = x^2$ and $g(y) = \sin y$. If we differentiate $\sin x^2$ we get $(\cos x^2) \cdot 2x$, from which we quickly conclude

$$\int x\cos x^2\, dx = \frac{\sin x^2}{2} + C.$$

Now consider $h(x) = x\cos x$. This could not have arisen by application of the Chain Rule. It is, in fact, one of the summands resulting from differentiating a product

$$\frac{d}{dx}(x\sin x) = x\cos x + \sin x.$$

But $\int \sin x \, dx = -\cos x + C$, whence if we add $\cos x$ to $x \sin x$, we see

$$\frac{d}{dx}(x \sin x + \cos x) = x \cos x + \sin x - \sin x = x \cos x$$

and we conclude

$$\int x \cos x \, dx = x \sin x + \cos x + C.$$

There is a technique of *integration by parts* by which one begins with a product such as $x \cos x$, assumes it to be one of the products $f'(x)g(x)$ or $f(x)g'(x)$ arising from application of the rule C3, and hopes that if, say, $f'(x)g(x)$ is difficult to integrate then $f(x)g'(x)$ will be easy.

Some products are not the result of the application of the Chain Rule and do not succumb to integration by parts. Students in the Calculus are tortured for months on end (or so it seemed when I was a student) integrating products of trigonometric functions, where trigonometric identities can be called into play to make the *integrand*, as the function being integrated is called, more amenable. For example, $y = \sin^2 x = (\sin x)(\sin x)$ is not the result of applying the Chain Rule. One can apply integration by parts:

$$f'(x) = \sin x \Rightarrow f(x) = -\cos x$$
$$g(x) = \sin x \Rightarrow g'(x) = \cos x$$

and

$$\int f'(x)g(x)dx = f(x)g(x) - \int f(x)g'(x)dx$$
$$= (-\cos x)(\sin x) - \int -(\cos x)\cos x \, dx$$
$$= -\sin x \cos x + \int \cos^2 x \, dx. \qquad (40)$$

Integrating $\cos^2 x$ by parts yields

$$\int \cos^2 x \, dx = \sin x \cos x + \int \sin^2 x \, dx,$$

which, when plugged into (40) results in the useless

$$\int \sin^2 x \, dx = -\sin x \cos x + \left(\sin x \cos x + \int \sin^2 x \, dx \right) = \int \sin^2 x \, dx.$$

However, (40) is not completely worthless because we have the trigonometric identity $\cos^2 x = 1 - \sin^2 x$. Plugging $1 - \sin^2 x$ in for $\cos^2 x$ in the integral of (40) yields

$$\int \sin^2 x \, dx = -\sin x \cos x + \int (1 - \sin^2 x) dx$$
$$= -\sin x \cos x + x - \int \sin^2 x \, dx,$$

and we can solve for the integral:

$$2 \int \sin^2 x \, dx = -\sin x \cos x + x,$$

i.e.,
$$\int \sin^2 x \, dx = \frac{x - \sin x \cos x}{2} + C,$$
when we add the *constant of integration*. The reader may wish to differentiate this last expression using the rules of TABLE 3 to verify the correctness of the result.

Integration emerges as not so much an algorithm like differentiation as a loose bag of tricks with some algorithmic features. Fortunately for us, we won't be needing to perform any tricky integrations in this book. If we do need an integral, there are plenty of tables of integrals available, expensive software that can do the integration for us, and, for those of us who have one, the *TI-89 Titanium*. If I should need to know a difficult integral I will simply inform the reader, who can then wonder which oracle I consulted to arrive at the result, and then perform the differentiation to see that my oracle was trustworthy.

7. A Brief Calculus Primer, IV: Series

Up till now our discussions of topics from the Calculus have been, if the reader will excuse the expression, of merely tangential interest. More immediately relevant is another topic from the Calculus—infinite sums. An infinite sum,
$$\sum_{k=1}^{\infty} a_k \quad \text{or} \quad \sum_{k=0}^{\infty} a_k,$$
is called an infinite *series*, the lists of terms $a_1, a_2, a_3, \ldots$ or $a_0, a_1, a_2, \ldots$ being called infinite *sequences*.

Before the age of Newton, infinite series made only sporadic appearances. The granddaddy of all infinite series is that based on geometric progressions. A *geometric progression* is a sequence obtained from an initial element a by repeatedly multiplying by a fixed ratio r:
$$a, ar, ar^2, \ldots, ar^n, \tag{41}$$
for a finite progression, and
$$a, ar, ar^2, \ldots, \tag{42}$$
for an infinite one.

Geometric progressions are among the easiest sequences to sum. For a finite such progression (41), form the sum
$$S_n = a + ar + ar^2 + \ldots + ar^n,$$
and notice
$$rS_n = ar + ar^2 + ar^3 + \ldots + ar^{n+1},$$
whence
$$a + rS_n = a + ar + ar^2 + ar^3 + \ldots + ar^{n+1} = S_n + ar^{n+1}.$$
Simple algebra yields
$$a - ar^{n+1} = (1 - r)S_n,$$

7. A BRIEF CALCULUS PRIMER, IV: SERIES

i.e.,
$$S_n = \frac{a - ar^{n+1}}{1-r} = a \cdot \frac{1 - r^{n+1}}{1-r}, \qquad (43)$$
provided $r \neq 1$. This sum was known already to the ancient Egyptians and was either independently discovered by or transmitted to the rest of the ancient civilisations of the old world.

The behaviour of the series formed by an infinite geometric progression is a matter of greater subtlety. Today we would define the infinite sum,
$$S = \sum_{k=0}^{\infty} ar^k \qquad (44)$$
to be the limit of the sequence of partial sums $S_0, S_1, S_2, \ldots$,
$$S = \lim_{n \to \infty} S_n = \lim_{n \to \infty} \sum_{k=0}^{n} ar^k,$$
provided this limit exists. And we would show that the limit exists for $|r| < 1$ and does not exist for $|r| \geq 1$.

We can verify this existence/nonexistence claim directly by rewriting (43),
$$S_n = a \cdot \frac{1 - r^{n+1}}{1-r} = a \cdot \frac{1}{1-r} - a \cdot \frac{r^{n+1}}{1-r},$$
and noting that the first term is a fixed constant not depending on n, and the second term is a constant times r^{n+1} and for $|r| < 1$ gets as small as one pleases by choosing n large enough, but large without bound by choosing n large if $|r| > 1$. Thus, for $|r| < 1$, the sequence $S_0, S_1, S_2, \ldots$ differs from $a/(1-r)$ by an amount that tends to 0 and the limit of the sequence is determined:
$$S = \lim_{n \to \infty} S_n = \frac{a}{1-r}.$$
And, for $|r| > 1$ no limit can exist: The term $a/(1-r)$ is negligible in comparison with the ever increasing size of $ar^{n+1}/(1-r)$ and we have
$$\lim_{n \to \infty} S_n = +\infty, \quad \text{if } a > 0, r > 1$$
$$\lim_{n \to \infty} S_n = -\infty, \quad \text{if } a < 0, r > 1,$$
while $S_0, S_1, S_2, \ldots$ oscillates ever more wildly if $r < -1$, e.g., for $a = 1, r = -2$, the terms ar^k are
$$1, -2, 4, -8, 16, -32, 64, \ldots$$
with partial sums
$$1, -1, 3, -5, 11, -21, 43, \ldots$$

The cases $r = \pm 1$ are special. For $r = 1$, we cannot use (43), but we don't need to:
$$S_n = a + a \cdot 1 + a \cdot 1^2 + \ldots + a \cdot 1^n = \underbrace{a + a + \ldots + a}_{n+1} = (n+1)a.$$
And $S_n \to \pm \infty$ as $n \to \infty$ according as a is positive or negative.

For $r = -1$, we have the partial sums
$$a, 0, a, 0, a, \ldots$$
and an alternation, a particularly unimaginative form of oscillation.

I emphasise that this is the modern treatment—*definition* of the sum as a limit *if* the limit exists. There were alternatives. Bernard Bolzano, whom we shall encounter in greater detail in Chapter 3, below, had three distinct views of what an infinite series was. In his work on the Binomial Theorem published in 1816, he espoused the modern definition of a series as the limit of its partial sums. In his later work on the theory of science, an infinite sum was to him an *accumulation*. Suppose, for example, we had infinitely many buckets with varying amounts of sand. The sum of these amounts would measure the total amount of sand we would have if we dumped all the buckets into one big pile. It wouldn't matter what order we dumped the buckets in, or if we first took some buckets and distributed their contents into several smaller buckets: we would still have the same total, for that is the nature of sums. This agrees with his earlier definition—provided all the terms in the series share the same sign.

In the meantime, in the early 1830s, Bolzano had yet another view of infinite sums, a description of how mathematicians dealt with them: A sum was an *infinite number expression* which mathematicians manipulated algebraically. He declared an infinite number expression to be *measurable* if one could fix its place among the rational numbers. This fixing was a sort of inadequately defined nod in the direction of the definition of a sum as a limit.

It is a bit hard to accept that mathematicians did not always stick faithfully to the notion that a series only had meaning as the limiting value of its partial sums, but such was the case.

> In writing to Goldbach on August 7, 1765, Euler refers to Bernoulli's argument that divergent series such as
> $$+1 - 2 + 6 - 24 + 120 - 720 + \ldots$$
> have no sum but says that these series have a definite *value*. He notes that we should not use the term "sum" because this refers to actual addition. He then states the general principle which explains what he means by a definite value. He points out that the divergent series comes from finite algebraic expressions and then says that the value of the series is the *value of the algebraic expression from which the series comes*.[17]

Every series arising from a geometric progression for $r \neq 1$ has such a value, which can be determined by the method used to sum a finite such progression. Let
$$S = a + ar + ar^2 + \ldots$$
Then
$$a + rS = a + r(a + ar + ar^2 + \ldots) = a + ar + ar^2 + \ldots = S,$$

---

[17]Morris Kline, *Mathematical Thought from Ancient to Modern Times*, Oxford University Press, New York, 1972, pp. 462 – 463.

and $a = (1-r)S$, whence
$$S = \frac{a}{1-r}. \tag{45}$$
For $|r| < 1$, this agrees with the limit. For $r = 1$, it yields no value. But for $|r| > 1$ or $r = -1$, it gives definite values: Let $a = 1$. For $r = 2, -2, -1$, we have

$$\sum_{k=0}^{\infty} 2^k = 1 + 2 + 4 + 8 + \ldots = \frac{1}{1-2} = -1$$

$$\sum_{k=0}^{\infty} (-2)^k = 1 - 2 + 4 - 8 + \ldots = \frac{1}{1-(-2)} = \frac{1}{3}$$

$$\sum_{k=0}^{\infty} (-1)^k = 1 - 1 + 1 - 1 + \ldots = \frac{1}{1-(-1)} = \frac{1}{2}.$$

If one thinks of these as sums, the results are paradoxical. The first is a series of positive numbers with a negative sum. The second offers a rather small positive sum for an infinitely oscillating sum. Moreover, grouping the terms (consolidating pairs of sand buckets á la Bolzano),

$$(1-2) + (4-8) + (16-32) + \ldots = -1 - 4 - 16 - \ldots = \frac{1}{3},$$

yields a positive sum for a series of negative numbers. And both the second and third offer proper fractions for integral sums. Moreover, grouping the terms in the third series results in different values:

$$(1-1) + (1-1) + (1-1) + \ldots = 0 + 0 + 0 + \ldots = 0$$
$$1 + (-1+1) + (-1+1) + \ldots = 1 + 0 + 0 + \ldots = 1.$$

I did not comment on it when discussing Newton's remarks, but he referred explicitly to generating series by division. Some mathematicians of the day preferred to generate and sum the geometric series in this way as in FIGURE 5, below.

$$\begin{array}{r}
1 + x + x^2 + \ldots \\
1-x \overline{\smash{)}\, 1 } \\
\underline{1 - x } \\
x \\
\underline{x - x^2 } \\
x^2 \\
\underline{x^2 - x^3} \\
x^3 \\
\text{etc.}
\end{array}
\qquad
\begin{array}{r}
1 + x + x^2 + \ldots \\
1-x \overline{\smash{)}\, 1 } \\
1 + x + x^2 + \ldots \\
-x - x^2 - \ldots \\
\underline{-x - x^2 - \ldots} \\
0
\end{array}$$

FIGURE 5.

A brief look at the history of series before and after Newton will give us some idea of the importance of Newton's Binomial Theorem.

One all but finds the geometric progression for $r = 1/4$ summed in the *Quadrature of the Parabola* by Archimedes (287 B.C. – 212 B.C.). One of the methods he used to find the area of a parabolic segment was to progressively fill up the segment with triangles, the areas of which are in geometric progression. He sums only finite portions of the progression, but in essence determines the limit of the infinite series, without reference to limits or to summing the infinite progression. In his classic history of infinite series, Richard Reiff describes the work thus:

> The application and consideration of an infinite process to the resolution of problems, namely for the determination of areas goes, as is known, back to Archimedes. In his quadrature of the parabola he applied the series
> $$1 + \frac{1}{4} + \frac{1}{4^2} + \ldots$$
> of which he showed that it can be brought arbitrarily close to a certain limit. Similar considerations play a rôle by his other rectifications and quadratures.[18]

The next episode in the European history of infinite series comes a millennium and a half later. Edward Grant introduces it as follows:

> The subject of infinite series, with its associated paradoxes, exerted a strong fascination for medieval natural philosophers and mathematicians. Many series were formulated and utilized by Oxford (Merton College) scholastics such as Swineshead, Dumbleton, Heytesbury, and others in the first half of the fourteenth century, and then by Parisian scholastics such as Nicole Oresme and, much later, Alvarus Thomas, to name only two.[19]

Of particular interest are the works of Oresme (*c.* 1320 – 1382) and Alvaro Thomaz (*fl.* 2nd half 15th century). As is often the case with original works, their approaches would today appear overly complicated. Oresme begins by assuming a certain quantity we shall call a and imagines removing a fixed proportion r ($0 < r < 1$) of it. Then one removes r of the remainder, then r of that remainder, ... Symbolically, he is considering a sequence

$$a_0 = a$$
$$a_{n+1} = a_n - ra_n = a_n(1-r).$$

Clearly $a_n = a(1-r)^n$. He claims[20], for $0 < r < 1$, the remainders tend to 0,

$$\lim_{n \to \infty} a_n = \lim_{n \to \infty} a(1-r)^n = 0.$$

---

[18] Richard Reiff, *Geschichte der unendlichen Reihen*, Verlag der H. Laupp'schen Buchhandlung, Tübingen, 1889, p. 4.

[19] Edward Grant (ed.), *A Source Book in Medieval Science*, Harvard University Press, Cambridge (Mass.), 1974, p. 132.

[20] Pertinent passages from Oresme's work on series can be found in English translation in Grant, *ibid.*, pp. 131 – 135. The discussion of this proof (p. 133) is, however, barely intelligible even in translation.

7. A BRIEF CALCULUS PRIMER, IV: SERIES

Thus, the total amount to be removed must equal all of a: What is removed is $r \times$ the remainder at a given stage:

$$ar + a(1-r)r + a(1-r)^2 r + \ldots = a$$
$$a + a(1-r) + a(1-r)^2 + \ldots = \frac{a}{r},$$

or, replacing $1-r$ by r,

$$a + ar + ar^2 + \ldots = \frac{a}{1-r},$$

agreeing with (45).

More impressive than his roundabout approach to summing the infinite geometric progression was Oresme's simple proof of the divergence of the *harmonic series*:

$$\sum_{k=1}^{\infty} \frac{1}{k} = +\infty.$$

Unlike his remarks on the geometric progression, Oresme is marvellously clear on this point:

> For example, let a one-foot quantity be assumed to which one-half of a foot is added during the first proportional part of an hour, then one-third of a foot in another [or next proportional part of an hour], then one-fourth [of a foot], then one-fifth, and so on into infinity following the series of [natural] numbers, I say that the whole would become infinite, which is proved as follows: There exist infinite parts of which any one will be greater than one-half foot and [therefore] the whole will be infinite. The antecedent is obvious, since $1/4$ and $1/3$ are greater than $1/2$, similarly [the sum of the parts] from $1/5$ to $1/8$ [is greater than $1/2$] and [also the sum of the parts] from $1/9$ to $1/16$, and so on into infinity.[21]

This is the proof still given in modern textbooks, but nowadays presented more explicitly:

$$1 > \frac{1}{2}$$
$$\frac{1}{2} \geq \frac{1}{2}$$
$$\frac{1}{3} + \frac{1}{4} > \frac{1}{4} + \frac{1}{4} = \frac{1}{2}$$
$$\frac{1}{5} + \frac{1}{6} + \frac{1}{7} + \frac{1}{8} > \frac{1}{8} + \frac{1}{8} + \frac{1}{8} + \frac{1}{8} = \frac{4}{8} = \frac{1}{2}$$
$$\vdots$$

---

[21] Grant, *op. cit.* p. 135. The bracketed insertions are Grant's.

More generally,

$$\frac{1}{2^k+1} + \frac{1}{2^k+2} + \ldots + \frac{1}{2^k+2^k} > \underbrace{\frac{1}{2^{k+1}} + \frac{1}{2^{k+1}} + \ldots + \frac{1}{2^{k+1}}}_{2^k} = \frac{2^k}{2^{k+1}} = \frac{1}{2}.$$

Thus

$$1 + \frac{1}{2} + \frac{1}{3} + \ldots \geq \frac{1}{2} + \frac{1}{2} + \frac{1}{2} + \ldots = +\infty.$$

Another often cited result of Oresme's is the summation of the series

$$\frac{1}{2} + \frac{2}{4} + \frac{3}{8} + \frac{4}{16} + \frac{5}{32} + \ldots = 2. \tag{46}$$

In publishing an account of a work, *Liber de triplici motu*, published in 1509 by Alvaro Thomaz, Heinrich Wieleitner described it as "a systematic presentation of the application of infinite series".[22] Mention of the word "application" helps clear up an odd feature of Thomaz's exposition. Whereas today we are given a and r and look to find the sum S of the infinite series, Thomaz started with r and S and solved for a. This harks back to the Merton scholars, mentioned in one of our citations from Grant's source book on page 98, above, and even farther back to the, as yet unmentioned in this book, paradoxes of Zeno of Elea ($c.$ 490 B.C. – $c.$ 425 B.C.). The simplest of these asserts that you cannot get from point A to point B. For, before you can do that you must reach the midpoint C. But before you can do that you must reach the point halfway between A and C, and the point halfway between it and A, etc. One must do infinitely many things in a finite amount of time, which is clearly impossible.

The task is not impossible if each successive task takes only half as much time as its predecessor. If the first task takes 1 minute to perform, the second half a minute, the third a quarter minute, etc., then all the tasks can be accomplished in

$$1 + \frac{1}{2} + \frac{1}{4} + \frac{1}{8} + \ldots = \frac{1}{1 - \frac{1}{2}} = 2 \text{ minutes}.$$

The Merton scholars are reported to have applied numerous series in discussing motion.[23] Oresme and Thomaz worked in that tradition, as is evident from Oresme's dividing the hour into proportional parts in discussing the harmonic series. This also probably explains his, to modern eyes strange, approach to discussing the sum of a geometric progression: He started with a given quantity a from which he subtracted proportional parts. Thomaz does the same.

---

[22]Heinrich Wieleitner, "Zur Geschichte der unendlichen Reihen im christlichen Mittelalter", *Bibliotheca Mathematica*, 3rd series, vol. 14 (1914), pp. 150 – 168; here: p. 151.

[23]I have not read them, preferring to discuss the more accessible work of Oresme and Thomaz. (I have Grant's source book with excerpts from Oresme and a copy of Wieleitner's discussion of Thomaz I downloaded from the internet. A number of excerpts from the Merton scholars can be found in Marshall Clagett's *The Science of Mechanics in the Middle Ages* (University of Wisconsin Press, Madison, 1959), but they contain little of interest on infinite series.)

7. A BRIEF CALCULUS PRIMER, IV: SERIES

For us, however, the interest is in the series themselves and summing them. As Wieleitner points out, Thomaz, lacking algebraic symbolism, cannot formulate and prove general results, but presents examples of each of several types, enough to allow the reader to see the general form of the solutions.

In addition to presenting the sum of an infinite geometric progression, Thomaz treats sums like Oresme's (46) as well as a few other types. If one multiplies (46) by 2 one gets a series of the form,

$$1 + 2r + 3r^2 + 4r^3 + \ldots = \sum_{k=0}^{\infty}(k+1)r^k, \qquad (47)$$

for $r = 1/2$. Wieleitner explains Thomaz's geometric presentation of the summation and then explains it algebraically, by separating like terms. If S denotes the sum (47), then

$$S = 1 + r + r^2 + r^3 + \ldots$$
$$+ r + r^2 + r^3 + \ldots$$
$$+ r^2 + r^3 + \ldots$$
$$+ r^3 + \ldots$$
$$= (1 + r + r^2 + r^3 + \ldots)(1 + r + r^2 + r^3 + \ldots)$$
$$= \frac{1}{1-r} \cdot \frac{1}{1-r} = \frac{1}{(1-r)^2}.$$

In Oresme's case, we have $r = 1/2$ and

$$1 + \frac{2}{2} + \frac{3}{4} + \frac{4}{8} + \ldots = 4, \qquad (48)$$

and dividing by 2 yields (46).

We can also sum the series in imitation of the way we summed the geometric progression. Write

$$S = \sum_{k=0}^{\infty}(k+1)r^k \quad \text{and} \quad G = \sum_{k=0}^{\infty} r^k.$$

Then

$$S - G = \sum_{k=0}^{\infty} kr^k = 0 + r + 2r^2 + 3r^3 + \ldots$$
$$= r(1 + 2r + 3r^2 + \ldots) = rS.$$

Thus $(1-r)S = G$, and

$$S = \frac{1}{1-r}G = \frac{1}{1-r} \cdot \frac{1}{1-r} = \frac{1}{(1-r)^2}.$$

2.7.1. Exercise. *Carry out the long division,*

$$1 + 2r + 3r^2 + 4r^3 + \ldots \,)\overline{1 \qquad},$$

as another method of determining the sum.

Another series Thomaz summed was

$$1 + \frac{3}{2} \cdot \frac{1}{2} + \frac{5}{4} \cdot \frac{1}{2^2} + \frac{9}{8} \cdot \frac{1}{2^3} + \cdots \qquad (49)$$

To do this, he simply subtracted

$$1 + \frac{1}{2} + \frac{1}{4} + \frac{1}{8} + \ldots = 2$$

to get

$$\frac{1}{2} \cdot \frac{1}{2} + \frac{1}{4} \cdot \frac{1}{2^2} + \frac{1}{8} \cdot \frac{1}{2^3} + \cdots,$$

i.e.,

$$\frac{1}{4} + \frac{1}{4^2} + \frac{1}{4^3} + \ldots = \frac{1}{4}\left(1 + \frac{1}{4} + \frac{1}{4^2} + \ldots\right) = \frac{1}{4} \cdot \frac{1}{1 - \frac{1}{4}} = \frac{1}{3}.$$

Thus (49) is $2 + 1/3 = 7/3$.

He couldn't handle every series exactly. For example,

$$1 + \frac{2}{1} \cdot \frac{1}{2} + \frac{3}{2} \cdot \frac{1}{2^2} + \frac{4}{3} \cdot \frac{1}{2^3} + \ldots = 2 + \ln 2 \qquad (50)$$

was beyond him. There are two good reasons for this. First, logarithms were unknown in his day. Second, the presence of a logarithm in the final result rules out an algebraic determination of the sum. However, he could squeeze the result between two bounds by comparing the series with

$$1 + \frac{1}{2} + \frac{1}{4} + \frac{1}{8} + \ldots = 2 \qquad (51)$$

and

$$1 + 2 \cdot \frac{1}{2} + 3 \cdot \frac{1}{4} + 4 \cdot \frac{1}{8} + \ldots = 4,$$

this last sum determined by (48), above. He was thus able to conclude the series in question to be trapped between 2 and 4. He could easily have done better by subtracting (51) to obtain

$$\frac{1}{2} \cdot \frac{1}{2} + \frac{1}{4} \cdot \frac{1}{3} \cdot \frac{1}{8} + \ldots < \frac{1}{2} + \frac{1}{4} + \frac{1}{8} + \ldots = 1,$$

to conclude the sum to lie between 2 and 3.

In a recent history of infinite series, Giovanni Ferraro notes summations of infinite geometric progressions in the work of François Viète and Grégoire de Saint-Vincent. The latter made a most impressive application of geometric progressions in his *Opus geometricum quadraturæ circuli et sectionum coni decem libris comprehensum* of 1647, proving a result that was recognised two years later by his friend and fellow Jesuit Alphonse Antonio de Sarasa as showing that the area beneath the hyperbola,

$$f(x) = \int_1^x \frac{dt}{t},$$

behaved like a logarithm.[24] It is in fact the natural logarithm of x, as we know from TABLE 4. We will use a variant of this observation in the next section.

A little less impressive, but still of interest, is a work by Pietro Mengoli, *Novæ quadraturæ arithmeticæ, seu de additione fractionum* (1650). In this work he summed such infinite series as

$$\sum_{k=1}^{\infty} \frac{1}{k(k+1)} = 1$$

$$\sum_{k=1}^{\infty} \frac{1}{k(k+2)} = \frac{3}{4}.$$

Moreover, he offered an alternate proof of Oresme's result on the divergence of the harmonic series. From

$$\frac{1}{n-1} + \frac{1}{n} + \frac{1}{n+1} = \frac{n(n+1) + (n-1)(n+1) + (n-1)n}{(n-1)n(n+1)}$$

$$= \frac{n^2 + n + n^2 - 1 + n^2 - n}{n(n^2 - 1)} = \frac{3n^2 - 1}{n(n^2 - 1)}$$

$$> \frac{3n^2 - 3}{n(n^2 - 1)} = \frac{3}{n},$$

he concluded for

$$S = 1 + \frac{1}{2} + \frac{1}{3} + \frac{1}{4} + \frac{1}{5} + \frac{1}{6} + \frac{1}{7} + \ldots,$$

that

$$S = 1 + \left(\frac{1}{2} + \frac{1}{3} + \frac{1}{4}\right) + \left(\frac{1}{5} + \frac{1}{6} + \frac{1}{7}\right) + \ldots$$

$$> 1 + \frac{3}{3} + \frac{3}{6} + \frac{3}{9} + \ldots = 1 + \left(1 + \frac{1}{2} + \frac{1}{3} + \ldots\right)$$

$$> 1 + S,$$

whence S cannot exist.[25]

For all the dexterity and cleverness exhibited, the work of Oresme, Thomaz, and Mengoli is very primitive and very limited. About the time Oresme was toying with his series, the Hindu mathematicians in India were dealing with infinite series representations of trigonometrical functions. It would take another century and a half for the Europeans to catch up. And Newton's Binomial Theorem would be an important tool in this.

2.7.2. Exercise. *Let $f_0 = f_1 = 1, f_{n+2} = f_n + f_{n+1}$. Express the sum of the series*

$$f_0 + f_1 r + f_2 r^2 + \ldots$$

as a rational function. [*Hint: Use long division.*]

---

[24]Giovanni Ferraro, *The Rise and Development of the Theory of Series Up to the Early 1820s*, Springer Science+Business Media, LLC, New York, 2008, pp. 5 – 7.
[25]*Ibid.*, pp. 8 – 9.

8. Power Series

From 1200 to 1600 Indian mathematics was largely carried out by what is called the Kerala school of mathematics, so named in honour of a region on the southwest coast of India. It was here, while Europeans were still pretty much tied to summing geometric progressions, that Indian mathematicians were using infinite series expansions of trigonometrical functions. The primary exponent of this was Madhava. Here is how George Gheverghese Joseph introduces him:

> **Madhava of Sangamagramma (c. 1340 – 1425)**
> Madhava was probably the greatest of the Indian medieval astronomer-mathematicians, but he has come to the fore only in recent years as a result of research into Kerala mathematics. It was Madhava who 'took the decisive step onwards from the finite procedures of ancient mathematics to treat their limit-passage to infinity, which is the kernel of modern classical analysis' (Rajagopal and Rangachari, 1978, p. 101).[26]
>
> ...Of his works that have survived, all are astronomical treatises; for his mathematical contributions we rely on reports by his contemporaries and successors. These contributions, which include infinite series expansions of circular and trigonometric functions and finite-series approximations, are discussed in a later section.[27]

After discussing Madhava's work on infinite series, he cautions

> In our rather brief look at Kerala mathematics, the name that keeps recurring is that of Madhava of Sangamagramma. His brilliance is generously acknowledged by those who came after him, and the effects of his teaching on the works of Paramesvara, Nilakantha, Jyesthadeva and others are there to see. It would be quite in keeping with Indian tradition, if, in holding him in such awe, his successors were to have credited him with more than his share of discoveries. Of his teachers we know nothing. Madhava's outstanding contributions, in the area of infinite-series expansions of circular and trigonometric functions, and finite-series approximations to them, predate European work on the subject by two to three hundred years.[28]

The results referred to of concern here are the infinite power series representations of the sine, cosine, and inverse tangent functions:

$$\sin x = x - \frac{x^3}{3!} + \frac{x^5}{5!} - \frac{x^7}{7!} + \ldots$$
$$\cos x = 1 - \frac{x^2}{2!} + \frac{x^4}{4!} - \frac{x^6}{6!} + \ldots$$

---

[26] The reference is to C.T. Rajagopal and M.S. Rangachari, "On an untapped source of medieval Keralese mathematics", *Archive for History of Exact Sciences* 18 (1978), pp. 89 – 102.

[27] George Gheverghese Joseph, *The Crest of the Peacock: Non-European Roots of Mathematics*, I.B. Tauris, London, 1991; paperback edition: Penguin Books, London, 1992. The quotation is from pp. 270 – 271 of the Penguin edition.

[28] *Ibid.*, p. 293.

$$\tan^{-1} x = x - \frac{x^3}{3} + \frac{x^5}{5} - \frac{x^7}{7} + \ldots, \text{ for } -1 \leq x \leq 1.$$

Joseph cites two 16th century sources crediting Madhava with the series for the inverse tangent, quoting Jyesthadeva's *Yuktibhasa* (c. 1550):

> The first term is the product of the given Sine and radius of the desired arc divided by the Cosine of the arc. The succeeding terms are obtained by a process of iteration when the first term is repeatedly multiplied by the square of the Sine and divided by the square of the Cosine. All the terms are then divided by the odd numbers $1, 3, 5, \ldots$. The arc is obtained by adding and subtracting [respectively] the terms of odd rank and those of even rank. It is laid down that the [Sine of the] arc or that of its complement whichever is smaller should be taken here [as the given Sine]. Otherwise the terms obtained by the above iteration will not tend to the vanishing magnitude.[29]

Joseph points out that the Indian sine was the radius times our sine, a common practice later continued among European astronomers calculating trigonometrical tables, hence the capitalisation of the function names. Thus the description given very clearly points to the identification

$$r\theta = \frac{r(r\sin\theta)}{1(r\cos\theta)} - \frac{r(r\sin\theta)^3}{3(r\cos\theta)^3} + \frac{r(r\sin\theta)^5}{5(r\cos\theta)^5} - \cdots,$$

i.e.,

$$\theta = \tan\theta - \frac{\tan^3\theta}{3} + \frac{\tan^5\theta}{5} - \cdots,$$

and if x denotes the tangent of θ,

$$\tan^{-1} x = \theta = x - \frac{x^3}{3} + \frac{x^5}{5} - \cdots$$

There is even the warning that the series does not converge if $x > 1$ and that, since $\tan(\pi/2 - \theta) = 1/\tan\theta$, if $x > 1$, one should invert it (switch the rôles of Sine and Cosine) to determine the complementary angle $\pi/2 - \theta$.

For the proofs, Joseph merely gives a reference[30] regarding these series, noting that the proofs are geometrical, but related to more familiar analytic proofs

---

[29] *Ibid.*, p. 290.

[30] T.A. Sarasvati, "Development of mathematical series in India", *Bulletin of the National Institute of Science* 21 (1963), pp. 320 – 343. [Since writing this, I have acquired two more recent publications. First, by Joseph himself is *A Passage to Infinity; Medieval Indian Mathematics from Kerala and Its Impact*, SAGE Publications, New Delhi, 2009. Chapter 7 of this work offers an accessible discussion of the derivation of the power series for sine and cosine as given by Jyesthadeva in his *Yuktibhasa*. My other acquisition is the *Yuktibhasa* itself: K.V. Sarma, K. Ramasubramanian, M.D. Srinivas, and M.S. Srivam, *Gaṇita-Yukti-Bhāṣā (Rationales in Mathematical Astronomy) of Jyeṣṭhadeva*, Hindustan Book Agency, New Delhi, 2008, and Springer. (Each of the two volumes I purchased at a discounted price is missing the page with publication data, so I can give no specific information on the city or date of the Springer issue.) The two volumes (I on mathematics and II on astronomy) each contain an English translation of the text by Sarma, mathematical commentary in English by the other three contributors, and the text in the original Malayalam language spoken in

by integration. As our concern in this book is the Binomial Theorem, I shall pass over these proofs in silence in favour of discussing their derivation from Newton's Binomial Theorem. Indeed, the achievements of Madhava in expanding trigonometrical functions into series is a digression from our main story, but it is one that deserves coverage. It is missing from the two histories of the theory of series as well as from the more general histories of the Calculus in my personal library.[31]

The European rediscovery of the infinite series representations of trigonometrical and other functions is much better documented and is closely tied to Newton's Binomial Theorem.

The type of infinite series representation of a function f we are considering is called a *power series* and is of the form,

$$f(x) = a_0 + a_1 x + a_2 x^2 + \ldots = \sum_{k=0}^{\infty} a_k x^k, \qquad (52)$$

or, more generally,

$$f(x) = a_0 + a_1(x-a) + a_2(x-a)^2 + \ldots = \sum_{k=0}^{\infty} a_k (x-a)^k. \qquad (53)$$

A series of the form (53) is called a *Taylor series*, or *Taylor series expansion* of f around the the point a; the expansion (52) is, of course, the Taylor series expansion around 0, but is also called a *Maclaurin series* or the *Maclaurin series expansion* of f. These are eponymous names: Brook Taylor (1685 – 1731) attempted to prove that every function had such a representation, and the modern theorem asserting that functions satisfying certain conditions do have such is called *Taylor's Theorem*. Colin Maclaurin (1698 – 1746) gave the most convenient determination of the coefficients of the power series expansion of a given function.

Just as the Finite Binomial Theorem is more than the collection of its instances, a power series representation is not just a collection of series summations of a common form. Before Newton discovered his Binomial Theorem, the identity,

$$\frac{1}{1-x} = 1 + x + x^2 + \ldots, \quad |x| < 1,$$

was not a power series expansion of the *function* $f(x) = 1/(1-x)$, but the fact that, whatever value r one chose for x, if $|r| < 1$, then the sum of the infinite series $1 + r + r^2 + \ldots$ was $1/(1-r)$. The notion of function was emerging and evolving. The word "function" was not yet a part of the mathematical vocabulary, and after Leibniz would introduce it, it would undergo a succession

---

Kerala. The two chapters 7 (of the translation and commentary, respectively) discuss sines, with §7.5 deriving the power series for the sine and versed sine (versin $\theta = 1 - \cos\theta$).]

[31] These histories of infinite series are Reiff, *op. cit.*, and Ferraro, *op. cit.* Reiff wrote his book half a century before the work of the Kerala school was publicised in the West, and, despite the title, Ferraro's primary concern was the evolution of ideas of series and functions in the 17th and 18th centuries.

of changes of meaning until it took on the modern general definition as a set of ordered pairs possessing certain properties.[32]

When Newton discovered his Binomial Theorem, the time was ripe for power series to emerge in Europe. Ever one to avoid the limelight, Newton kept quiet about his discovery until Nicolaus Mercator (c. 1619 – 1687) published the expansion,
$$\ln(1+x) = x - \frac{x^2}{2} + \frac{x^3}{3} - \cdots$$
About the same time, James Gregory (1638 – 1675) had discovered the power series expansion of $\tan^{-1} x$. The series for the logarithm is today called *Mercator's series*, while that for $\tan^{-1} x$ is often called the *Gregory series*, though Joseph justifiably refers to it as the *Madhava-Gregory series*.[33] By the time of Mercator's publication (1668) and Gregory's discovery (1667), Newton had also discovered these series.[34]

Mercator, who should not be confused with the famous map-maker Gerardus Mercator, preceded Newton in publication. This was in 1668 in a book bearing a then customarily long title, *Logarithmotechnica; sive Methodus construendi logarithmos nova, accurata et facilis* [*Logarithmotechnica; with a new accurate and easy method of constructing logarithms*]. Today, the book is generally referred to by the first word in its title. Referring to the work of Saint-Vincent and de Sarasa he noted that
$$\ln(1+x) = \int_0^x \frac{dx}{1+x} \left(= \int_0^x \frac{dt}{1+t} \right),$$
(if one prefers[35]) and that long division yielded
$$\frac{1}{1+x} = 1 - x + x^2 - x^3 + \cdots$$
Then, essentially[36] integrating term-by-term, he concluded
$$\int_0^x \frac{dx}{1+x} = x - \frac{x^2}{2} + \frac{x^3}{3} - \frac{x^4}{4} + \cdots$$

---

[32]I do not wish to get into a discussion of these matters, so will instead refer the reader to Jacqueline Stedall's source book, *Mathematics Emerging: A Sourcebook 1540 – 1900* (Oxford University Press, New York, 2008), the 9th chapter of which contains some indications of the evolution of the concept.

[33]Joseph, *op. cit.*, p. 289.

[34]And conversely, Gregory rediscovered Newton's binomial series. He left behind no proof or explanation of his discovery, but only a letter describing the result. *Cf.* Stedall, *op. cit.*, pp. 198 – 201, Ferraro, *op. cit.*, pp. 20 – 24, especially p. 23, or Herman H. Goldstine, *A History of Numerical Analysis from the 16th through the 19th Century*, Springer-Verlag, New York, Inc., 1977, pp. 75 – 76. And, more impressively, Henry Briggs had derived the binomial series for the single exponent $q = 1/2$ already in 1624 in his *Arithmetica Logarithmica*, as reported by Goldstine, p. 19. Neither of these discoveries, however, had the impact that Newton's did.

[35]Or if one possesses a *TI-89 Titanium* and wishes to check my integration with it: Entering ∫(1/(1+x),x,0,x) results in an error message, while ∫(1/(1+t),t,0,x) will yield ln(1+x) or a numerical value if x has been assigned a value.

[36]"Essentially" = it was a bit more complicated than this.

In 1667, Newton also drew the same conclusion. Newton, however, had the choice of expanding $1/(1+x) = (1+x)^{-1}$ either by long division or by reference to his Binomial Theorem.

The expansion of the inverse tangent depended on a similar representation. Let $\theta = \tan^{-1} x$, i.e., $x = \tan \theta$. Then
$$x = \tan(\tan^{-1} x)$$
and the Chain Rule yields
$$1 = \frac{d\tan}{d\theta}(\tan^{-1} x) \cdot \frac{d\tan^{-1} x}{dx}. \tag{54}$$
But
$$\frac{d\tan\theta}{d\theta} = \frac{d}{d\theta}\left(\frac{\sin\theta}{\cos\theta}\right) = \frac{\cos\theta\cos\theta - \sin\theta(-\sin\theta)}{\cos^2\theta} = \frac{\cos^2\theta + \sin^2\theta}{\cos^2\theta} = \frac{1}{\cos^2\theta}. \tag{55}$$
Now, if $x = \sin\theta/\cos\theta$ then $x\cos\theta = \sin\theta$, but $\sin^2\theta + \cos^2\theta = 1$, whence
$$x^2\cos^2\theta + \cos^2\theta = 1,$$
i.e., $(1+x^2)\cos^2\theta = 1$ and
$$1 + x^2 = \frac{1}{\cos^2\theta}.$$
Plugging this into (55) yields
$$\frac{d\tan\theta}{d\theta} = 1 + x^2,$$
and plugging this into (54) yields
$$1 = (1+x^2)\frac{d\tan^{-1} x}{dx},$$
i.e.,
$$\frac{d\tan^{-1} x}{dx} = \frac{1}{1+x^2}$$
and
$$\tan^{-1} x = \int_0^x \frac{dx}{1+x^2} + C$$
for some constant C. Using the value $\tan^{-1} 0 = 0$ determines $C = 0$ and we can now once again expand $1/(1+x^2) = (1+x^2)^{-1}$ into an infinite series and integrate term-by-term to derive the Madhava-Gregory series.

2.8.1. Exercise. *Carry out the details of this derivation.*

So far Newton's Binomial Theorem hasn't really been needed, as one can use long division to obtain the power series to be integrated. Newton, however, also considered
$$\sin^{-1} x = \int_0^x \frac{dx}{\sqrt{1-x^2}} = \int_0^x (1-x^2)^{-1/2} dx$$
$$= \int_0^x \sum_{k=0}^\infty \binom{-1/2}{k}(-x^2)^k dx$$

$$= \int_0^x \left(1 + \left(-\frac{1}{2}\right)(-x^2) + \frac{\left(-\frac{1}{2}\right)\left(-\frac{3}{2}\right)}{2 \cdot 1}(-x^2)^2 \right.$$

$$\left. + \frac{\left(-\frac{1}{2}\right)\left(-\frac{3}{2}\right)\left(-\frac{5}{2}\right)}{3 \cdot 2 \cdot 1}(-x^2)^3 + \ldots \right) dx$$

$$= \int_0^x \left(1 + \frac{x^2}{2} + \frac{3x^4}{8} + \frac{5x^6}{16} + \ldots\right) dx$$

$$= x + \frac{x^3}{6} + \frac{3x^5}{40} + \frac{5x^7}{112} + \ldots$$

To be honest, Newton's actual derivation was a little less direct.[37] Let $0 \leq x \leq 1$ and $\theta = \sin^{-1} x$. Then, referring to FIGURE 6, below, θ is the angle

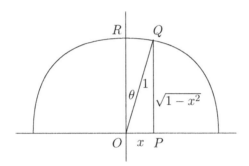

FIGURE 6.

$\angle QOR$ and it equals twice the area of the sector QOR,

$$\frac{\theta}{2\pi r} = \frac{\text{Area}(QOR)}{\pi r^2}, \quad \text{for } r = 1.$$

But this area is the difference between the area of the segment $OPQR$ and the triangle OPQ:

$$\theta = 2\left(\text{Area}(OPQR) - \frac{1}{2}x\sqrt{1-x^2}\right).$$

And back in section 2 we saw the first area to be (by equation (32))

$$x - \frac{x^3}{6} - \frac{x^5}{40} - \frac{x^7}{112} - \frac{x^9}{1152} - \ldots$$

and

$$\sqrt{1-x^2} = 1 - \frac{x^2}{2} - \frac{x^4}{8} - \frac{x^6}{16} - \frac{5x^8}{128} - \ldots,$$

---

[37] I don't read Latin and am relying on C.H. Edwards, Jr., *The Historical Development of the Calculus*, Springer-Verlag New York, Inc., New York, 1979, pp. 205 – 206, for this proof.

whence
$$0 = 2\left(x - \frac{x^3}{6} - \frac{x^5}{40} - \frac{x^7}{112} - \frac{x^9}{1152} - \cdots\right) - x\left(1 - \frac{x^2}{2} - \frac{x^4}{8} - \frac{x^6}{16} - \frac{5x^8}{128} - \cdots\right)$$
and a little algebra yielded
$$0 = x + \frac{x^3}{6} + \frac{3x^5}{40} + \frac{5x^7}{112} + \frac{43x^9}{1152} + \cdots$$

Having got the power series expansion of $f(x) = \sin^{-1} x$, he was now able to derive the power series for $g(x) = \sin x$. This is a computational chore comparable in the amount of pleasure it affords to multiplication and division of multidigit numbers or digit-by-digit root extraction.

Conceptually, the matter is simple. We have a series representation of the form
$$z = a_1 x + a_2 x^2 + a_3 x^3 + \cdots = \sum_{k=1}^{\infty} a_k x^k \tag{56}$$
for $\sin^{-1} x$, and assume $\sin x$ to have a representation
$$x = b_1 z + b_2 z^2 + b_3 z^3 + \cdots = \sum_{k=1}^{\infty} b_k z^k. \tag{57}$$

Basically, one has but to plug the value given by (57) for x into (56) and solve for $b_1, b_2, b_3, \ldots$ This, of course, is not proof that there is such a series representation for $\sin x$ or that the series so obtained will converge anywhere, but if such a series exists, the procedure ought to find it. Indeed it will, but the proof is a matter of some depth.

Note that the series (56) and (57) have no constant terms. This is because $\sin^{-1} 0 = 0$ and $\sin 0 = 0$, and the constant term c_0 of a Maclaurin series representation,
$$f(x) = c_0 + c_1 x + c_2 x^2 + \cdots,$$
is just $f(0)$. That $a_0 = 0$ is a great computational convenience. And it is not a serious restriction. If f is an invertible function with $f(0) \neq 0$, then $F(z) = f(z) - f(0)$ will be invertible with $F(0) = 0$, its Maclaurin series (56) will have an inverse, F^{-1} with series (57), and $f^{-1}(z) = F^{-1}(z - f(0))$ will be expressed as a Taylor series around $f(0)$.

There is an additional condition that is more than a convenience and must be imposed. It is that $a_1 \neq 0$. The algebraic necessity of this will be apparent shortly. Another reason is this. As will be verified in the next section, $a_1 = f'(0)$ and one shows in the standard Calculus course that $f'(a) \neq 0$ is a sufficient condition that f be invertible in some interval containing a. It is possible that f be so invertible even when the derivative is 0, e.g., $f(x) = x^3$. However, the coefficient of the linear term of the Maclaurin expansion of f^{-1} would have to be $(f^{-1})'(0) = 1/f'(f(0)) = 1/a_1$, by the Chain Rule. But if $a_1 = 0$ then f^{-1} has no derivative at 0 and hence no Maclaurin expansion.

The method of solving for $b_1, b_2, b_3, \ldots$ is very simple. Find the successive powers of z:
$$z = a_1 x + a_2 x^2 + a_3 x^3 + \cdots$$

$$z^2 = a_1^2 x^2 + 2a_1 a_2 x^3 + (2a_1 a_3 + a_2^2)x^4 + \ldots$$
$$z^3 = a_1^3 x^3 + 3a_1^2 a_2 x^4 + (3a_1^2 a_3 + 3a_1 a_2^2)x^5 + \ldots$$
$$\vdots$$

Then one plugs these into (57) to get

$$x = b_1(a_1 x + a_2 x^2 + a_3 x^3 + \ldots)$$
$$+ b_2(a_1^2 x^2 + 2a_1 a_2 x^3 + (2a_1 a_3 + a_2^2)x^4 + \ldots)$$
$$+ b_3(a_1^3 x^3 + 3a_1^2 a_2 x^4 + (3a_1^2 a_3 + 3a_1 a_2^2)x^5 + \ldots) + \ldots,$$

and matches coefficients:

$$1 = b_1 a_1$$
$$0 = b_1 a_2 + b_2 a_1^2$$
$$0 = b_1 a_3 + 2b_2 a_1 a_2 + b_3 a_1^3$$
$$\vdots$$

One can then successively solve for $b_1, b_2, b_3, \ldots$:

$$b_1 = \frac{1}{a_1}$$
$$b_2 = -\frac{b_1 a_2}{a_1^2} = -\frac{a_2}{a_1^3}$$
$$b_3 = \frac{-b_1 a_3 - 2b_2 a_1 a_2}{a_1^3} = \frac{-\dfrac{a_3}{a_1} + 2\dfrac{a_1 a_2^2}{a_1^3}}{a_1^3} = \frac{2a_2^2 - a_1 a_3}{a_1^5}$$
$$\vdots$$

The next coefficient is

$$b_4 = -\frac{5a_1 a_2 a_3 - 5a_2^3 - a_1^2 a_4}{a_1^7}.$$

We can see from all of this just why it was so important to assume (56) to have no constant term. When $a_0 = 0$ each successive series for $z, z^2, z^3, \ldots$ starts with a higher power of x, whence in the final sum $b_1 z + b_2 z^2 + \ldots$ each x^k will occur in only finitely many terms $z, z^2, \ldots, z^k$. This means that the equations for solving for $b_1, b_2, b_3, \ldots$ will all be finite, b_k will not occur before the $(k-1)$-th equation, allowing one to solve for them successively, as we have done.

For the function at hand, namely $z = \sin^{-1} x$, the computations already given do not go very far. The coefficients given are

$$a_1 = 1, \ a_2 = 0, \ a_3 = \frac{1}{6}, \ a_4 = 0, \ a_5 = \frac{3}{40}, \ a_6 = 0, \ a_7 = \frac{5}{112}, \ a_8 = 0, \ldots$$

Using these we can read off the values

$$b_1 = 1, \ b_2 = 0, \ b_3 = -\frac{1}{6}, \ b_4 = 0,$$

using the formulæ cited. We have more work to do to get more terms—and more terms are desirable because they make for greater accuracy in numerical application of the series and because, if there is a discernible pattern to the coefficients, more terms make it more easily discovered. Calculating more coefficients, we see that

$$\sin x = x - \frac{x^3}{6} + \frac{x^5}{120} - \frac{x^7}{5040} + \dots,$$

which Newton was quick to recognise as

$$x - \frac{x^3}{3!} + \frac{x^5}{5!} - \frac{x^7}{7!} + \dots = \sum_{k=0}^{\infty} \frac{(-1)^k x^{2k+1}}{(2k+1)!}.$$

In like manner, Newton inverted

$$f(x) = \ln\left(\frac{1}{1-x}\right) = -\ln(1-x) = x + \frac{x^2}{2} + \frac{x^3}{3} + \frac{x^4}{4} + \dots$$

to obtain

$$1 - e^{-x} = x - \frac{x^2}{2} + \frac{x^3}{6} - \frac{x^4}{24} + \dots$$

Solving for e^{-x} yields

$$e^{-x} = 1 - x + \frac{x^2}{2} - \frac{x^3}{6} + \frac{x^4}{24} - \dots$$

and replacing x by $-x$:

$$e^x = 1 + x + \frac{x^2}{2} + \frac{x^3}{6} + \frac{x^4}{24} - \dots$$

$$= \frac{1}{0!} + \frac{x}{1!} + \frac{x^2}{2!} + \frac{x^3}{3!} + \frac{x^4}{4!} + \dots = \sum_{k=0}^{\infty} \frac{x^k}{k!}$$

It would be a good exercise for the reader to carry out one of these *reversions* as Newton (or his translators) called it. However, it is rather tricky. One has to keep track of which combinations of the a_i's go into the making up of individual b_k's, and the expression of b_k in terms of its predecessors introduces minus signs, a traditional stumbling block for students learning the Calculus. The computer, or programmable calculator, is not subject to the sort of mistakes that the student (or the author) is prone to, whence automating the procedure is a good idea.

For one knowing a little beginning Linear Algebra, this is a fairly straightforward programming task once one decides how to represent the data. To determine $b_1, b_2, \dots, b_n$ we need the series for $z, z^2, \dots, z^n$, each out to the x^n term. The constant terms of these series are all 0, whence we can collect the coefficients in an $n \times n$ matrix, each column being the coefficients of $x, x^2, \dots, x^n$ of one of these series. The first column will have the entries $a_1, a_2, \dots, a_n$. And if the i-th column is $c_1, c_2, \dots, c_n$, the j-th element of the $(i+1)$-th column will be

$$d_j = a_1 c_{j-1} + a_2 c_{j-2} + \dots + a_{j-1} c_1.$$

The program will essentially generate the matrix one column at a time, then perform a Gaussian elimination on the matrix augmented by a column with entries $1, 0, 0, \ldots, 0$ representing the series $x = x + 0x^2 + 0x^3 + \ldots$. The coefficients $b_1, b_2, \ldots, b_n$ of the inverse series will be the final column of the augmented matrix after it has been put into reduced row echelon form (rref).

On the next page, I present the programs for the *TI-83 Plus* and the *TI-89 Titanium* side-by-side for comparison. The former assumes $a_1, a_2, \ldots, a_n$ to have been stored in a list ∟ACOEF. The latter assumes the list to be entered as an argument to the program. The last line of the program for the *TI-83 Plus*

```
PROGRAM:POWREVRT



:dim(∟ACOEF)→N
:{N,N+1}→dim([A])
:Fill(0,[A])
:For(J,1,N)
:∟ACOEF(J)→[A](J,1)
:End
:For(I,2,N)
:For(J,I,N)
:sum(seq([A](K,I−1)∗
   ∟ACOEF(J−K),K,1,J−1)→[A](J,I)
:End
:End
:1→[A](1,N+1)
:rref([A])→[A]
:Matr▶list([A],N+1,∟BCOEF)

:∟BCOEF▶Frac
```

```
:powRevrt(acoef)
:Func
:Local n,mat,j,i,k
:dim(acoef)→n
:newMat(n,n+1)→mat

:For j,1,n
:acoef[j]→mat[j,1]
:EndFor
:For i,2,n
:For j,i,n
:sum(seq(mat[k,i−1]∗
   acoef[j−k],k,1,j−1)→mat[j,i]
:EndFor
:EndFor
:1→mat[1,n+1]
:rref(mat)→mat
:Return seq(mat[i,n+1],i,1,n)
:EndFunc
```

converts the list, insofar as it is possible, into ordinary fractions to enable one to more easily recognise any pattern to the coefficients. The line also produces output which appears on the home screen in horizontally scrollable form. For the *TI-89 Titanium*, running the program powRevrt() in EXACT mode produces the fractions, and the fact that powRevrt() is a function places the scrollable result in the history portion of the homescreen.

Before running the first of these programs one might want to insert some commands deleting unneeded variables ∟ACOEF, N, [A], J, I, K immediately before the final line.

2.8.2. Exercise. *Apply one of these programs to the series given for* $\sin^{-1} x$, $\ln(1/(1-x))$, $\ln(1+x)$, *and* $\tan^{-1} x$ *to obtain the first terms of the series for* $\sin x$, $1 - e^{-x}$, $e^x - 1$, *and* $\tan x$. *If you have a TI-89 Titanium, use* EXACT *mode and carry out the expansion at least as far as* x^{10}. *Then enter* factor(ans(1)) to see if the pattern to the coefficients is discernible. Apply the program to the series obtained to verify that you get the original series.

With all this activity, the reader may not have noticed that we still do not have a series representation for $\cos^{-1} x$ or $\cos x$. The Maclaurin series for $\cos^{-1} x$ is an easy matter:
$$\cos^{-1} x + \sin^{-1} x = \frac{\pi}{2},$$
whence
$$\cos^{-1} x = \frac{\pi}{2} - x + \frac{x^3}{6} - \frac{x^5}{120} + \frac{x^7}{5040} - \cdots,$$
a series with a non-zero constant term. The series reversion program will not apply directly. However, we do have the Taylor expansion
$$\cos x = \sin\left(\frac{\pi}{2} - x\right) = \frac{\pi}{2} - x - \frac{1}{6}\left(\frac{\pi}{2} - x\right)^3 + \frac{1}{120}\left(\frac{\pi}{2} - x\right)^5 - \cdots$$
One can use the identity
$$\cos^2 x = 1 - \sin^2 x \tag{58}$$
to find $\cos x$ by root extraction, or by writing
$$\cos x = b_0 + b_1 x + b_2 x^2 + \cdots,$$
plugging this into (58), and solving successively for $b_0, b_1, b_2, \ldots$ Alternatively, one can note that
$$\cos x = \frac{\sin x}{\tan x}$$
and obtain the coefficients by long division.

2.8.3. Exercise. *Find the first few terms of the Maclaurin expansion of $\cos x$ by one of these methods.*

As with the reversion of series, an exercise such as this last one offers another excuse to program our calculators. As the first n terms of the sum, difference, product, or quotient of two infinite series depend only on the first n terms of the individual series, such automation is feasible. For the sums and differences the task is, in fact, trivial. On the *TI-83 Plus* one assumes the first n coefficients of the two power series to be stored in lists ∟ACOEF and ∟BCOEF, respectively, and enters the programs of the table below. On the *TI-89 Titanium*, the lists are input as variables acoef, bcoef to functions.

PROGRAM:POWSUM	PROGRAM:POWDIFF
:∟ACOEF+∟BCOEF→∟CCOEF	:∟ACOEF−∟BCOEF→∟CCOEF
:powSum(acoef,bcoef)	:powDiff(acoef,bcoef)
:Func	:Func
:Return acoef + bcoef	:Return acoef−bcoef
:EndFunc	:Endfunc

These are pathetically stupid programs. POWSUM and POWDIFF replace perfectly good built-in functions by non-functional programs. powSum(,) and

8. POWER SERIES

powDiff(,) are functions, but are equally unnecessary. We could remedy this situation by not assuming the two lists to have the same length and ignoring the excess terms of the longer list. For example, for POWSUM and powSum(,) we modify the programs as in the table below.

PROGRAM:POWSUM	:powSum(acoef,bcoef)
	:Func
	:Local, m,n,k,p1,p2
:dim(∟ACOEF)→M	:dim(acoef)→m
:dim(∟BCOEF)→N	:dim(bcoef)→n
:min({M,N})→K	:min({m,m})→k
:∟ACOEF→∟P1	:acoef→p1
:K→dim(P1)	:k→dim(p1)
:∟BCOEF→∟P2	:bcoef→p2
:K→dim(P2)	:k→dim(p2)
:∟P1+∟P2→∟CCOEF	:Return p1+p2
	:EndFunc
:DelVar M	
:DelVar N	
:DelVar K	
:DelVar ∟P1	
:DelVar ∟P2	
:∟CCOEF	

The product is also quite simple to program. Note that, if

$$\left(\sum_{n=0}^{\infty} a_k x^k\right)\left(\sum_{k=0}^{\infty} b_k x^k\right) = \left(\sum_{k=0}^{\infty} c_k x^k\right),$$

then

$$c_k = a_0 b_k + a_1 b_{k-1} + \ldots + a_k b_0.$$

If we have lists representing the first n coefficients of two power series, we obtain the k-th coefficient of the product by taking the first k elements of the two lists, one in proper order, one in reverse, then multiplying the lists and adding them as in the program below.

Finding the quotient is a bit more challenging. First, note that for $\left(\sum b_k x^k\right)/\left(\sum a_k x^k\right)$ to be representable as a power series, the first nonzero coefficient of $\sum a_k x^k$ must occur at least as early as the earliest such of $\sum b_k x^k$. If one makes the simplifying assumption that $a_0 \neq 0$, one can ignore this problem. After that one has to choose a strategy. Does one simulate long division, i.e., does one program the long division algorithm, or does one view the product

$$\left(\sum b_k x^k\right) = \left(\sum a_k x^k\right)\left(\sum c_k x^k\right)$$

as a system of simultaneous linear equations and program the calculator to solve for $c_0, c_1, \ldots, c_{n-1}$ in terms of $a_0, a_1, \ldots, a_{n-1}$ and $b_0, b_1, \ldots, b_{n-1}$?

```
PROGRAM:POWPROD                :powProd(acoef)
                               :Func
                               :Local n,ccoef,k,aux1,aux2
:dim(∟ACOEF)→N                 :dim(acoef)→n
:For(K,1,N)                    :For k,1,n
:seq(∟ACOEF(I),I,1,K)→∟AUX1    :seq(acoef[i],i,1,k)→aux1
:seq(∟BCOEF(I),I,K,1,⁻1)→∟AUX2 :seq(bcoef[i],i,k,1,⁻1)→aux2
:sum(∟AUX1∗∟AUX2)→∟CCOEF(K)    :sum(aux1∗aux2)→ccoef[k]
:End                           :EndFor
                               :Return ccoef
                               :EndFunc

:DelVar N
:DelVar K
:DelVar I
:DelVar ∟AUX1
:DelVar ∟AUX2
:∟CCOEF
```

2.8.4. Exercise. *Write a program POWQUOT on the TI-83 Plus or a function powQuot(,) on the TI-89 Titanium that will take two lists of the same length representing the first n terms of two power series $\sum a_k x^k$ ($a_0 \neq 0$) and $\sum b_k x^k$ and produce the first n terms of the quotient series $(\sum b_k x^k)/(\sum a_k x^k)$. How can one apply this to the problem of finding the initial terms of the Maclaurin expansion of $\cos x$ from those of $\sin x$ and $\tan x$ for which $a_0 = b_0 = 0$?*

I have only a few final points to make before moving on. The first is the observation that, since $\tan(\pi/4) = 1$, one has $\tan^{-1} 1 = \pi/4$, whence the series

$$\frac{\pi}{4} = 1 - \frac{1}{3} + \frac{1}{5} - \frac{1}{7} + \frac{1}{9} - \ldots \tag{59}$$

This is not the most rapidly converging series, hence its value is more for one's edification than for practical application. But it is pleasing and it does show that even the simplest series can be hard to sum—there is no way $\pi/4$ would appear after some algebraic manipulation like those in the preceding section.

The name of Leibniz is often associated with (59). Leibniz was about a decade behind Newton in developing the Calculus and independently derived the various series we've been considering. Unlike the case with the algorithms for differentiation and integration, he quickly learned of Newton's earlier work on series, whence their contact via Oldenburg and Newton's description of his discovery of his Binomial Theorem.

The second point again concerns the speed of convergence. Mercator's series,

$$\ln(1+x) = x - \frac{x^2}{2} + \frac{x^3}{3} - \frac{x^4}{4} + \ldots,$$

is also rather slow to converge, especially if x is close to 1. Noting that every real number $y \geq 1$ can be written in the form
$$y = \frac{1+x}{1-x} \quad \text{for some } 0 < x < 1,$$
Gregory and Edmond Halley (1656 – 1742) used Mercator's series to derive the more rapidly convergent series for $y \geq 1$,[38]

$$\ln y = \ln \frac{1+x}{1-x} = \ln(1+x) - \ln(1-x)$$
$$= \left(x - \frac{x^2}{2} + \frac{x^3}{3} - \frac{x^4}{4} + \cdots\right) - \left(-x - \frac{x^2}{2} - \frac{x^3}{3} - \frac{x^4}{4} - \cdots\right)$$
$$= 2x + \frac{2x^3}{3} + \frac{2x^5}{5} + \cdots = \sum_{k=0}^{\infty} \frac{2x^{2k+1}}{2k+1}. \tag{60}$$

For example, to calculate $\ln 2$ by Mercator's series,
$$\ln 2 = \ln(1+1) = 1 - \frac{1}{2} + \frac{1}{3} - \frac{1}{4} + \cdots, \tag{61}$$
enter
$$\{1, ^{-}1/2, 1/3, ^{-}1/4, 1/5, ^{-}1/6, 1/7, ^{-}1/8, 1/9, ^{-}1/10\} \to L_1,$$
and cumSum(L_1) on your *TI-83 Plus* (or the corresponding expressions on the *TI-89 Titanium* running in APPROXIMATE mode) to get

$$1, .5, .8\overline{3}, .58\overline{3}, .78\overline{3}, .61\overline{6}, .7595238095, .6345238095, .7456349206, .6456349206,$$

which still leaves us in doubt as to the first digit of $\ln 2$. Or, one can solve
$$2 = \frac{1+x}{1-x},$$
and plug the solution $x = 1/3$ into (60):

seq((2/(2K+1))(1/3)^(2K+1),K,0,9)$\to L_2$.

One can then enter cumSum(L_2) to obtain

$$.\overline{6}, .6913580247, .6930041152, .6931347573, .6931460474,$$
$$.6931470738, .6931471703, .6931471795, .6931471805, .6931471805,$$

which is very close. Entering $\ln 2$ yields $.6931471806$, and checking for hidden digits on the calculator yields $.69314718055994$. Convergence of the Gregory-Halley series is much more rapid.

Incidentally, the series (50) of the immediately preceding section,
$$1 + \frac{2}{1} \cdot \frac{1}{2} + \frac{3}{2} \cdot \frac{1}{2^2} + \frac{4}{3} \cdot \frac{1}{2^3} + \cdots,$$
which Thomaz couldn't sum, is now easily summable. Subtracting
$$1 + \frac{1}{2} + \frac{1}{4} + \cdots = 2,$$

---

[38] For $0 < y < 1$, one simply takes $\ln y = -\ln(1/y)$.

yields
$$1 + \frac{1}{2} + \frac{1}{2 \cdot 4} + \frac{1}{3 \cdot 8} + \ldots = -\ln\left(1 - \frac{1}{2}\right) = -\ln\left(\frac{1}{2}\right) = \ln 2,$$
whence the sum is $2 + \ln 2$ as mentioned earlier.

9. Taylor's Theorem

With so many functions admitting expansion into power series, it wasn't long before general results were formulated by various authors. Ferraro's history of infinite series begins its discussion of Taylor series with the words,

> The Taylor series... is one of the most important series in mathematics. It bears the name of Brook Taylor, who first published it in *Methodus incrementorum* [1715]. However, similar results were probably known to Gregory and certainly to Newton, Johann Bernoulli, Leibniz and de Moivre.[39]

Richard Reiff, in his classic history of infinite series, describes Taylor's book as follows:

> Bearing the date 1717, in 1716 the Methodus Incrementorum directa et inversa appeared, a book which his contemporaries accused of lack of clarity. Therein T a y l o r treated finite differences and their limits, the infinitely small differences; he sought to arrive at the differential calculus via the theory of differences...
>
> Of interest and particularly characteristic as an example of his methods is the derivation of what he called the Fundamental Theorem of Analysis...[40]

There appears to be a consensus nowadays of assigning the date 1715 to Taylor's book, the title of which, *Methodus incrementorum directa et inversa* [*Method of direct and inverse differences*], refers to the Calculus of Finite Differences. The result on which the "Fundamental Theorem of Analysis" rests concerns interpolation.

Linear interpolation between two successive values in a table goes back a long time, at least as far back as Hipparchus (*fl.* after 127 B.C.) and his early trigonometric tables. Liú Zhuó (544 – 610) used quadratic interpolation in the 6th or 7th century, as did Brahmagupta in India, Wáng Xún (1235 – 1281) and Guō Shǒujìng (1231 – 1316) used cubic interpolation in the 13th century, and Zhū Shìjié (*fl.* 1280 – 1303) was up to the 4th degree by 1303. By Taylor's day Europeans were generating many tables, primarily of trigonometric and logarithmic functions, and had begun developing techniques of interpolation to fill in missing entries in these tables. A central result, which Taylor took as the starting point of his derivation of the "Fundamental Theorem" was the *Newton*

---

[39] Ferraro, *op. cit.*, p. 87. I have taken the liberty of deleting the citation of the formula and the cross-references to other chapters of Ferraro's book.

[40] Reiff, *op. cit.*, p. 80.

9. TAYLOR'S THEOREM

Forward Difference Formula, also known as the *Newton-Gregory Interpolation Formula*.[41]

Somewhat de-emphasised in elementary education today, linear interpolation was drilled into those of us who received our primary education in the pre-calculator era when tables dominated hand calculation. Tables, however, can only hold a limited amount of data and one has to interpolate for untabulated values. In elementary school, this was done by simple proportions. Consider TABLE 6, below. Suppose we wanted the value of y for $x = .023$, which is not

TABLE 6.

x	.01	.02	.03	.04	.05
y	.1	.1414	.1732	.2000	.2236

in the table. We could reason that x is $3/10$ of the way from .02 to .03 and thus y should be about $3/10$ of the way from .1414 to .1732:

$$y \approx .1414 + .3(.1732 - .1414) \tag{62}$$
$$\approx .1414 + .3(.0318)$$
$$\approx .1414 + .00954 = .15094$$
$$\approx .1509.$$

Old tables often included a list of the differences Δy along with the lists of x and y values to facilitate interpolation. Thus TABLE 6 might appear as TABLE 7, below. The presence of such a row allows one to skip the line (62) in the

TABLE 7.

x	.01	.02	.03	.04	.05
y	.1	.1414	.1732	.2000	.2236
Δy	.0414	.0318	.0268	.0236	

calculation above. As this is a simple enough task, the amount of effort put into performing all the subtractions for the possibly hundreds or even thousands of entries in a table seems a bit much. While the table itself might save the end user hours of labour, the list of differences will save a negligible amount of time. One thing it can do is facilitate the finding of second differences—the differences of differences—thus requiring only a single subtraction to find $\Delta^2 y = \bigl(f(x+2h) - f(x+h)\bigr) - \bigl(f(x+h) - f(x)\bigr)$ (h being the increment of the x's), which would otherwise require three subtractions.

---

[41] James Gregory made a number of such independent discoveries, with credit, in line with the biblical *Matthew Effect* ("Them that's got shall get; them that's not shall lose."), going to the more illustrious Newton. In the present case, however, the result is at least a generation older, having been used by Henry Briggs (1561 – 1630) in calculating logarithms and before him by Thomas Harriot (c. 1560 – 1621), who, however, left the technique unpublished.

Why second differences? Well, if we recast the simple use of proportions into algebraic language, we see that it yields a linear interpolation formula. In our case, the proportion,
$$\frac{y - .1414}{x - .02} = \frac{.1732 - .1414}{.03 - .02}$$
yields the linear function
$$y = .1414 + \frac{.0318}{.01}(x - .02)$$
connecting the points $\langle .02, .1414 \rangle$ and $\langle .03, .1732 \rangle$ of the table and approximating the tabulated function in the interval $[.02, .03]$. Higher order differences,
$$\Delta_h^{i+1} f(x) = \Delta_h^i f(x+h) - \Delta_h^i f(x),$$
yield higher order interpolating functions.

2.9.1. Theorem (Newton Forward Difference Formula). *Let f be a function defined for $a, a+h, a+2h, \ldots, a+nh$. Let $x = a + ih$ for $i \in \{0, 1, 2, \ldots, n\}$. Then*
$$f(x) = f(a) + \sum_{j=1}^{n} \frac{\Delta_h^j f(a)}{h^j} \cdot \frac{(x-a)(x-a-h)\cdots(x-a-(j-1)h)}{j!}. \quad (63)$$

Taylor did not provide a proof of this Theorem, but offered a table which made evident why the Theorem was true and what the induction hypothesis should be to provide a proof. I confess that my lack of knowledge shines through here. I've only read the excerpts of Taylor that appear in the source books[42], so I cannot say how familiar Taylor was with formal inductive proofs. John Crossley's brief discussion of the history of induction in his *The Emergence of Number*[43], Chapter II, sections 3 to 6, offers an outline of the history. Following the discussion of some inductive reasoning of the 16th century that approaches mathematical induction, he discusses that of the 17th century thus:

> In the following century, the seventeenth, we *do* find quite explicit use of the word "induction", though not always in the neat and precise forms given by Pascal and Bernoulli (see below). Wallis discusses the value of
> $$\frac{0 + 1 + 4 + 9 + \ldots + n^2}{n^2 + n^2 + n^2 + n^2 + \ldots + n^2}$$
> in his *Arithmetica infinitorum* [1699], p. 336). For successive values of n he gets smaller and smaller values and concludes that eventually the value vanishes (*evaniturus fit*), i.e. becomes zero. However, the sense in which Wallis uses "induction" is that of scientists rather

---

[42] Dirk J. Struik, *A Source Book in Mathematics, 1200 – 1800*, Harvard University Press, Cambridge (Mass.), 1969, pp. 328 – 333; and Stedall, *op. cit.*, pp. 201 – 206. Both sources give facsimile reproductions of Taylor's original Latin presentation along with English translation. Struik also translates the archaic notation into a more familiar modern mathematical language; Stedall sticks to the original, but explains how to perform this step.

[43] Crossley, John, *The Emergence of Number*, World Scientific, Singapore, 1987.

than of mathematicians. That is to say, he relies on repeated "experiments" giving the same result rather than a mathematical (viz. logical) proof of the result...

We next turn to Pascal, who meets the most stringent requirements when one is seeking the first to recognize, define and formally use the principle of mathematical induction. This is exemplified in Pascal's triangle,...

Pascal, however, in his *Traité du triangle arithmétique* [1665] of 1654 not only uses an inductive proof, he even draws attention to it as a *method* of proof...

About the same time, Fermat used a method of proof which we now know is equivalent to induction...

With Jacob Bernoulli, however, we again find the principle of induction clearly enunciated. The *Acta Eruditorum* of 1686 includes an excerpt from a letter...

But having given this specific example, he goes on to note that the *method* is quite general...

Thus by 1686 the idea of mathematical induction is both explicit and used. In the succeeding century, though not very widely used as a name, induction was employed though so, too, was Wallis's "induction" (scientific or experimental induction as we say). It was not until the latter half of the nineteenth century that it started to play a central rôle. The name "mathematical induction" appears to be due to De Morgan in 1838.[44]

Taylor, writing in 1715, lived in this period shared by experimental and mathematical induction, so his presentation can be viewed in either way. Today, such a presentation would be considered as evidence the author thought the ritual of a formal mathematical induction obvious and he couldn't be bothered to carry it out. For those early days of mathematical induction, when Pascal and Bernoulli both felt it necessary to spell things out, this interpretation is less clear cut.

But, I digress! Taylor derived Theorem 2.9.1 by setting up a table something like TABLE 8, below. He did not, however, present it in this form, but listed

TABLE 8.

x	y	$\Delta_h y$	...	$\Delta_h^{n-1} y$	$\Delta_h^n y$
a	$f(a+h)$	$\Delta_h f(a+h)$	...	$\Delta_h^{n-1} f(a+h)$	
$\vdots$	$\vdots$				
$a+(n-1)h$	$f(a+(n-1)h)$	$\Delta_h f(a+(n-1)h)$			
$a+nh$	$f(a+nh)$				

---

[44]Crossley, *op. cit.*, pp. 43 – 46.

the calculated values obtained by replacing
$$\Delta_h^j f(a + (i+1)h)$$
by the sum of the value above it and that immediately to the right of that number:
$$\Delta_h^j f(a + ih) + \Delta_h^{j+1} f(a + ih).$$
If one takes the first row as given, then makes the substitutions described to obtain the second row, then uses these substituted expressions, and so forth, one obtains a table like TABLE 9, below.

TABLE 9.

y	$\Delta_h y$
$f(a)$	$\Delta_h f(a)$
$f(a) + \Delta_h f(a)$	$\Delta_h f(a) + \Delta_h^2 f(a)$
$f(a) + 2\Delta_h f(a) + \Delta_h^2 f(a)$	$\Delta_h f(a) + 2\Delta_h^2 f(a) + \Delta_h^3 f(a)$
$f(a) + 3\Delta_h f(a)$ $+3\Delta_h^2 f(a) + \Delta_h^3 f(a)$	$\Delta_h f(a) + 3\Delta_h^2 f(a)$ $+3\Delta_h^3 f(a) + \Delta_h^4 f(a)$
$\vdots$	

$\Delta_h^2 y$	$\Delta_h^3 y$
$\Delta_h^2 f(a)$	$\Delta_h^3 f(a)$
$\Delta_h^2 f(a) + \Delta_h^3 f(a)$	$\Delta_h^3 f(a) + \Delta_h^4 f(a)$
$\Delta_h^2 f(a) + 2\Delta_h^3 f(a) + \Delta_h^4 f(a)$	$\Delta_h^3 f(a) + 2\Delta_h^4 f(a) + \Delta_h^5 f(a)$
$\Delta_h^2 f(a) + 3\Delta_h^3 f(a)$ $+3\Delta_h^4 f(a) + \Delta_h^5 f(a)$	$\Delta_h^3 f(a) + 3\Delta_h^4 f(a)$ $+3\Delta_h^5 f(a) + \Delta_h^6 f(a)$
$\vdots$	

Proof of Theorem 2.9.1. It will simplify matters if we first consider the case in which $a = 0$ and $h = 1$:
$$f(x) = f(0) + \sum_{j=1}^{n} \Delta^j f(0) \frac{x(x-1)\cdots(x-j+1)}{j!}$$
$$= \sum_{j=0}^{n} \binom{x}{j} \Delta^j f(0),$$

where $\Delta = \Delta_1$ and $\Delta^0 f(0) = f(0)$. In this case, $x = a + ih = i$ and we must prove for $i = 0, 1, \ldots, n$,

$$f(i) = \sum_{j=0}^{n} \binom{i}{j} \Delta^j f(0).$$

The proof is by induction, using the stronger induction hypothesis suggested by comparing the entries of TABLES 8 and 9:

$$\sum_{j=0}^{n} \binom{i}{j} \Delta^{j+k} f(0) = \Delta^k f(i) \tag{64}$$

for $i + k \leq n$.

For $i = 0$,

$$\sum_{j=0}^{n} \binom{i}{j} \Delta^{j+k} f(0) = \binom{0}{0} \Delta^{0+k} f(0) = \Delta^k f(0).$$

For the inductive step,

$$\sum_{j=0}^{n} \binom{i+1}{j} \Delta^{j+k} f(0) = \sum_{j=0}^{n} \left(\binom{i}{j} + \binom{i}{j-1}\right) \Delta^{j+k} f(0)$$

$$= \sum_{j=0}^{n} \binom{i}{j} \Delta^{j+k} f(0) + \sum_{j=0}^{n} \binom{i}{j-1} \Delta^{j+k} f(0)$$

$$= \Delta^k f(i) + \sum_{j=0}^{n-1} \binom{i}{j} \Delta^{j+k+1} f(0)$$

by the induction hypothesis and a change in index,

$$= \Delta^k f(i) + \sum_{j=0}^{n} \binom{i}{j} \Delta^{j+k+1} f(0),$$

since $i + 1 \leq n \Rightarrow i < n \Rightarrow \binom{i}{n} = 0$,

$$= \Delta^k f(i) + \Delta^{k+1} f(i),$$

by induction hypothesis,

$$= \Delta^k f(i+1),$$

by definition of Δ^{k+1}.

Letting $k = 0$ in (64),

$$\sum_{j=0}^{n} \binom{i}{j} \Delta^j f(0) = f(i).$$

Thus we have proven (63) for the special case in which $a = 0$ and $h = 1$. The general case reduces to the special one via a simple substitution. Given f, define
$$g(y) = f(a + hy),$$
so that $g(0) = f(a)$ and $\Delta_1^k g(y) = \Delta_h^k f(a + hy)$ and for $x = a + hi$,

$$f(x) = g(i) = \sum_{j=0}^n \binom{i}{j} \Delta_1^j g(0)$$

$$- \sum_{j=0}^n \binom{(x-a)/h}{j} \Delta_h^j f(a)$$

$$= \sum_{j=0}^n \frac{\frac{x-a}{h}\left(\frac{x-a}{h} - 1\right) \cdots \left(\frac{x-a}{h} - j + 1\right)}{j!} \Delta_h^j f(a)$$

$$= \sum_{j=0}^n \frac{\Delta_h^j f(a)}{h^j} \cdot \frac{(x-a)(x-a-h) \cdots (x-a-(j-1)h)}{j!}, \quad (65)$$

as was to be shown. $\square$

At the cost of a little abstraction, we can give a more memorable, equally rigorous proof that connects explicitly with the Binomial Theorem. One introduces the *operators* Δ_h, E_h, and I on functions as follows:

$$(\Delta_h(f))(x) = f(x+h) - f(x)$$
$$(E_h(f))(x) = f(x+h)$$
$$(I(f))(x) = f(x).$$

For each h, the collection of operators generated from Δ_h, E_h, and I by closing under multiplication by a scalar, addition, and composition turns out to be a *commutative ring*, an algebraic structure in which the Finite Binomial Theorem can be proven to hold.[45] Now, $E_h = I + \Delta_h$:

$$E_h f(x) = f(x+h) = f(x) + \big(f(x+h) - f(x)\big) = If(x) + \Delta_h f(x)$$

and if we choose $h = (x-a)/n$ for some n, then

$$f(x) = E_{x-a} f(a).$$

But

$$E_{x-a} = E_{nh} = E_h^n = (I + \Delta_h)^n = \sum_{j=0}^n \binom{n}{j} \Delta_h^j$$

$$= \sum_{j=0}^n \binom{(x-a)/h}{j} \Delta_h^j,$$

from which (63) can be read off as in the last part of the proof of Theorem 2.9.1.

---

[45]If the reader is skeptical, he can take this argument as a heuristic instead of a rigorous proof.

9. TAYLOR'S THEOREM

2.9.2. Exercise. *Extend* TABLE 7 *to include rows for* Δ^2 *and* Δ^3. *Apply the Newton Forward Difference Formula to obtain quadratic and cubic interpolation functions, and compare the results when interpolating values for* $x = .023$. [*To compare these with the "correct" value, I note that in* TABLE 6 *the y-values were obtained by rounding the square roots of the x-values to four places.*]

From Newton's Forward Interpolation Formula, Taylor derived the power series representation of a function through what Felix Klein called "a transition to the limit of extraordinary audacity".[46] Assume f is defined on an interval containing a and x and divide the interval $[a, x]$ into n pieces of length $h = (x-a)/n$. By (65),

$$f(x) = \sum_{j=0}^{n} \frac{(x-a)(x-a-h)\cdots(x-a-(j-1)h)}{j!} \cdot \frac{\Delta_h^j f(a)}{h^j}.$$

As $n \to \infty$ we have $h \to 0$ and

$$f(x) = \sum_{j=0}^{\infty} \lim_{h \to 0} \frac{(x-a)(x-a-h)\cdots(x-a-(j-1)h)}{j!} \cdot \lim_{h \to 0} \frac{\Delta_h^j f(a)}{h^j}$$

$$= \sum_{j=0}^{\infty} \frac{(x-a)^j}{j!} f^{(j)}(a)$$

$$= \sum_{j=0}^{\infty} \frac{f^{(j)}(a)}{j!}(x-a)^j.$$

It hardly needs mention that this is no proof. Certainly

$$\lim_{h \to 0} \frac{(x-a)(x-a-h)\cdots(x-a-(j-1)h)}{j!} = \frac{(x-a)^j}{j!}$$

for each j. And it can be shown, if f has sufficiently many derivatives, that

$$\lim_{h \to 0} \frac{\Delta_h^j f(a)}{h^j} = f^{(j)}(a),$$

for each j. Thus each term

$$\frac{(x-a)(x-a-h)\cdots(x-a-(j-1)h)}{j!} \cdot \frac{\Delta_h^j f(a)}{h^j}$$

tends to

$$\frac{(x-a)^j}{j!} f^{(j)}(a),$$

but do they all tend to these limits rapidly enough so that the infinite accumulation of errors,

$$\sum_{j=0}^{\infty} \left(\frac{(x-a)(x-a-h)\cdots(x-a-(j-1)h)}{j!} \cdot \frac{\Delta_h^j f(a)}{h^j} - \frac{(x-a)^j}{j!} f^{(j)}(a) \right)$$

---

[46] Cf. Struik, op. cit., p. 332.

$$= \sum_{j=0}^{\infty} \varepsilon_j(h)$$

tends to 0 with h? Obviously, for some functions it will and for others it won't.

Without some accounting for the occasional success of his derivation, Taylor's proof is nothing more than a heuristic device, a method of determining the coefficients $f^{(j)}(a)/j!$ of the power series expansion $\sum a_j(x-a)^j$ of f around a. If all we want is a heuristic argument, we can do much better than this.

Identifying I with 1 and writing D for the differentiation operator,

$$D = \lim_{h \to 0} \frac{\Delta_h}{h},$$

we have, for $h = (x-a)/n$,

$$E_{x-a} = E_h^n = (1+\Delta_h)^n$$
$$\to (1+hD)^{(x-a)/h}$$
$$= \sum_{j=0}^{\infty} \binom{(x-a)/h}{j}(hD)^j,$$

by Newton's Theorem(!) and application of this and a little algebra to f yields the desired

$$f(x) = \sum_{j=0}^{\infty} \frac{f^{(j)}(a)}{j!}(x-a)^j.$$

Of course, this again requires an audacious passage to the limit. We can get around this by recalling the formula,

$$\lim_{n \to \infty}\left(1+\frac{x}{n}\right)^n = e^x.$$

Letting $h = (x-a)/n$ and recalling $f(x) = E_{nh}f(a)$, we have

$$E_{x-a} = E_{nh} = (1+\Delta_h)^n = \left(1+\frac{(x-a)\Delta_h}{hn}\right)^n$$
$$\to \left(1+\frac{(x-a)D}{n}\right)^n$$
$$\to e^{(x-a)D}.$$

I.e.,

$$E_{x-a} = e^{(x-a)D} = \sum_{j=0}^{\infty}\frac{(x-a)^j D^j}{j!},$$

and

$$f(x) = E_{x-a}f(a) = \sum_{j=0}^{\infty}\frac{f^{(j)}(a)}{j!}(x-a)^j.$$

The best heuristic determination of the coefficients of the power series expansion, one that is still given in textbooks today is that of Colin Maclaurin

9. TAYLOR'S THEOREM

in his *Treatise on fluxions*[47] [1742]. He assumes the existence of an expansion,

$$f(x) = a_0 + a_1 x + a_2 x^2 + \ldots,$$

and proceeds to determine $a_0, a_1, a_2, \ldots$ successively. Obviously $f(0) = a_0$. Differentiating term-by-term as if the power series were an oversized polynomial of infinite degree, one has

$$f'(x) = a_1 + 2a_2 x + 3a_3 x^2 + \ldots,$$

whence $f'(0) = a_1$. Differentiating again,

$$f''(x) = 2a_2 + 3 \cdot 2a_3 x + 4 \cdot 3a_4 x^2 + \ldots,$$

and $f''(0) = 2a_2$. Continuing,

$$f^{(n)}(0) = n! a_n,$$

i.e., $a_n = f^{(n)}(0)/n!$.

Maclaurin dealt with the series expansion around $a = 0$. To obtain the Taylor expansion around general a, simply write $g(x) = f(x+a)$ and observe

$$f(x) = g(x - a) = \sum_{j=0}^{\infty} \frac{g^{(j)}(0)}{j!} (x-a)^j = \sum_{j=0}^{\infty} \frac{f^{(j)}(a)}{j!} (x-a)^j$$

by the Chain Rule.

Taylor's and Maclaurin's derivations do not provide proofs, but, at least once one has proven that power series can be differentiated term-by-term, Maclaurin's argument tells us what such a series must be. For example, for $f(x) = (1+x)^\mu$ note that

$$f'(x) = \mu(1+x)^{\mu-1}$$
$$f''(x) = \mu(\mu-1)(1+x)^{\mu-2}$$
$$\vdots$$
$$f^{(n)}(x) = \mu(\mu-1) \cdots (\mu-n+1) x^{\mu-n},$$

whence

$$f'(0) = \mu, \quad f''(0) = \mu(\mu-1), \quad \ldots, \quad f^{(n)}(0) = \mu(\mu-1) \cdots (\mu-n+1)$$

and

$$f(x) = (1+x)^\mu = \sum_{j=0}^{\infty} \frac{\mu(\mu-1) \cdots (\mu-j+1)}{j!} (1+x-1)^j$$

$$= \sum_{j=0}^{\infty} \binom{\mu}{j} x^j,$$

exactly as Newton claimed.

---

[47]The relevant passage is reproduced in Struik, *op. cit.*, pp. 338 – 340, and again in Stedall, *op. cit.*, p. 207.

2.9.3. Exercise. *Assuming the expansions exist, find the Maclaurin series representing the functions:*

i. $f(x) = \sin x$
ii. $f(x) = \sin^{-1} x$
iii. $f(x) = \cos x$
iv. $f(x) = \tan^{-1} x$
v. $f(x) = \ln(1+x)$.

Comparing the results of this exercise with the derivations of these series given earlier in the Chapter shows that, while the results are still lacking proofs, something has been gained. We still do not know when the equation,
$$f(x) = a_0 + a_1 x + a_2 x^2 + \ldots,$$
is valid, but we know how to find $a_0, a_1, a_2, \ldots$ easily.

We needn't consider only transcendental functions. The following Example cleans up one loose end from Chapter 1 (page 48, above).

2.9.4. Example. *Let*
$$P(x) = a_0 + a_1 x + \ldots + a_n x^n$$
be a polynomial of degree n. Then, for each k, $a_k = f^{(k)}(0)/k!$, i.e.,
$$P(x) = P(0) + \frac{P'(0)}{1!} x + \frac{P''(0)}{2!} x^2 + \ldots + \frac{P^{(n)}(0)}{n!} x^n.$$

Maclaurin's proof of Taylor's Theorem works perfectly in this case as there is no problem differentiating finite sums termwise. The question of the termwise differentiability of infinite sums is much thornier and will be taken up in Chapter 3, section 8 (Theorem 3.8.14 on page 211 and its ensuing discussion).

In the meantime I suggest one more exercise.

2.9.5. Exercise. *Find the first few terms of the Maclaurin series for the rational function,*
$$f(x) = \frac{1}{1 - x - x^2},$$
i. *by applying Maclaurin's method, and*
ii. *by long division.*
Which method is easier in this case.

10. Newton's Method and the Mean Value Theorem

I shall finish this chapter with some material that harks back to the numerical work of Chapter 1, but explicitly invokes the Calculus. These are Newton's Method for approximating the zeros of functions and the Mean Value Theorem. The latter is an important theoretical tool in the theory of the Calculus, a hint of which importance will be seen in the next chapter, but is also a useful tool in error estimation as we will see later in the present section.

Newton's Method, or the Newton-Raphson Method or the Tangent Method, as it is also called, is, in the form taught today, primarily the work of four individuals—Newton himself, Joseph Raphson (1678 – 1765), Jean Raymond

10. NEWTON'S METHOD AND THE MEAN VALUE THEOREM

Pierre Mouraille (1720 – 1808), and Cauchy. Newton worked out the original method, explaining it algebraically; Raphson reformulated it as a simple recursion; Mouraille interpreted it geometrically, first noted its limitations, raised the question of the extent of its applicability, and gave conditions under which the procedure was guaranteed success; and Cauchy determined how quickly the sequence of approximations converges. The story, with some of its aftermath, is nicely told via the commentary on and excerpts from the texts of these men in Chabert's source book[48]. Insofar as the discussion digresses from our main theme, I shall be brief in my summary.

Newton's point of departure was again to find the area under an algebraic curve, $F(x,y) = 0$, where F is a polynomial in the two variables x and y. His solution was to express y as a power series in x, $y = f(x)$, finding the terms one at a time. The technique being fairly complicated, he began his exposition by treating the simpler problem of solving an equation $F(x, 0) = 0$, i.e., of finding the root of a polynomial. The polynomial he chose to use for this demonstration,

$$x^3 - 2x - 5 = 0, \tag{66}$$

being of sufficiently general form, became the standard example used by mathematicians to test their own procedures to illustrate any particular benefits these methods might afford.

Entering

Y₁=X^3−2X−5

in the equation editor and using ZStandard in the ZOOM window to graph the function, we see that the polynomial crosses the x-axis exactly once, at some value slightly greater than 2. Thus, it is no surprise that Newton begins his search for a root to the equation there, assuming the root to be $x = 2 + y$ for some small error y. Now, writing $P(x) = x^3 - 2x - 5$, we make the substitution $x = y + 2$ to obtain a new polynomial,

$$Q(y) = y^3 + 6y^2 + 10y - 1.$$

[The reader who finds the algebra tiresome will find programs for the *TI-83 Plus* that will do the work for him in the first section of the Appendix. One possessing the *TI-89 Titanium* need only successively enter

x^3−2x−5→p(x)
expand(p(y+2))→q(y)
q(y)

to read the coefficients of the polynomial Q—provided, of course, x and y have not been assigned specific values.]

Now, y should be small, whence y^2 and y^3 should be negligible and we can take $Q(y) \approx 10y - 1$, i.e., $y = 1/10$. Again, $1/10$ is only an approximation to the error and we may write $y = 1/10 + z$ and estimate the error of the error

---

[48] Jean-Luc Chabert (ed.), *A History of Algorithms; From the Pebble to the Microchip*, Springer-Verlag, Heidelberg, 1999, chapter 6 on "Newton's Methods", pp. 169 – 197.

by solving $R(z) = 0$ for

$$R(z) = Q\left(z + \frac{1}{10}\right) = z^3 + \frac{63z^2}{10} + \frac{1123z}{100} + \frac{61}{1000}.$$

Again we can ignore the higher degree terms and approximate z by solving

$$\frac{1123z}{100} + \frac{61}{1000} = 0,$$

i.e.,

$$z \approx \frac{-61}{11230}.$$

Altogether, we thus far have as an approximation to the root of P,

$$x + y + z = 2 + \frac{1}{10} - \frac{61}{11230} = \frac{11761}{5615} \approx 2.094568121104,$$

which is a bit large, but not too shabby: the actual root lies between 2.09455148 and 2.09455149.

In spirit, Newton's Method is similar to the Chinese-Horner Method. One starts with an estimate of the root of a polynomial and makes a substitution to determine a polynomial whose root is the error. One then estimates the root of the new polynomial by choosing the negative of the ratio of the constant coefficient to the linear one exactly as Qín instructs us to. The difference is that Qín always produces a lower bound yielding only non-negative correction terms in the final sum and he gets exactly one additional digit accuracy in each successive step. Newton's sequence,

$$2,\ 2.1,\ 2.094568,\ \ldots$$

did not start out well in this respect. Certainly 2.1 is closer to the root than 2.0, but 1 is not the correct digit. The next level, where Qín would get 2.09, Newton is already correct to the 4th decimal place. And the next entry in the Newtonian series is a slightly large 2.09455148699, correct to 8 decimals, while Qín would claim only 3. While each step for Qín adds a single secured digit, each iteration of Newton will double the number of secured digits for this polynomial. Qín requires many more steps to achieve a fixed high accuracy, while Newton requires multiplications and divisions of successively larger multidigit numbers. For one calculating by hand, it is not clear where the actual advantage lies, but on a calculator like the *TI-89 Titanium*, where the multiplications and divisions of multidigit numbers are built-in, the advantage seems to lie with Newton's approach.

I relegate such considerations to the Appendix, preferring here to consider the Method itself rather than its application.

To illustrate the technique in two variables, consider again Newton's original problem of finding the area of a circular segment, i.e., the area under the curve $y = \sqrt{1 - x^2}$. We want to express y as a power series in x. To this end consider the polynomial equation

$$y^2 + x^2 - 1 = 0. \tag{67}$$

10. NEWTON'S METHOD AND THE MEAN VALUE THEOREM

Now x must be less than 1 in absolute value, so we may consider x^2 as small and ignore it, yielding the solution to $y^2 - 1 = 0$, i.e., $y = 1$ as our first approximation to y as a root of (67).

y is not actually equal to 1, but has an error, say u: $y = 1 + u$. Substituting, (67) becomes
$$(1 + u)^2 + x^2 - 1 = 0,$$
i.e.,
$$u^2 + 2u + x^2 = 0. \tag{68}$$
Because u is small, u^2 is very small and we can ignore it to get $2u + x^2 = 0$, i.e.,
$$u = -\frac{x^2}{2}.$$
Again this is not exact and $u = v - x^2/2$ for some error v. Plugging this into (68) yields
$$\left(v - \frac{x^2}{2}\right)^2 + 2\left(v - \frac{x^2}{2}\right) + x^2 = 0,$$
i.e.,
$$v^2 - vx^2 + \frac{x^4}{4} + 2v = 0.$$
Ignoring the very very small terms v^2 and vx^2, we have
$$2v + \frac{x^4}{4} = 0, \quad \text{i.e., } v = -\frac{x^4}{8}.$$
Thus far we have
$$y = 1 + u + v = 1 - \frac{x^2}{2} - \frac{x^4}{8},$$
the first three terms of Newton's power series expansion for $\sqrt{1 - x^2}$.

2.10.1. Exercise. *Derive the binomial series for*
$$y = (1 + x)^{2/3}$$
by applying this method to the polynomial equation
$$y^3 - x^2 - 2x - 1 = 0.$$

Once one is aware of the nature of the Taylor expansion of a function, Newton's Method does not seem that efficient for finding power series representations. But it is still of numerical use in finding approximate numerical solutions.

In 1690, Joseph Raphson published a method that at first appeared to differ from Newton's, but which is actually the same. According to Raphson, one would start with an approximation a to the root and choose
$$x = a - \frac{f(a)}{f'(a)}$$
as the next approximation. Implicitly he defined a sequence $x_0, x_1, x_2, \ldots$ of successive approximations by the recursion
$$x_0 = a$$

$$x_{n+1} = x_n - \frac{f(x_n)}{f'(x_n)}.$$

Newton's and Raphson's methods are closely related, but not identical, the difference being that Raphson bypasses the polynomial substitutions. Newton starts with $x_0 = a$ and calculates a sequence of estimates $y_0(= x_0), y_1, y_2, \ldots$ of the errors using

$$y_{n+1} = \frac{-P_n(y_n)}{P'_n(y_n)}$$

where $P_n(y) = P_{n-1}(y_{n-1} + y)$. But

$$P'_n(y) = \frac{dP_n}{dy}(y) = \frac{dP_{n-1}}{dy}(y_{n-1} + y) = P'_{n-1}(y_{n-1} + y).$$

Working backward, for Q_n denoting either P_n or P'_n,

$$Q_n(y) = Q_{n-1}(y_{n-1} + y)$$
$$= Q_{n-2}(y_{n-2+y_n-1} + y) \ldots$$
$$\vdots$$
$$= Q_0(x_0 + y_1 + \ldots + y_{n-1} + y)$$
$$= Q_0(x_1 + y_2 + y_3 + \ldots + y_{n-1} + y)$$
$$\vdots$$
$$= Q_0(x_{n-1} + y)$$
$$= Q_0(x_n).$$

Thus,

$$y_{n+1} = \frac{-P(x_n)}{P'(x_n)}$$

and

$$x_{n+1} = x_n + y_{n+1} = x_n - \frac{P(x_n)}{P'(x_n)},$$

i.e., Newton and Raphson agree at each step. Newton's Method, being essentially the Chinese algorithm minus the restriction to improve the estimate from below a single digit at a time, is initially better motivated; while Raphson's Method, by omitting the unnecessary substitution step is computationally more efficient. A little geometry, however, provides proper motivation for Raphson's algorithm.

Newton's Method is also called the Tangent Method because it can be formulated geometrically as a sort of variant of the Secant Method. The Secant Method is just a geometric rebranding of linear interpolation. In FIGURE 7, below, we illustrate this method. Given two points a, b for which $f(a) < 0 < f(b)$, we can approximate the zero of f by drawing the secant line connecting the points $\langle a, f(a) \rangle$ and $\langle b, f(b) \rangle$ and seeing where it crosses the x-axis. Newton's Method is a one-point variant. Given a near the root, one uses the tangent line

10. NEWTON'S METHOD AND THE MEAN VALUE THEOREM

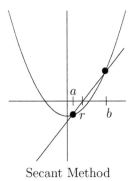
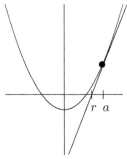

Secant Method   Tangent Method

FIGURE 7.

through $\langle a, f(a) \rangle$ in place of a secant line, taking the intersection of said line with the x-axis as the next approximation: The equation of the tangent line is

$$\frac{y - f(a)}{x - a} = f'(a),$$

i.e.,

$$y - f(a) = f'(a)(x - a)$$

i.e.,

$$x - a = \frac{y - f(a)}{f'(a)},$$

the x-intercept of which is given by

$$x - a = \frac{-f(a)}{f'(a)}.$$

Thus, if x is an attempt to approximate the root, $x - a = -f(a)/f'(a)$ is Newton's familiar error estimate.

Thus Raphson's Method, under the names "Newton's Method" and "the Newton-Raphson Method" has superceded the procedure as Newton actually described it.

Chabert summarises:

> For a long time, the two methods, Newton's and Raphson's, were set out in distinctly different ways. Newton put the accent on choosing initial values sufficiently close in order to ensure linear approximations, whereas Raphson used intervals, which were obtained by choosing an initial value which was greater or less than the root, according to the type of equation being considered. Lagrange... showed that "these two methods are fundamentally the same, but presented differently".
>
> This process, which had been used by Newton and Raphson only for algebraic equations with rational coefficients, was used by Thomas Simpson for equations with irrational and transcendental coefficients

(*Essays on Mathematics*, 1740). However, the initial conditions necessary to guarantee convergence in these cases was [*sic*] not properly considered.[49]

Another quarter century would pass before the convergence question would begin to be seriously addressed. In his *Traité de la Résolution des Equations en général* [*Treatise on the resolution of equations in general*] of 1768, emphasising the geometry Mourraille first demonstrated some of the bad behaviour of the Newton-Raphson Method. The excerpt in Chabert's source book deals with a generic 4th degree polynomial and is stated qualitatively. With today's graphing calculators, it is easy to examine concrete examples quantitatively.[50]

Let us begin with the polynomial

$$P(x) = x^4 - 10x^3 + 34x^2 - 44x + 17,$$

the derivative of which is

$$P'(x) = 4x^3 - 30x^2 + 68x - 44.$$

In the equation editor enter

Y_1=X^4−10X^3+34X^2−44X+17
Y_2=4X^3−30X^2+68X−44
Y_3=X−Y_1(X)/Y_2(X).

Graphing only Y_1 we see that it has only two zeros, one near 2/3 and one near 1.8. Using the calculator's built-in zero-finding function, one quickly improves the estimates of the two zeros to .6678102335 and 1.769707421.

One can, of course, also obtain these estimates fairly quickly using the Newton-Raphson iteration. If one first enters

1→A

and then repeatedly enters

Y_3(A)→A,

one will see scrolling before one's eyes the list of numbers

1, 0, .3863636364, .5920960786, .6602225921,

.6677227941, .6678102217, .6678102335, ...,

the last entry repeating from here on. This yields the first root. Repeating the exercise with 2 as an initial approximation to the second root,

2→A
Y_3(A)→A,

one will generate the sequence

2, 1.75, 1.769691781, 1.769707421, ...,

---

[49] *Ibid.*, p. 178.

[50] In 1770, another edition of the book appeared under the slightly altered title, *Traité de la Résolution des Equations invariable*. I've not seen the original, but this second version can be accessed online—or most of it can: the copy I downloaded seems to omit every other page of figures from the end of the book. In this version he does not follow up his qualitative geometric discussion with any specific examples.

10. NEWTON'S METHOD AND THE MEAN VALUE THEOREM

the last entry again repeating forever.

Thus far the Newton-Raphson iteration has performed as advertised. However, if we start with 2.5 as our initial approximation to the larger root,

2.5→A
Y_3(A)→A,

and repeatedly press ENTER, the sequence that scrolls on the screen is

2.5, .1875, .4936073314, .6339076162, .6661587838,

.6678060373, .6678102335, ...,

the last entry repeating. The larger approximation to the larger root has resulted in a sequence converging to the smaller root.

The failure to converge to the desired root can be even worse. Enter

Y_4=Y_3(Y_3(X))−X

and graph this function in DOT mode using the window

Xmin=3
Xmax=4
Xscl=.1
Ymin=⁻.5
Ymax=.5
Yscl=.1.

You will see there to be several zeros to this function, including one between 3.1 and 3.2 near 3.18, one between 3.4 and 3.5 near 3.44, and two between 3.8 and 3.9 near 3.81 and 3.85, respectively. Use the built-in zero-finder of the calculator to find these in succession, exiting to the home screen after each zero is located to store the result X in the variables A, B, C, D, respectively. These values are (showing only the displayed digits)

A: 3.179172001
B: 3.437592487
C: 3.814895286
D: 3.855059635.

If you now enter

Y_3({A,B,C,D}),

you will get

{3.814895286, 3.855059635, 3.179172001, 3.437592487}.

That is, starting at the points A, B, C, D, the Newton-Raphson Method, instead of converging to a root of the polynomial, repeatedly maps A to C and back and B to D and back.

The calculator can give us a nice picture of this. On the *TI-83 Plus*[51], set the window to

---

[51] The instructions for the *TI-89 Titanium* differ slightly. Most notably, that calculator's PtOn command does not accept the argument 2 choosing a box to represent a point and one must indicate the points on the curve by drawing small circles around them.

```
Xmin=3
Xmax=4.2
Xscl=.1
Ymin=⁻1
Ymax=2
Yscl=.1 ,
```
and make sure that only the graph of Y_1 is turned on. Exit to the home screen and enter

Pt−On(A,Y_1(A),2):Pt−On(C,Y_1(C),2) .

[Use both commands on the same line separated by a colon to avoid having to exit to the home screen again to execute the second command.] This will place little boxes on the curve above the points $x =$ A and $x =$ C. Now draw the tangent lines. This can be done in two ways. The first is to enter the equations for the tangent lines in the equation editor. The following will graph the two simultaneously:

$Y_5=Y_1$({A,C})+Y_2({A,C})(X−{A,C}) .

Alternatively, one can use DRAW commands. From the home screen one enters

Tangent(Y_1,A):Tangent(Y_1,C) ,

or from the graphics screen, choose TRACE and make sure Y_1 is selected, choose Tangent from the DRAW menu, and then enter A. Repeat for C. Finally, one can add vertical lines by using the Vertical command from the DRAW menu. Either enter

Vertical A:Vertical C

in the home screen, or from the graphics screen press the TRACE button, enter A, choose and enter Vertical from the DRAW menu, and repeat for C. Then hit CLEAR to remove the information at the bottom of the screen and see an uncluttered view of a rather cluttered diagram.

2.10.2. Exercise. *Consider the rational function*
$$f(x) = \frac{x}{x^2+1}.$$
This has a unique zero at 0.
i.   *Find $f'(x)$ and $g(x) = x - f(x)/f'(x)$.*
ii. *What is $g(1)$?*
iii. *Solve for $g(g(x)) = x$ on the calculator. How does g behave at these values?*
iv. *Let $x_0 > 1$ and define $x_{n+1} = g(x_n)$. How does the sequence $x_0, x_1, x_2, \ldots$ behave?*
v.   *Let $x_0 = .6$ and define $x_{n+1} = g(x_n)$. How does the sequence $x_0, x_1, x_2, \ldots$ behave?*

2.10.3. Exercise. *Examine the behaviour of the Newton-Raphson iteration for the curves $y = x^3 - 2x - 5$ and $y = x^3 - 2x + 2$.*

Mourraille did more than this. He also noted qualitatively that various conditions guaranteed convergence and he even explained qualitatively where convergence would be more-or-less rapid. But he did not give the information, useful in practice, on how rapid convergence was. Jean Baptiste Joseph Fourier (1768 – 1830) considered this problem in 1818, but his work contained an error. By the time his correction was posthumously published in 1831, Cauchy had already published two solutions, a Calculus-free solution in the third appendicial note to his famous *Cours d'analyse algébrique* [*Course in algebraic analysis*] of 1821, and a more general, more definitive, solution in his *Leçons sur le calcul différentiel* [*Lectures on the differential calculus*] of 1829.

Cauchy's analysis of the convergence of the iterates rested on a central result of the Differential Calculus called the *Mean Value Theorem*. With roots stretching back to 14th century Oxford and Paris and making a more impressive appearance in the Kerala mathematics of the 16th century before assuming its current position of prominence at the end of the 18th and beginning of the 19th centuries at the hands of Lagrange, Bolzano, and, most notably, Cauchy, the Mean Value Theorem could easily occupy a chapter of its own. Its lineage is not as venerable as that of the Binomial Theorem, but its importance has outlived that of the latter. One can no longer say with Bolzano that "it would hardly be an exaggeration to say that almost the whole of the so-called differential and integral calculus (higher analysis) rests on" the Binomial Theorem[52], but one can still repeat the phrase with respect to the Mean Value Theorem with some justification.

The modern statement of the Mean Value Theorem reads as follows:

2.10.4. Theorem (Mean Value Theorem). *Let $a < b$ and let the function f be continuous on $[a, b]$ and differentiable on (a, b). There is a point $c \in (a, b)$ such that*
$$f'(c) = \frac{f(b) - f(a)}{b - a}.$$

Geometrically, the Mean Value Theorem asserts that there is some point c strictly between a and b such that the tangent line to the curve $y = f(x)$ through $\langle c, f(c) \rangle$ is parallel to the secant line connecting $\langle a, f(a) \rangle$ and $\langle b, f(b) \rangle$. Its truth is fairly obvious on intuitive grounds: If one slides the secant line along a line perpendicular to it (PQ in FIGURE 8, below), without rotation, it will eventually have no intersection with the curve. At the last point P of intersection, the translated secant is tangent to the curve.

Geometric intuition, however, is not a substitute for proof. Before the rigorisation of analysis led by Bolzano, Cauchy, *et alia*, it was believed that all functions were differentiable at all but a few isolated points and that the Taylor series of a function equalled the function wherever it converged. These beliefs turned out to be false, Bolzano producing, albeit not publishing, a counterexample to the former in the early 1830s and Cauchy giving a counterexample to the latter in his *Résumé des leçons données a l'école royale polytechnique sur le calcul infinitésimal*[*Summary of the lectures on the infinitesimal calculus given at the royal polytechnical school*], commonly referred to as the *Calcul infinitésimal* in 1823.[53] Thus, a rigorous proof is required. Such a proof requires

---

[52]*Cf.* page 147, below, for Bolzano's remark.
[53]I refer the advanced reader to my own *Adventures in Formalism* (College Publications, London, 2012) as a convenient source for both results. Bolzano's function is described and a

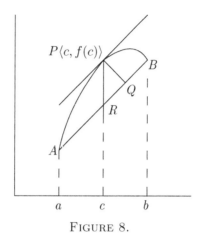

FIGURE 8.

a deeper look into the properties of the real numbers and of continuous functions than are established in the basic Calculus course and the now standard proof involves a bit of hand waving in assuming without proof that a continuous function on a closed bounded interval assumes a maximum and a minimum value:

2.10.5. Theorem (Extreme Value Theorem). *Let $a < b$ and let the function f be continuous on $[a,b]$. There are $c, d \in [a,b]$ such that, for all $x \in [a,b]$,*
$$f(c) \leq f(x) \leq f(d).$$

In one's mathematical education, the Extreme Value Theorem is stated and used in the standard Calculus course, but first proven in the more advanced theory course. Likewise, historically it was assumed until Bolzano and Cauchy provided demonstrations in their respective foundational works on the Calculus. For us, it will suffice to know its truth and not the reasons for this truth.

Its use here is in the observation, first exploited by Pierre de Fermat, that the tangent to a curve at a local extremum is horizontal. I.e., if f assumes its maximum or minimum for the interval $[a,b]$ at an interior point $c \in (a,b)$, then $f'(c) = 0$. A quick corollary to this observation is a special case of the Mean Value Theorem eponymously called *Rolle's Theorem* after Michel Rolle who proved nothing of the kind[54].

---

proof of the nowhere differentiability of a slightly more amenable variant is given in Chapter III, section 8.2 of that book. Also included, in III.8.3, therein is a full English translation of Karl Weierstrass's later presentation of such a function. Cauchy's example concerning Taylor series is broken into several steps and presented as Exercise 6.6 (pp. 184 – 185) at the end of Chapter II of that work.

[54]Rolle proved the Theorem for polynomials algebraically using an algebraic description of the derivative. Rolle was a critic of the Calculus and did not believe in derivatives. For details of the history I refer the read to June Barrow-Green, "From cascades to calculus: Rolle's theorem", in: Eleanor Robson and Jacqueline Stedall (eds.), *The Oxford Handbook of the History of Mathematics*, Oxford University Press, Oxford, 2009.

2.10.6. Theorem (Rolle's Theorem). *Let $a < b$ and let the function f be continuous on $[a, b]$ and differentiable on (a, b). If $f(a) = f(b)$, there is some $c \in (a, b)$ such that $f'(c) = 0$.*

For, either f is constant and $f'(c) = 0$ for all $c \in [a, b]$, or it is not and f assumes a value larger or smaller than $f(a) = f(b)$ somewhere in the interior and hence has an extremum (maximum or minimum) at some $c \in (a, b)$ where f' must be 0.

One would like to conclude the Mean Value Theorem as a Corollary to Rolle's Theorem by simply rotating the curve in FIGURE 8 until AB is horizontal. However, $f(x)$ may not be as well-behaved as in my picture and, if it oscillates, the curve may cease to be a function: The now vertical line passing through Q might intersect the curve in several points. The function mapping Q to the intersecting point of maximum distance from AB may no longer be continuous...

One could consider the distance from P to Q as a function of c and maximise that, noting that its derivative at c must be 0, and then determine $f'(c)$ from that. However, the algebra involved can be replaced by a simple geometric observation. Notice that if P' is any other point on the curve, and Q' and R' are the points on AB obtained by dropping a perpendicular from P' to AB and a vertical from P' to AB, respectively, then the triangle $P'Q'R'$ is similar to PQR, whence $P'Q'$ is maximised when $P'R'$ is maximised. And the algebraic determination of the length of $P'R'$ is easy. One simply subtracts the function defining the secant line AB from f. The equation of the line connecting $A = \langle a, f(a) \rangle$ to $B = \langle b, \langle f(b) \rangle$ is

$$y = f(a) + \frac{f(b) - f(a)}{b - a}(x - a).$$

Subtracting this from f gives the function

$$g(x) = f(x) - f(a) - \frac{f(b) - f(a)}{b - a}(x - a),$$

which is also continuous on $[a, b]$ and differentiable on (a, b). But

$$g(a) = f(a) - f(a) - \frac{f(b) - f(a)}{b - a}(a - a) = 0 - \frac{f(b) - f(a)}{b - a} 0 = 0,$$

and

$$g(b) = f(b) - f(a) - \frac{f(b) - f(a)}{b - a}(b - a) = f(b) - f(a) - \big(f(b) - f(a)\big) = 0,$$

whence $g(a) = g(b) = 0$ and Rolle's Theorem tells us $g'(c) = 0$ for some $c \in (a, b)$. But

$$g'(x) = f'(x) - \frac{f(b) - f(a)}{b - a},$$

whence

$$f'(c) - \frac{f(b) - f(a)}{b - a} = 0, \text{ i.e. } f'(c) = \frac{f(b) - f(a)}{b - a}.$$

One can follow the proof of the Mean Value Theorem with theoretical consequences or applications. Some simple theoretical consequences are collected in the following corollary.

2.10.7. Corollary. *Let $a < b$ and let f be continuous on $[a,b]$ and differentiable on (a,b).*
i. if $f'(x) = 0$ for all $x \in (a,b)$ then f is constant;
ii. if $f'(x) > 0$ for all $x \in (a,b)$ then f is strictly increasing on $[a,b]$;
iii. if $f'(x) < 0$ for all $x \in (a,b)$ then f is strictly decreasing on $[a,b]$.

Proof. The proofs all being similar, I present only the proof of ii. Suppose by way of contradiction, that f is not strictly increasing on $[a,b]$. Then there are $a', b' \in [a,b]$ such that $a' < b'$ and $f(a') \geq f(b')$. By the Mean Value Theorem, there is some $c \in (a', b') \subseteq [a,b]$ such that

$$f'(c) = \frac{f(b') - f(a')}{b - a} \leq \frac{0}{b - a} = 0,$$

contrary to the assumption that $f'(c) > 0$. □

By Corollary 2.10.7.i, if $f' = g'$ in an interval, then $f - g$ is constant and the assertion on page 90, above, that the antiderivative of a function is unique up to a constant has been rigorously justified.

That the converse of Corollary 2.10.7.i is true, i.e., that the derivative of a constant function is 0, we have already checked by direct computation. The converses to parts ii and iii of the Corollary are not true in general, but are nearly so. A strictly increasing continuous and differentiable function can have its derivative equal to 0 infinitely often, though not on any subinterval (as the function would be constant there).

2.10.8. Example. *Let $f(x)$ be the area under the curve $y = g(x)$ between $x = -.5$ and $x = .5$, where*

$$g(x) = \begin{cases} |x \sin \frac{1}{x}|, & x \neq 0 \\ 0, & x = 0. \end{cases}$$

g is a well-known continuous function, whence the Fundamental Theorem of Calculus tells us that f is defined and differentiable and $f'(x) = g(x)$. $g(x)$ is strictly increasing, yet $f'(x) = 0$ infinitely often, namely at $x = 0$ and at each point $x = \pm 1/(k\pi)$ for $k = 0, 1, 2, \ldots$

2.10.9. Exercise. *The function g of the preceding example is easy to graph to verify visually its continuity, its nonnegativity, and the locations of its zeros. It is, however, not easy to integrate to determine and graph f in order to visualise that function. Thus, define $g(x)$ on $[0,1]$ by setting*

$$g(x) = \frac{1}{2^{k+2}} - \left| x - \left(1 - \frac{1}{2^k} + \frac{1}{2^{k+2}}\right) \right|$$

on the subinterval $[1 - 1/2^k, 1 - 1/2^{k+1}]$ for $k = 0, 1, 2, \ldots$ and setting $g(1) = 0$. Show that

$$f(x) = \int_0^x g(y)\, dy$$

is strictly increasing on $[0,1]$ and $f'(x) = g(x) = 0$ infinitely often. Express $f(x)$ explicitly on appropriate intervals and graph f on your calculator. [Note.

One can graph g fairly easily, if a bit slowly, on the TI-83 Plus by first choosing the window,

    Xmin=⁻.1
    Xmax=1.1
    Xscl=1
    Ymin=⁻.1
    Ymax=.4
    Yscl=.1,

then entering

    {0,1,2,3,4}→L₁

on the home screen, then entering

    Y₁=1/2^(L₁+2)−abs(X−1+1/2^L₁−1/2^(L₁+2))
    Y₂=Y₁(X)∗(Y₁(X)≥0),

and finally graphing only Y₂.]

More pertinent to our ongoing discussion is the use of the Mean Value Theorem in estimating errors.

Consider again Newton's polynomial $P(x) = x^3 - 2x - 5$ and an initial approximation a to the root r. By the Mean Value Theorem

$$\frac{P(r) - P(a)}{r - a} = P'(c)$$

for some c between a and r. It follows that

$$r - a = \frac{P(r) - P(a)}{P'(c)} = \frac{-P(a)}{P'(c)}.$$

Newton's Method estimates the error $r - a$ as $-P(a)/P'(a)$, while the Mean Value Theorem says it is exactly $-P(a)/P'(c)$ for some c. We don't know what c is, but we can use this fact and knowledge of P and P' to give bounds on $r - a$.

$P'(x) = 3x^2 - 2$ and $P''(x) = 6x > 0$ for $x > 0$. By Corollary 2.10.7.ii, P' is strictly increasing for $x \geq \sqrt{2/3}$, whence the minimum value of $P'(x)$ on any interval $[b, c] \subseteq [\sqrt{2/3}, \infty)$ is $P'(b)$. Thus, if $a \geq 2$,

$$|r - a| = \frac{|P(a)|}{|P'(c)|} < \frac{|P(a)|}{P'(2)} = \frac{|P(a)|}{10}.$$

Starting with $a = 2$,

$$|r - 2| \leq \frac{|P(2)|}{10} = \frac{|8 - 4 - 5|}{10} = \frac{1}{10}$$

and, since $2 < r$, we have

$$2 < r < 2.1. \tag{69}$$

Using 2.1 as our next estimate for r,

$$|r - 2.1| < \frac{P(2.1)}{P'(2)} < \frac{.061}{10} = .0061,$$

and $2.1 - .0061 < r < 2.1$, i.e.,
$$2.0939 < r < 2.1. \tag{70}$$

Choosing 2.0939 as our next estimate,
$$|r - 2.0939| < \frac{P(2.0939)}{P'(2.0939)} = .000651720612$$
$$< .000652.$$

Thus
$$2.0939 < r < 2.094552. \tag{71}$$

Choosing 2.094552 as our estimate,
$$|r - 2.094552| < \frac{P(2.094552)}{P'(2.0939)} = .0000005188383524$$
$$< .00000052,$$

whence
$$2.09455148 < r < 2.094552. \tag{72}$$

The procedure we are following is not much different from Newton's. At each stage it starts with an interval $[b, c]$, alternately choosing a to be b or c, and narrowing the interval by estimating
$$|r - a| < \frac{P(a)}{P'(b)}.$$

As the endpoints approach each other, they provide better and better upper and lower bounds for the root, either one of which or their average serving as a good estimate of the root.

We can offer a more sophisticated analysis as follows. Let $g(x) = x - P(x)/P'(x)$ be the improved estimate yielded by Newton's Method. And let
$$x_0 = 2, \quad x_{n+1} = g(x_n)$$
be the Newton-Raphson iteration. We can estimate how close x_n is to r as follows. First notice that
$$g(r) = r - P(r)/P'(r) = r - 0 = r.$$

Now
$$|x_0 - r| = |2 - r| \le \frac{|P(x_0)|}{P'(2)} = \frac{|P(2)|}{P'(2)} = \frac{1}{10}$$
as before. And
$$x_{n+1} - r = g(x_n) - r = g(x_n) - g(r) = g'(c)(x_n - r),$$
for some c between x_n and r. But
$$g'(x) = 1 - \frac{P'(x)P'(x) - P(x)P''(x)}{(P'(x))^2}$$
$$= \frac{(P'(x))^2 - (P'(x))^2 + P(x)P''(x)}{(P'(x))^2}$$

10. NEWTON'S METHOD AND THE MEAN VALUE THEOREM   143

$$= \frac{P(x)P''(x)}{(P'(x))^2} = \frac{6xP(x)}{(P'(x))^2}.$$

Now, P is strictly increasing, whence $|P(x)|$ is maximised on, say $[2.09, 2.1]$ by $\max\{|P(2.09)|, |P(2.1)|\} = \max\{.050671, .061\} = .061$, so on $[2.09, 2.1]$,

$$|g'(x)| = \frac{|P(x)| \cdot 6x}{(P'(x))^2} < \frac{.061 \cdot 6 \cdot 2.1}{(P'(2.09))^2}$$

$$< \frac{.7686}{11.1043^2} < \frac{.7686}{123.305}$$

$$< .0063 < .007.$$

Thus

$$|x_{n+1} - r| < .007|x_n - r| < .007^2|x_{n-1} - r| < \ldots < .007^{n-1}|2.1 - r|$$
$$< .007^{n-1}(.0061).$$

And we see that Newton's iterates will successively improve accuracy by two figures. In point of fact, the number of secured digits actually doubles. In the *Calcul différentiel* Cauchy demonstrated this by applying the *Mean Value Theorem of the Second Order*.

2.10.10. Theorem (Mean Value Theorem of the Second Order). *Let $a < b$ and let the functions f and f' be continuous on $[a, b]$ and f' differentiable on (a, b). There is some $c \in (a, b)$ such that*

$$f(b) = f(a) + f'(a)(b - a) + \frac{f''(c)}{2}(b - a)^2.$$

Proof. One mimics the proof of the Mean Value Theorem by cleverly choosing the auxiliary function[55]

$$h(x) = f(b) - f(x) - (b - x)f'(x) - \left(\frac{b - x}{b - a}\right)^2 \big(f(b) - f(a) - (b - a)f'(a)\big).$$

This vanishes at $x = a$ and $x = b$, whence for some $c \in (a, b)$ one has $h'(c) = 0$. But

$$h'(x) = -f'(x) + f'(x) - (b - x)f''(x)$$
$$- 2\left(\frac{b - x}{b - a}\right)\frac{-1}{b - a}\big(f(b) - f(a) - (b - a)f'(a)\big)$$
$$= -(b - x)f''(x) + 2\frac{b - x}{(b - a)^2}\big(f(b) - f(a) - (b - a)f'(a)\big),$$

---

[55]The function

$$g(x) = f(x) - f(a) - (x - a)f'(a) - \left(\frac{x - a}{b - a}\right)^2 \big(f(b) - f(a) - (b - a)f'(a)\big)$$

is a more direct analogue to the function g used in reducing the Mean Value Theorem to Rolle's Theorem. One does get $g(a) = g(b) = 0$, whence some c between a and b for which $g'(c) = 0$. However, when one calculates $g'(c)$, $f''(c)$ does not occur. But $f'(c) - f'(a)$ does and one can apply the Mean Value Theorem to replace this by $f''(\theta)(c - a)$ for some θ between c and a. (*Exercise.* Carry out the steps and complete this variant of the proof of the Theorem.)

whence
$$(b-c)f''(c) = 2\frac{b-c}{(b-a)^2}(f(b) - f(a) - (b-a)f'(a))$$

$$\frac{(b-a)^2}{2}f''(c) = f(b) - f(a) - (b-a)f'(a)$$

and
$$f(b) = f(a) + (b-a)f'(a) + \frac{(b-a)^2}{2}f''(c). \qquad \square$$

Letting f be Newton's polynomial P, $b = r$, and a any point $x \in [2, 2.1]$, we have
$$0 = P(r) = P(x) + P'(x)(r-x) + \frac{P''(c)}{2}(r-x)^2,$$

whence
$$\frac{-P(x)}{P'(x)} = r - x + \frac{P''(c)}{2P'(x)}(r-x)^2,$$

i.e.,
$$g(x) = x - \frac{P(x)}{P'(x)} = r + \frac{P''(c)}{2P'(x)}(r-x)^2,$$

and
$$g(x) - r = \frac{P''(c)}{2P'(x)}(r-x)^2.$$

Now
$$\frac{P''(c)}{2P'(x)} = \frac{6c}{2P'(x)} = \frac{3c}{P'(x)} < \frac{3 \cdot 2.1}{P'(2)} = \frac{6.3}{10} = .63,$$

whence
$$|x_0 - r| < .1$$
$$|x_1 - r| = |g(x_0) - r| < .63(.1)^2 < .0063$$
$$|x_2 - r| = |g(x_1) - r| < .63(.0063)^2 < .00004$$
$$|x_3 - r| = |g(x_2) - r| < .63(.00004)^2 < .000000001,$$

etc. At each step, the number of zeros immediately to the right of the decimal point in the error doubles at least. Thus, once we have two digits secured with 2.09 the next approximation will be correct to 4 decimals, then 8, then 16, ... after rounding to the 4th, 8th, 16th ... places.

Our discussion of Newton's Method and the Mean Value Theorems, which seems to be drawing us away from binomial matters, actually brings us full circle as some of the binomial based algorithms for root extraction that we met in Chapter 1 coincide with Newton's Method.

2.10.11. Exercise. *Let D be a positive number and consider, for positive integers $n \geq 1$, the polynomial $P(x) = x^n - D$. Use Newton's Method to derive formula (14) of page 16 anew. Apply the Mean Value Theorem to show again that the procedure gives successively better approximations to $\sqrt[n]{D}$. How good are the approximations for $n = 2, 3, 5$?*

With this we take leave of numerical work. The reader who enjoys playing with the calculator is invited to peruse the Appendix where we carry such matters further. For now, it is time to switch gears. Initially an aid to numerical work, both in the Finite and Newton's versions, the Binomial Theorem became, as we saw with Taylor's proof, the foundation on which the use of power series rested. But, by the end of the 18th century, dissatisfaction with the foundations of the Calculus led to Lagrange's attempt to replace a limit-based Calculus by an algebra of power series. The need for a solid basis for the Binomial Theorem and Taylor's Theorem was keenly felt and attempts in this direction were made. That is the story to be told in our next chapter.

CHAPTER 3

The Binomial Theorem Proven

1. Prefatory Remarks

In 1816 Bernard Bolzano published a short monograph bearing a long title: *Der binomische Lehrsatz und als Folgerung aus ihm der polynomische, und die Reihen, die zur Berechnung der Logarithmen und Exponentialgrößen dienen, genauer als bisher erwiesen*[1]. He began the preface to this work with the words

> The *binomial theorem* is usually quite rightly considered as one of the most important theorems in the whole of analysis. For not only can the *polynomial theorem*[2] and all the formulae used for the calculation of *logarithmic* and *exponential quantities* be derived from it by easy arguments, as will be shown here, but upon it also rests, as a further consequence, the widely applicable *Taylor's theorem* which can in no way be proved strictly, unless the binomial theorem is presupposed. It would hardly be an exaggeration to say that almost the whole of the so-called differential and integral calculus (higher analysis) rests on this theorem. However, the main theorems in every science have the property that it is usually their presentation in particular that is most beset by difficulties, and this is also the case with the binomial theorem. Therefore it is not surprising that the list of mathematicians who have tried to discover a completely strict proof of the theorem is already so long. *Newton's* immortal name stands at the top of the list. Following him—to mention only those known to *me*—*Colson, Horsley, Th. Simpson, Robertson, Sewell, Landen, Clairaut, Aepinus, Castillon, L'Huillier, Lagrange, Kästner, Euler, Segner, Scherfer, Klügel, Karsten, Busse, Pfaff, Rothe, Hindenburg, Kaussler, Schultz, Pasquich, Rösling, Jungius, Fischer* and *Krause, Crelle, Nordmann*, amongst many others.

> However, grateful as we must be that so much really excellent work on our theorem has already been achieved through the efforts of all these men, there are certainly still a few things that they have left

---

[1] C.W. Enders Buchhandlung, Prague, 1816. An English translation, "The Binomial Theorem and as a Consequence from it the Polynomial Theorem and the Series which serve for the Calculation of Logarithmic and Exponential Quantities proved more strictly than before", appears as an article in Steve Russ (ed.), *The Mathematical Works of Bernard Bolzano*, Oxford University Press, Oxford, 2004.

[2] I.e., the Multinomial Theorem.

148   3. THE BINOMIAL THEOREM PROVEN

> to be gleaned by later scholars [*Bearbeiter*]. I may be permitted to state here generally, and without mentioning anyone by name, what I find missing in previous presentations of this theorem.[3]

There follow, without specific attribution, some of the strategies that had been applied and the shortcomings inherent in them. The main body of the work then proceeded with a careful exposition of his own proof of the Binomial Theorem and related matters.

Just four years earlier, another, far more modest and less successful, work by one of those "amongst many others" was published. This was Charles Hutton's short tract[4] on the Binomial Theorem. It starts with a brief summary of the European history of the Finite Binomial Theorem, mention of Newton's version, the proof by John Landen, a mention of Euler's proof, and then his own proof. His introduction to the history begins

> 4. The truth of this series was not demonstrated by Newton, but only inferred by way of induction[5]. Since his time however, several attempts have been made to demonstrate it, with various success, and in various ways; of which however those are justly preferred, which proceed by pure algebra, and without the help of fluxions[6]. And such has been esteemed the difficulty of proving the general case, independent of the doctrine of fluxions, that many eminent mathematicians to this day account the demonstration not fully accomplished, and still a thing greatly to be desired. Such a demonstration I think is here effected.[7]

According to Hutton, the proof by Landen and one by Leonhard Euler "are the principle demonstrations and investigations that have been given of this important theorem"[8] and I am compelled to discuss these here.

2. Landen's Attempted Proof

John Landen (1719 – 1790) is not one of the big names in the history of mathematics. He gets an honorary mention as one of the first to concern himself with the foundation of the real number line, opting for an axiomatic/algebraic approach. And he attempted an algebraic proof of the Binomial Theorem for fractional exponents. He published the proof in his *Discourse concerning the*

---

[3]Russ, *op. cit.*, p. 157.

[4]"Tract XII. Of the Binomial Theorem. With a demonstration of the truth of it in the general case of fractional exponents", in: Charles Hutton, *Tracts on Mathematical and Philosophical Subjects; Comprising among numerous important articles, ...* vol. I, London, 1812.

[5]Meant here is not the Principle of Mathematical Induction, which is a valid method of proof, but general inductive reasoning.

[6]I.e., without appeal to differentiation.

[7]Hutton, *op. cit.*, pp 229 – 230.

[8]*Ibid.*, p. 237.

2. LANDEN'S ATTEMPTED PROOF

Residual Analysis[9] in 1758 and in the *Residual Analysis* in 1764. The proof assumes, for positive integers m, n, that
$$(1+x)^{\frac{m}{n}} = 1 + ax + bx^2 + cx^3 + \ldots \tag{73}$$
for some sequence $a, b, c, \ldots$ of coefficients and determines them essentially by differentiating both sides and using the new equation
$$\frac{m}{n}(1+x)^{\frac{m}{n}-1} = a + 2bx + 3cx^2 + 4dx^3 + \ldots \tag{74}$$
along with (73) to determine the coefficients $a, b, c, \ldots$ in order. He offers two variants of the proof, one purely algebraic and one using the fluxions.

Landen's algebraic treatment starts with a ratio,
$$\frac{(1+x)^{\frac{m}{n}} - (1+y)^{\frac{m}{n}}}{(1+x) - (1+y)} = \frac{(1+x)^{\frac{m}{n}} - (1+y)^{\frac{m}{n}}}{x - y},$$
which the substitutions $u = (1+x)^{1/n}, v = (1+y)^{1/n}$ convert to
$$\frac{u^m - v^m}{u^n - v^n} = \frac{(u-v)(u^{m-1} + u^{m-2}v + \ldots + uv^{m-2} + v^{m-1})}{(u-v)(u^{n-1} + u^{n-2}v + \ldots + uv^{n-2} + v^{n-1})}$$
$$= \frac{u^{m-1} + u^{m-2}v + \ldots + uv^{m-2} + v^{m-1}}{u^{n-1} + u^{n-2}v + \ldots + uv^{n-2} + v^{n-1}}$$
$$= \frac{u^{m-1}\left(1 + \frac{v}{u} + \left(\frac{v}{u}\right)^2 + \ldots + \left(\frac{v}{u}\right)^{m-1}\right)}{u^{n-1}\left(1 + \frac{v}{u} + \left(\frac{v}{u}\right)^2 + \ldots + \left(\frac{v}{u}\right)^{n-1}\right)}.$$

Partially reintroducing x and y converts this last into
$$\frac{(1+x)^{\frac{m}{n}} - (1+y)^{\frac{m}{n}}}{x-y} = \frac{(1+x)^{\frac{m-1}{n}}\left(1 + \frac{v}{u} + \left(\frac{v}{u}\right)^2 + \ldots + \left(\frac{v}{u}\right)^{m-1}\right)}{(1+x)^{\frac{n-1}{n}}\left(1 + \frac{v}{u} + \left(\frac{v}{u}\right)^2 + \ldots + \left(\frac{v}{u}\right)^{n-1}\right)}$$
$$= (1+x)^{\frac{m}{n}-1} \frac{\left(1 + \frac{v}{u} + \left(\frac{v}{u}\right)^2 + \ldots + \left(\frac{v}{u}\right)^{m-1}\right)}{\left(1 + \frac{v}{u} + \left(\frac{v}{u}\right)^2 + \ldots + \left(\frac{v}{u}\right)^{n-1}\right)}. \tag{75}$$

But from (73) we have
$$(1+x)^{\frac{m}{n}} - (1+y)^{\frac{m}{n}} = (1 + ax + bx^2 + cx^3 + \ldots) - (1 + ay + by^2 + cy^3 + \ldots)$$
$$= a(x-y) + b(x^2 - y^2) + c(x^3 - y^3) + \ldots,$$

---

[9]The passage (pp. 4 – 7) of this work presenting the proof is reprinted in: Dirk J. Struik (ed.), *A Source Book in Mathematics, 1200 – 1800*, Harvard University Press, Cambridge (Mass.), 1969, pp. 386 – 388; and Jacqueline Stedall (ed.), *Mathematics Emerging; A Sourcebook 1540 – 1900*, Oxford University Press, New York, 2008, pp. 398 – 401. Struik modernises the notation, while Stedall offers a facsimile reproduction of Landen's original, including a bit more of his introduction.

150   3. THE BINOMIAL THEOREM PROVEN

whence

$$\frac{(1+x)^{\frac{m}{n}} - (1+y)^{\frac{m}{n}}}{x-y} = a + b(x+y) + c(x^2 + xy + y^2) + \ldots \quad (76)$$

Combining (75) and 76 and restoring the rest of the x's and y's we have

$$(1+x)^{\frac{m}{n}-1} \frac{\left(1 + \left(\frac{1+x}{1+y}\right)^{\frac{1}{n}} + \left(\frac{1+x}{1+y}\right)^{\frac{2}{n}} + \ldots + \left(\frac{1+x}{1+y}\right)^{\frac{m-1}{n}}\right)}{\left(1 + \left(\frac{1+x}{1+y}\right)^{\frac{1}{n}} + \left(\frac{1+x}{1+y}\right)^{\frac{2}{n}} + \ldots + \left(\frac{1+x}{1+y}\right)^{\frac{n-1}{n}}\right)}$$

$$= a + b(x+y) + c(x^2 + xy + y^2) + \ldots \quad (77)$$

Landen now claimed that this holds for any y, so we can choose $y = x$, making $(1+x)/(1+y) = 1$ and converting this last equation into

$$(1+x)^{\frac{m}{n}-1} \cdot \frac{m}{n} = a + 2bx + 3cx^2 + 4dx^3 + \ldots \quad (78)$$

Multiplying this by $1 + x$ yields

$$\frac{m}{n}(1+x)^{\frac{m}{n}} = a + (a + 2b)x + (2b + 3c)x^2 + (3c + 4d)x^3 + \ldots \quad (79)$$

And comparison with (73) yields

$$a = \frac{m}{n} \cdot 1 = \frac{m}{n} = \binom{m/n}{1}$$

$$a + 2b = \frac{m}{n} \cdot a, \quad \text{i.e.,} \quad b = \frac{\frac{m}{n} - 1}{2} \cdot \frac{m}{n} = \binom{m/n}{2}$$

$$2b + 3c = \frac{m}{n} \cdot b, \quad \text{i.e.,} \quad c = \frac{\frac{m}{n} - 2}{3} \cdot b = \frac{\frac{m}{n}\left(\frac{m}{n} - 1\right)\left(\frac{m}{n} - 2\right)}{3 \cdot 2} = \binom{m/n}{3},$$

etc.

There are several points to criticise about this proof. The first is the unproven *assumption* that $(1+x)^{m/n}$ can be written as a power series (73). That this is possible—for *some* values of x—is not an easy thing to prove, even for students with a background in the Calculus up to and including convergence criteria. Hutton would recognise this and attempt to prove $(1+x)^{m/n}$ could be written in the form (73) by assuming a more general form of power series in which the exponents might not be integral and algebraically determining these exponents. This is ostensibly a weaker assumption, but it is an assumption nonetheless.

The second weak spot in the proof concerns the derivations of (75) – (78). The presence of the expression $x - y$ in the denominators of the left-hand-sides of (75) and (76) implicitly requires y not to equal x, whence (77) has only been demonstrated for $y \neq x$. Yet, he assumes $y = x$ in deriving (78) from (77).

The derivation is reminiscent of those fallacies one learns in beginning Algebra. For example: Assume $x = y$ and multiply both numbers by x to obtain

$$x^2 = xy.$$

Subtract y^2 from both sides:

$$x^2 - y^2 = xy - y^2.$$

Factor both sides,

$$(x-y)(x+y) = (x-y)y,$$

and divide by $x - y$ to get

$$x + y = y.$$

Setting $x = y = 1$, this yields $1 + 1 = 1$, i.e., $2 = 1$, a contradiction. The problem, of course, is that in dividing by $x - y$, one is dividing by 0, which cannot be done.

Landen's division, however, concerned a difference quotient and the technique was a familiar way of calculating derivatives: Find $f(x) - f(y)$ and divide by $x - y$. Simplify the ratio until no remnant of the division occurs and set $y = x$. The result is the derivative. One might as well replace the algebraic manipulations and differentiate (73) to obtain (74) directly. In doing so, one is differentiating a power series term-by-term, which can be done, but the fact that it can requires a proof. As with the representability of $(1 + x)^{m/n}$ by a power series, the termwise differentiability of a power series is not a trivial result.

And, finally, the determination of the coefficients $a, b, c, \ldots$ by comparing (73) and (79) requires one to know the uniqueness of the power series of a function. That is, if a function f satisfies

$$f(x) = a_0 + a_1 x + a_2 x^2 + \ldots \quad \text{and} \quad f(x) = b_0 + b_1 x + b_2 x^2 + \ldots$$

it follows that $a_0 = b_0, a_1 = b_1, a_2 = b_2, \ldots$ And this needs to be proven.

Landen's argument is not without value. While it has gaps and cannot be said to prove the result, it has a certain heuristic value. For, though it does not provide its own justification, it does yield the correct values for the coefficients $a, b, c, \ldots$ of the power series expansion of $(1 + x)^{m/n}$. In this, it is like Maclaurin's "proof" of Taylor's Theorem; it provides an algorithm if not a proof.

3. Euler's Attempted Proof

Leonhard Euler (1707 – 1783), one of the greatest mathematicians of all time and certainly the most prolific, made several attempts to prove the Binomial Theorem. I shall describe his proof of 1744.[10]

---

[10]Knowing no Latin, I have not consulted Euler's original paper, but rely on the sketch given in Klaus Volkert, *Geschichte der Analysis*, Bibliographisches Institut, Mannheim, 1988, p. 147.

For each rational number p, Euler considers the series

$$[p] = \sum_{k=0}^{\infty} \binom{p}{k} x^k \qquad (80)$$

and intends to show $[p] = (1+x)^p$. The key to his proof is the following lemma.

3.3.1. Lemma. *For all rational numbers p, q, $[p] \cdot [q] = [p+q]$.*

This is almost no more than an observation:

$$\begin{aligned}[p] \cdot [q] &= \left(\sum_{k=0}^{\infty} \binom{p}{k} x^k\right)\left(\sum_{k=0}^{\infty} \binom{q}{k} x^k\right) \\ &= \sum_{k=0}^{\infty} \left(\sum_{i=0}^{k} \binom{p}{i}\binom{q}{k-i}\right) x^k, \end{aligned} \qquad (81)$$

after collecting like terms. One needs only verify the identity

$$\binom{p}{0}\binom{q}{k} + \binom{p}{1}\binom{q}{k-1} + \cdots + \binom{p}{k}\binom{q}{0} = \binom{p+q}{k} \qquad (82)$$

to see that the right-hand side of (81) is $[p+q]$. This is not as easy as it looks.

For p and q positive integers, a simple combinatorial argument applies. The binomial coefficient $\binom{p+q}{k}$ is the number of ways one can choose a committee of k from a collection of p men and q women. We can do this by choosing all the committee members from the q women. This can be done in $\binom{p}{0}\binom{q}{k}$ ways. Or we can choose 1 man and $k-1$ women in any of $\binom{p}{1}\binom{q}{k-1}$ ways. Or ... Putting it all together we see that (82) holds—for p, q positive integers.

For general rational p and q identity (82), which is called *Vandermonde's Theorem* after Alexandre Theophile Vandermonde (1735 – 1796) who is best known for his work in the theory of determinants[11], is not that easy to prove in general. A few special cases are quickly verified:

$k = 0$. Observe

$$\binom{p}{0}\binom{q}{0} = 1 \cdot 1 = 1 = \binom{p+q}{0}.$$

$k = 1$. Observe

$$\binom{p}{0}\binom{q}{1} + \binom{p}{1}\binom{q}{0} = 1 \cdot q + p \cdot 1 = q + p = p + q = \binom{p+q}{1}.$$

$k = 2$. Observe

$$\begin{aligned}\binom{p}{0}\binom{q}{2} + \binom{p}{1}\binom{q}{1} + \binom{p}{2}\binom{q}{0} &= 1 \cdot \frac{q(q-1)}{2 \cdot 1} + pq + \frac{p(p-1)}{2 \cdot 1} \cdot 1 \\ &= \frac{q^2 - q + 2pq + p^2 - p}{2} \\ &= \frac{(q+p)^2 - q - p}{2} = \frac{(q+p)(q+p-1)}{2}\end{aligned}$$

---

[11] He is particularly well-known for the *Vandermonde determinant*, which did not, however, originate with him.

$$= \frac{(p+q)(p+q-1)}{2} = \binom{p+q}{2}.$$

A proof by induction on k is possible, but requires a page and a half of dense notational scribbling. Thus, I shall omit the details, referring the reader elsewhere for a proof.[12]

If we accept Vandermonde's Theorem, i.e., identity (82), as proved, we have the Lemma and it is a short step to proving the Binomial Theorem. Assume first that p is positive and $p = m/n$ where m, n are positive integers, and observe that the Lemma yields

$$[p]^n = [p] \cdot [p] \cdots [p] = [np] = [m] = \sum_{k=0}^{\infty} \binom{m}{k} x^k$$

$$= \sum_{k=0}^{m} \binom{m}{k} x^k = (1+x)^m,$$

the last inference by the Finite Binomial Theorem. Thus

$$[p] = ((1+x)^m)^{1/n} = (1+x)^{m/n} = (1+x)^p.$$

If p is negative, note that the Lemma yields

$$[p] \cdot [|p|] = [p + |p|] = [0] = 1,$$

whence

$$[p] = \frac{1}{[|p|]} = \frac{1}{(1+x)^{|p|}} = (1+x)^p.$$

Thus, Euler claims to have proven Newton's Binomial Theorem for all rational exponents.

There are two gaps in this proof. One is the assumption that $[p]$ converges for appropriate values of x and the second is that the *Cauchy product* of two convergent series,

$$\left(\sum_{k=0}^{\infty} a_i\right)\left(\sum_{k=0}^{\infty} b_i\right) = \sum_{k=0}^{\infty} \left(\sum_{i=0}^{k} a_i b_{k-i}\right),$$

is again a convergent series and, in fact, equals the product of the two series. Both of these gaps would be famously filled by Bolzano in 1816, Cauchy in 1821, and Niels Henrik Abel in a classical paper of 1826.

4. Hutton's Attempted Proof

Hutton followed his exposition of Landen's attempt and mention of Euler's with his own attempt at a proof. This proof, he says, dates from almost 30 years before the publication of his *Tracts*, thus from the early 1780s. Francis Maseres had made an attempt at proving the theorem by appeal to a variant of the Multinomial Theorem raising an infinite series to a finite power. "But,

---

[12] For my own exposition, cf. Craig Smoryński, *Adventures in Formalism*, College Publications, 2012, pp. 136 – 138 for the inductive proof, and Exercises 6.8 and 6.9 on pp. 186 – 187 for slicker proofs. The latter exercise is borrowed from W.L. Ferrar, *A Text-Book of Convergence*, Oxford University Press, Oxford, 1938, pp. 94 – 95.

not being quite satisfied with his own demonstration, as not expressing the law of continuation of the terms which are not actually set down, he was pleased to urge me[13] to attempt a more complete and satisfactory demonstration of the general case of roots, or fractional exponents." [14] This Hutton did.

As usual, Hutton ignored the problem of convergence, assuming at the outset the existence of a series

$$(1+x)^{1/n} = 1 + Ax + Bx^2 + Cx^3 + \ldots \tag{83}$$

Raising this to the n-th power,

$$(1 + Ax + Bx^2 + Cx^3 + \ldots)^n, \tag{84}$$

should result in $1 + x$. But it would also result in an infinite series,

$$1 + ax + bx^2 + cx^3 + \ldots, \tag{85}$$

with $a, b, c, \ldots$ expressed in terms of $A, B, C, \ldots$ and *multinomial coefficients*[15]. From the assumed identities,

$$a = 1, \ b = c = d = \ldots = 0,$$

one can successively solve for $A, B, C, \ldots$

Finding the individual coefficients $A, B, C, \ldots$ is not difficult, requiring only rudimentary algebraic skills, a tiny bit of combinatorial skill, and perhaps some computational agility to avoid the pitfalls of elementary errors. For example, to determine A, we ask how one can form a linear term in expanding (84). The terms in the expansion, before collecting like terms, are n-fold products, one factor coming from each of the n factors $1 + Ax + Bx^2 + \ldots$ Clearly exactly one of these must be Ax and all the others must be 1. But there are n ways of choosing a single factor of (84) from which to take Ax. ax will thus be the n-fold sum of terms Ax:

$$ax = nAx.$$

Thus, $a = nA$ and since $a = 1$, we have

$$A = \frac{1}{n} = \binom{1/n}{1}.$$

To determine B, note that a term in the expansion of (84) in which x occurs to degree 2 consists either of a product of two Ax and $n - 2$ 1 terms or one Bx^2 and $n - 1$ 1 terms. There are $\binom{n}{2}\binom{n-2}{n-2} = \binom{n}{2}$ terms of the first form and $\binom{n}{1}\binom{n-1}{n-1} = \binom{n}{1}$ terms of the second. Collecting like terms will yield

$$bx^2 = \binom{n}{2} A^2 x^2 + \binom{n}{1} Bx^2,$$

i.e.,

$$0 = \frac{n(n-1)}{2} \cdot \frac{1}{n^2} + nB,$$

---

[13] I.e., Hutton.
[14] Hutton, *op. cit.*, p. 237.
[15] I.e., the coefficients analogous to the binomial coefficients when a multinomial is raised to a power.

4. HUTTON'S ATTEMPTED PROOF

since $b = 0$ and $A = 1/n$. Thus

$$nB = -\frac{n-1}{n} \cdot \frac{1}{2},$$

i.e.,

$$B = \frac{1}{n} \cdot \frac{1-n}{n} \cdot \frac{1}{2} = \frac{\frac{1}{n}\left(\frac{1}{n}-1\right)}{2 \cdot 1} = \binom{1/n}{2}.$$

Things get algebraically more complicated at the next step. Cubic terms arise by choosing one factor of (84) to choose Cx^3 from and $n-1$ factors to choose 1 from; or one factor of (84) to choose Bx^2 from, one to choose Ax from, and $n-2$ to choose 1 from; or three factors to choose Ax from and $n-3$ to choose 1 from. From this one concludes

$$c = \binom{n}{1}C + \binom{n}{1}B\binom{n-1}{1}A + \binom{n}{3}A^3.$$

Plugging in the known values $c = 0$, $A = 1/n$, $B = (1-n)/(2n^2)$, we have

$$0 = nC + n\frac{1-n}{2n^2}(n-1)\frac{1}{n} + \frac{n(n-1)(n-2)}{3 \cdot 2 \cdot 1} \cdot \frac{1}{n^3}$$

$$-C = \frac{n-1}{2n^2} \cdot \frac{1-n}{n} + \frac{n-1}{3 \cdot 2 \cdot 1 \cdot n^2} \cdot \frac{n-2}{n}$$

$$= \frac{n-1}{3 \cdot 2 \cdot 1 \cdot n^2}\left(\frac{3-3n}{n} + \frac{n-2}{n}\right) = \frac{n-1}{3 \cdot 2 \cdot 1 \cdot n^2}\left(\frac{1-2n}{n}\right)$$

$$C = \frac{(1-n)(1-2n)}{3 \cdot 2 \cdot 1 \cdot n^3} = \frac{\frac{1}{n}\left(\frac{1}{n}-1\right)\left(\frac{1}{n}-2\right)}{3 \cdot 2 \cdot 1} = \binom{1/n}{3}.$$

This is a good start:

$$(1+x)^{1/n} = 1 + \binom{1/n}{1}x + \binom{1/n}{2}x^2 + \binom{1/n}{3}x^3 + \text{ higher degree terms.}$$

But the calculation of successive terms looks to require a bit of algebraic skill which Maseres may have lacked to see the way clearly to establish the general term. This is where Hutton takes over.

3.4.1. Exercise. *Find the coefficient D of x^4 in the series (83) by the method so far used in determining $A, B,$ and C.*

Hutton begins by listing the terms in a nicely organised, but slightly puzzlingly typeset display of the coefficients a, b, c, d, e that I will spare the reader. From it certain patterns emerge:

$a: \quad \dfrac{n}{1}A$

$b: \quad \dfrac{n}{1} \cdot \dfrac{n-1}{2}A^2 + \dfrac{n}{1}B$

$c: \quad \dfrac{n}{1} \cdot \dfrac{n-1}{2} \cdot \dfrac{n-2}{3}A^3 + \dfrac{n}{1} \cdot \dfrac{n-1}{1}AB + \dfrac{n}{1}C$

$d:\quad \dfrac{n}{1}\cdot\dfrac{n-1}{2}\cdot\dfrac{n-2}{3}\cdot\dfrac{n-3}{4}A^4 + \dfrac{n}{1}\cdot\dfrac{n-1}{2}\cdot\dfrac{n-2}{1}A^2B + \dfrac{n}{1}\cdot\dfrac{n-1}{1}AC+$
$\qquad +\dfrac{n}{1}\cdot\dfrac{n-1}{2}B^2 + \dfrac{n}{1}D$

$e:\quad \dfrac{n}{1}\cdot\dfrac{n-1}{2}\cdot\dfrac{n-2}{3}\cdot\dfrac{n-3}{4}\cdot\dfrac{n-4}{5}A^5 + \dfrac{n}{1}\cdot\dfrac{n-1}{2}\cdot\dfrac{n-2}{3}\cdot\dfrac{n-3}{1}A^3B+$
$\qquad +\dfrac{n}{1}\cdot\dfrac{n-1}{2}\cdot\dfrac{n-2}{1}A^2C + \dfrac{n}{1}\cdot\dfrac{n-1}{2}\cdot\dfrac{n-2}{1}AB^2 + \dfrac{n}{1}\cdot\dfrac{n-1}{1}AD+$
$\qquad +\dfrac{n}{1}\cdot\dfrac{n-1}{1}BC + \dfrac{n}{1}E$

He now spots the pattern:

$a:\quad \dfrac{nA}{1}$

$b:\quad \dfrac{2nB + (n-1)Aa}{2}$

$c:\quad \dfrac{3nC + (2n-1)Ba + (n-2)Ab}{3}$

$d:\quad \dfrac{4nD + (3n-1)Ca + (2n-2)Bb + (n-3)Ac}{4}$

$e:\quad \dfrac{5nE + (4n-1)Da + (3n-2)Cb + (2n-3)Bc + (n-4)Ad}{5}.$

Plugging $a = 1$ and $b = c = d = \ldots = 0$ into the fractions on the right-hand-sides of these yields:

$a:\quad \dfrac{nA}{1}$

$b:\quad \dfrac{2nB + (n-1)A}{2}$

$c:\quad \dfrac{3nC + (2n-1)B}{3}$

$d:\quad \dfrac{4nD + (3n-1)C}{4}$

$e:\quad \dfrac{5nE + (4n-1)D}{5}.$

And from these we see

$$1 = nA, \quad \text{i.e., } A = \dfrac{1}{n} = \binom{1/n}{1}$$

$$0 = 2nB + (n-1)A, \quad \text{i.e., } B = -\dfrac{n-1}{2n}A = \dfrac{1-n}{2n}\cdot\dfrac{1}{n} = \dfrac{\tfrac{1}{n}-1}{2}\cdot\dfrac{\tfrac{1}{n}}{1} = \binom{1/n}{2}$$

$$0 = 3nC + (2n-1)B, \quad \text{i.e., } C = \dfrac{1-2n}{3n}\cdot\dfrac{\tfrac{1}{n}\left(\tfrac{1}{n}-1\right)}{2\cdot 1} = \binom{1/n}{3},$$

and so on.

4. HUTTON'S ATTEMPTED PROOF

This is as much detail as Hutton offers. It does not appear he has made that great an advance over Maseres in that it would seem he obtains the first pattern on a case-by-case basis and thus he has not handled the general transitional case. But with the pattern he does clarify what is to be proven, if not how he proved it or whether he accepted it by empirical induction.

We can state a bit more easily what is to be proven by modernising notation. We consider a power series,

$$1 + A_1 x + A_2 x^2 + A_3 x^3 + \ldots, \tag{86}$$

the coefficient of the k-th power of x being written A_k. And we assume its n-th power, for some positive integer n, is

$$1 + a_1 x + a_2 x^2 + a_3 x^3 + \ldots \tag{87}$$

Hutton claims one can determine successive coefficients of (87) by the recurrence

$$a_{k+1} = \frac{1}{k+1} \Big((k+1)n A_{k+1} + (kn-1) A_k a_1 + \big((k-1)n - 2 \big) A_{k-1} a_2 + $$
$$+ \ldots + (2n - k - 1) A_2 a_{k-1} + (n-k) A_1 a_k \Big) \tag{88}$$

$$= \frac{1}{k+1} \sum_{i=0}^{k} \big((k+1-i)n - i \big) A_{k+1-i} a_i, \tag{89}$$

where we write $a_0 = 1$ for the coefficient of the x^0 term in (87). Assuming (87) is just $1 + x$, so that $a_1 = 1$ and $a_i = 0$ for $i \geq 2$, the recurrence yields

$$0 = a_{k+1} = \frac{1}{k+1} \Big((k+1)n A_{k+1} + (kn-1) A_k \Big), \quad \text{for } k \geq 1,$$

i.e.,

$$A_{k+1} = -\frac{kn-1}{(k+1)n} A_k = \frac{1-kn}{(k+1)n} A_k = \frac{\frac{1}{n} - k}{k+1} A_k,$$

and assuming as induction hypothesis that

$$A_k = \binom{1/n}{k} = \frac{\frac{1}{n} \left(\frac{1}{n} - 1 \right) \cdots \left(\frac{1}{n} - k + 1 \right)}{k(k-1) \cdots 1},$$

we conclude

$$A_{k+1} = \frac{\frac{1}{n} \left(\frac{1}{n} - 1 \right) \cdots \left(\frac{1}{n} - k + 1 \right) \left(\frac{1}{n} - k \right)}{(k+1) k (k-1) \cdots 1} = \binom{1/n}{k+1},$$

as desired.

The task at hand is, thus, to derive (88) or (89) for an arbitrary power series (86) and its n-th power (87). Now a simple, direct algebraic verification of (88) eludes me and I can only establish it using differentiation as follows. Letting $A_0 = a_0 = 1$, define

$$f(x) = A_0 + A_1 x + A_2 x^2 + \ldots = \sum_{k=0}^{\infty} A_k x^k$$

$$f(x)^n = a_0 + a_1 x + a_2 x^2 + \ldots = \sum_{k=0}^{\infty} a_k x^k$$

Then

$$f(x)^{n+1} = f(x)^n f(x) = \left(\sum_{k=0}^{\infty} a_k x^k\right)\left(\sum_{k=0}^{\infty} A_k x^k\right) = \sum_{k=0}^{\infty}\left(\sum_{i=0}^{k} a_i A_{k-i}\right) x^k.$$

Differentiating this term-by-term, we have

$$\frac{df(x)^{n+1}}{dx} = \sum_{k=0}^{\infty} k\left(\sum_{i=0}^{k} a_i A_{k-i}\right) x^{k-1} = \sum_{k=1}^{\infty} k\left(\sum_{i=0}^{k} a_i A_{k-i}\right) x^{k-1}$$

$$= \sum_{k=0}^{\infty} (k+1)\left(\sum_{i=0}^{k+1} a_i A_{k+1-i}\right) x^k. \tag{90}$$

But we also have, by the Chain Rule,

$$\frac{df(x)^{n+1}}{dx} = (n+1)f(x)^n f'(x)$$

$$= (n+1)\left(\sum_{k=0}^{\infty} a_k x^k\right)\left(\sum_{k=0}^{\infty} k A_k x^{k-1}\right)$$

$$= (n+1)\left(\sum_{k=0}^{\infty} a_k x^k\right)\left(\sum_{k=0}^{\infty} (k+1) A_{k+1} x^k\right)$$

$$= \sum_{k=0}^{\infty} (n+1)\left(\sum_{i=0}^{k} a_i (k+1-i) A_{k+1-i}\right) x^k. \tag{91}$$

Equating like coefficients of (90) and (91), we see

$$(k+1)\sum_{i=0}^{k+1} a_i A_{k+1-i} = (n+1)\sum_{i=0}^{k} a_i (k+1-i) A_{k+1-i}.$$

Isolating the term involving a_{k+1} yields

$$(k+1)a_{k+1} A_0 = \sum_{i=0}^{k} \Big((n+1)(k+1-i) - (k+1)\Big) a_i A_{k+1-i}$$

$$(k+1)a_{k+1} = \sum_{i=0}^{k} \Big((k+1-i)n + k+1-i - (k+1)\Big) a_i A_{k+1-i}$$

$$a_{k+1} = \frac{1}{k+1}\sum_{i=0}^{k} \Big((k+1-i)n - i\Big) a_i A_{k+1-i},$$

4. HUTTON'S ATTEMPTED PROOF

which is (89).

With this we have completed Hutton's completion of Maseres's attempted proof. As usual, the proof assumed a lot that was left unproven—the existence of a power series for $(1+x)^{1/n}$, the convergence and form of the product of two power series, and the term-by-term differentiability of a power series. If one is willing to make such assumptions, the Binomial Theorem can be more readily derived. I will finish this section with two such derivations.

First, for those with a strong background in the Calculus, there is the standard technique of finding a power series solution to a differential equation. Let $y = (1+x)^{m/n}$ and observe

$$y' = \frac{dy}{dx} = \frac{m}{n}(1+x)^{\frac{m}{n}-1} = \frac{my}{n(1+x)}.$$

Writing $y = A_0 + A_1 x + A_2 x^2 + \ldots$ and using the initial value $y(0) = 1$ to conclude $A_0 = 1$, we have

$$A_1 + 2A_2 x + 3A_3 x^2 + \ldots = \frac{m(1 + A_1 x + A_2 x^2 + \ldots)}{n(1+x)}.$$

$$\frac{n}{m}(1+x)(A_1 + 2A_2 x + 3A_3 x^2 + \ldots) = 1 + A_1 x + A_2 x^2 + \ldots$$

$$\frac{n}{m}\left(A_1 + (2A_2 + A_1)x + (3A_3 + 2A_2)x^2 + \ldots\right) = 1 + A_1 x + A_2 x^2 + \ldots, \tag{92}$$

using only the distributive law and not the convergence of the Cauchy product of two power series. And, of course, (92) yields, on comparison of coefficients,

$$\frac{n}{m}A_1 = 1, \quad \frac{n}{m}(2A_2 + A_1) = A_1, \quad \frac{n}{m}(3A_3 + 2A_2) = A_2, \quad \ldots$$

and one finishes up exactly as Landen did.

A proof that is popular in modern textbooks proceeds as follows. Let $p = m/n$ and define $[p]$ as Euler did in (80):

$$[p] = \sum_{k=0}^{\infty} \binom{p}{k} x^k, \tag{80}$$

and consider

$$f(x) = \frac{[p]}{(1+x)^p}.$$

Termwise differentiation of $[p]$ yields

$$\frac{d[p]}{dx} = \sum_{k=0}^{\infty} k \binom{p}{k} x^{k-1}$$

$$= \sum_{k=1}^{\infty} k \frac{p(p-1)\cdots(p-k+1)}{k(k-1)\cdots 1} x^{k-1}$$

$$= p \sum_{k=1}^{\infty} \binom{p-1}{k-1} x^{k-1} = p \sum_{k=0}^{\infty} \binom{p-1}{k} x^k$$

$$= p[p-1].$$

Thus

$$f'(x) = \frac{p[p-1](1+x)^p - [p]p(1+x)^{p-1}}{(1+x)^{2p}}$$
$$= \frac{p(1+x)^{p-1}}{(1+x)^{2p}}\left((1+x)[p-1] - [p]\right). \qquad (93)$$

But

$$(1+x)[p-1] = \sum_{k=0}^{\infty}\binom{p-1}{k}x^k + \sum_{k=0}^{\infty}\binom{p-1}{k}x^{k+1}$$
$$= \binom{p-1}{0}x^0 + \sum_{k=1}^{\infty}\binom{p-1}{k}x^k + \sum_{k=0}^{\infty}\binom{p-1}{k}x^{k+1}$$
$$= x^0 + \sum_{k=1}^{\infty}\binom{p-1}{k}x^k + \sum_{j=1}^{\infty}\binom{p-1}{j-1}x^j,$$

where $j = k+1$,

$$= \binom{p}{0}x^0 + \sum_{k=1}^{\infty}\binom{p-1}{k}x^k + \sum_{k=1}^{\infty}\binom{p-1}{k-1}x^k,$$

relabelling again,

$$= \binom{p}{0}x^0 + \sum_{k=1}^{\infty}\left(\binom{p-1}{k} + \binom{p-1}{k-1}\right)x^k$$
$$= \binom{p}{0}x^0 + \sum_{k=1}^{\infty}\binom{p}{k}x^k = [p].$$

Thus $(1+x)[p-1] = [p]$ and the right-hand-side of (93) is 0, i.e., $f'(x) = 0$. Thus f is constant, $f(x) = f(0)$. But

$$f(0) = \frac{\sum_{k=0}^{\infty}\binom{p}{k}0^k}{(1+0)^p} = \frac{\binom{p}{0}0^0}{1^p} = \frac{1 \cdot 1}{1} = 1.$$

Thus

$$\frac{[p]}{(1+x)^p} = 1,$$

i.e.,

$$(1+x)^p = [p] = \sum_{k=0}^{\infty}\binom{p}{k}x^k.$$

5. Peacock's Attempted Metaphysical Proof

When Hutton's *Tracts* were published in 1812, the Binomial Theorem was only a short wait away from rigorous proofs by Bolzano (1816), Augustin Louis Cauchy (1821), and Abel (1826). I shall discuss these shortly, but first I wish to desert the chronology and discuss a completely bogus proof due to George Peacock (1791 – 1858), published in his *A Treatise on Algebra*[16]. A founding member of the Cambridge Analytical Society, and thus something of an intellectual activist, Peacock founded a Symbolical Algebra to be distinguished from the Arithmetical Algebra of nonnegative quantities. This would allow him to refer freely to negative numbers which some of his contemporaries, like Maseres and William Frend, felt were not quantities at all

The point is that nonnegative numbers are actual quantities, the results of counting or measuring. Negative numbers had no meaning: they stood for nothing and were often said to be "less than nothing", a phrase their opponents hurled as an epithet. Yet one used them as if they existed and their use was convenient and effective. They were convenient in that, for example, they unified disparate cases. When everything in sight had to be positive, the equations,
$$Ax^2 + Bx = C \quad \text{and} \quad Ax^2 + C = Bx,$$
were of different character and the solutions had to be found separately. If C, respectively B, is allowed to be negative, they are both of one form,
$$Ax^2 + B'x + C' = 0.$$
And, negative numbers are effective in that their use does not produce wrong answers.

Peacock sought to justify the use of negatives by treating them as ideal objects, nonexistent entities that we pretend exist and manipulate formally according to fixed rules. Thus, for him, Arithmetical Algebra was the algebraic manipulation of genuine quantities and Symbolical Algebra was the manipulation of symbols, which may or may not denote actual quantities, according to fixed rules. The establishment of these rules and their justification he termed the *Principle of the Permanence of Equivalent Forms*. Chapter XV of the second edition of his *A Treatise on Algebra* begins with his

> FORMAL STATEMENT OF THE PRINCIPLE OF THE PERMANENCE
> OF EQUIVALENT FORMS.
>
> 630. In the exposition of the fundamental operations of addition, subtraction, multiplication and division in Symbolical Algebra, we have adopted the corresponding rules of operation in Arithmetical Algebra, extending them to all values of the symbols involved, as well as to those additional derivative forms[17] which are the necessary results of that extension: and we have subsequently endeavoured to

---

[16]Cambridge University Press, Cambridge, 1830; 2nd edition in two volumes, 1842 and 1845.

[17]Peacock here inserts the footnote: Such are $+a$ and $-a$, and other forms which are thus derived, and which are not recognised in Arithmetical Algebra.

162   3. THE BINOMIAL THEOREM PROVEN

give to those extended operations and to their results, such an interpretation as was consistent with the conditions which they were required to satisfy. In the further developement of this science we shall continue to be guided by the same principle, making the results of defined operations, or the rules for forming them, the basis of the corresponding operations and results in Symbolical Algebra, and also of the interpretation of the meaning which must be given to them, whenever such interpretation is practicable.

631. This principle, which is thus made the foundation of the operations and results of Symbolical Algebra, has been called "The principle of the permanence of equivalent forms", and may be stated as follows:

"*Whatever algebraical forms are equivalent, when the symbols are general in form but specific in value, will be equivalent likewise when the symbols are general in value as well as in form.*"

It will follow from this principle, that all the results of Arithmetical Algebra will be results likewise of Symbolical Algebra: and the discovery of equivalent forms in the former science, possessing the requisite conditions, will be not only their discovery in the latter, but the *only* authority for their existence: for there are no definitions of the operations in Symbolical Algebra, by which such equivalent forms can be determined.[18]

Algebraical forms were expressions built up using the given operations. Thus, for example, $x(y+z)$ is an algebraical form. The equivalence of forms is valid identity of forms, e.g.,

$$x(y+z) = xy + xz.$$

The rules of Symbolical Algebra are just those identities valid in Arithmetical Algebra, i.e., valid in the nonnegative reals. When the forms are composed only of finite combinations of $+$ and $\cdot$, one generally only needs the familiar identities—the commutative laws, the associative laws, etc. For, if P, Q are polynomials with nonnegative coefficients, an "equivalence of forms" $P = Q$ holds in the nonnegative reals if and only if it holds in all the reals. This is not too difficult to prove. But when one starts dividing, taking square roots, forming infinite sums, and exponentiating in constructing forms, the matter becomes more problematic. A good example is afforded by the so-called Grandi series: Simple symbolical manipulation yields

$$1 - 1 + 1 - 1 + \ldots = (1-1) + (1-1) + \ldots = 0 + 0 + \ldots = 0$$
$$1 - 1 + 1 - 1 + \ldots = 1 + (-1+1) + (-1+1) + \ldots = 1 + 0 + 0 + \ldots = 1$$

and thus $0 = 1$. Does Peacock's principle allow us to conclude $0 = 1$ to hold in the reals?

---

[18]George Peacock, *A Treatise on Algebra, Vol. II. On Symbolical Algebra and its Applications to the Geometry of Position*, 2nd edition, Cambridge University Press, Cambridge, 1845, p. 59.

5. PEACOCK'S ATTEMPTED METAPHYSICAL PROOF

Obviously the principle requires a great deal of care in its formulation and application. As a heuristic principle it can be quite useful; as a method of proof I reject it as neither a proven principle nor an obvious axiom.

Anyway, our point to discussing Peacock's principle is to consider his application of it to derive the Binomial Theorem. The first hint of this derivation comes in the first edition of his *Treatise* at that point where he introduces the principle:

> 131. Our great object in the very lengthened discussion which we have just concluded, has been to point out the distinction between the science of Algebra when considered with reference to its own principles, and when considered with reference to its applications, and to shew in what manner and to what extent the assumptions which regulate the combinations of general and arbitrary symbols in Algebra were suggested, and their interpretation limited by other and subordinate sciences: the principles which determine the connection between these sciences being once established, we shall be fully prepared to consider to what extent we can consider equivalent forms suggested or investigated upon the principles of a subordinate science, as equally true when expressed in general symbols.
>
> Thus the principles of Arithmetical Algebra lead to the equation
> $$a^n \times a^m = a^{n+m},$$
> when n and m were whole numbers: it was the conversion of this conclusion in one science into an assumption in the other, which lead to the same equation,
> $$a^n \times a^m = a^{n+m},$$
> when n and m were general symbols.[19]
>
> If, however, we had commenced with the *assumption* that there existed some *equivalent* form for $a^n \times a^m$, when n and m were general symbols; and if we had discovered and *proved* that this form in Arithmetical Algebra was a^{n+m}, where n and m were such quantities as Arithmetical Algebra recognizes, then we might infer that such likewise must be the *equivalent* form in Symbolical Algebra: for this form can undergo no change, according to the assumptions which we have made, from any change in the nature of its symbols, and must therefore continue the same when the symbols are numbers: if, therefore, we discover this form in any one case, we discover it for all others.
>
> 132. Let us again recur to this principle or law of the *permanence of equivalent forms*, and consider it when stated in the form of a

---

[19]The German mathematician Hermann Hankel (1839 – 1873) formulated his own *Principle of the Permanence of Formal Laws* in 1867. In his *Zur Geschichte der Mathematik im Alterthum und Mittelalter* (B.G. Teubner, Leipzig, 1874), pp. 350 – 351, he credits Nicole Oresme as the first to apply the principle in using the exponential law $(a^m)^n = a^{mn}$ to define exponentiation with fractional exponents.

direct and *converse* proportion.

"*Whatever form is Algebraically equivalent to another, when expressed in general symbols, must be true, whatever those symbols denote.*"

Conversely, if we discover an equivalent form in Arithmetical Algebra or any other subordinate science, when the symbols are general in form though specific in their nature, the same must be an equivalent form, when the symbols are general in their nature as well as in their form.

The direct proposition must be true, since the laws of the combinations of symbols by which such equivalent forms are deduced, have been assumed without any reference to their specific nature, and the forms themselves, therefore, are equally independent.

The converse proposition must likewise be true for the following reasons:

If there be an equivalent form when the symbols are general in form and in their nature also, it must coincide with the form discovered and proved in the subordinate science, where the symbols are general in form but specific in their nature: for in passing from the first to the second, no change in its form can take place by the first proposition.

Secondly, we may assume such an equivalent form in general symbols, since the laws of the combinations of symbols are assumed in such a manner as to coincide strictly with the corresponding laws in subordinate sciences, such as Arithmetical Algebra: the conclusions, therefore, so far as their form is concerned, are necessarily the same in both; and the Algebraical *equivalence* which exists in one case must exist likewise in the other.

133. The principle which is expressed in these propositions, which we have named *the law of the permanence of algebraical forms*, is one of the greatest importance, and merits the most profound and careful consideration: it points out the proper object of those demonstrations in Algebra, which have reference to the research of equivalent forms, and shews why they may be safely generalized, even though they may be obtained by the aid of specific values of the symbols: if a general equivalent form be assumed to exist, it is clearly sufficient if we can discover it in any one of its states of existence, corresponding to the different specific values of its general symbols: and if we commence by detecting its existence for specific values of the symbols, we may generalize the symbols, since the same form continues to be equivalent for all algebraical operations.

134. Thus, we may assume the existence of an equivalent form for $(1+x)^n$, when n is a general symbol, and we may discover it when n is a whole number: or we may commence by the discovery of the

equivalent form of $(1+x)^n$, when n is a whole number, and subsequently assume the existence of it, when n is a general symbol: in the first case, assuming its existence, it is necessarily the form discovered: in the second case, the law we have mentioned and the reasoning by which it is established, would shew that the form discovered is algebraically equivalent, when n is a general symbol.

135. It is no objection to the use of this principle, that in consequence of its extreme abstraction and generality, it requires a great and painful effort of the mind to apprehend thoroughly the evidence of its truth, unless it can be shewn that there are other and more easily intelligible, though less general, methods of arriving at similar conclusions; but a very little consideration will shew, that the province of demonstration is extremely limited when general symbols are employed, since we must reject altogether the aid derived from the specific values of the symbols: we are thus confined to the assumed laws of their combination and incorporation with each other, and consequently demonstration can only extend to cases which those laws comprehend: it is for this reason that there can be no demonstration, independently of additional assumptions, of the existence of an equivalent form for $(1+x)^n$, when n is a general symbol.[20]

Peacock's discussion of the Binomial Theorem in the first edition of his treatise rambles a bit, but the proof as given in the second edition is direct and to the point:

CHAPTER XXI.
THE BINOMIAL THEOREM AND ITS APPLICATIONS.

679. In Chap. VIII (Art. 486...), we have proved, when the index n is a whole number, that

$$(1+x)^n = 1 + nx + n(n-1)\frac{x^2}{1.2} + n(n-1)(n-2)\frac{x^3}{1.2.3} + \ldots :$$

and it will be seen, from an examination of this series, and of the law of its formation, that the powers of x and their divisors are independent of n, and that the coefficients of

$$x, \frac{x^2}{1.2}, \frac{x^3}{1.2.3}, \ldots\ldots, \frac{x^r}{1.2\ldots r},$$

are n, $n(n-1)$, $n(n-1)(n-2)$, $\ldots$, $n(n-1)\ldots(n-r+1)$,

being, for the $(1+r)^{\text{th}}$ term, the continued product of the descending series of natural numbers from n to $n-r+1$.

680. This series for $(1+x)^n$ is perfectly general in its form, though n is specific in its value, and it will continue therefore, by "the principle of the permanence of equivalent forms" (Art. 631) to be

---

[20] George Peacock, *A Treatise on Algebra*, Cambridge University Press, Cambridge, 1830, pp. 103 – 106.

equivalent to $(1+x)^n$, when n is general in value as well as in form: and it will consequently admit, in virtue of this equivalence, of being immediately translated into the whole series of propositions respecting indices and their interpretation, which are given in Chapter XVI.

681. Thus "the general principle of indices" (Art. 635) shews that
$$(1+x)^n(1+x)^{n'} = (1+x)^{n+n'}$$
for all values of n and n', and consequently the product of the series for $(1+x)^n$ or
$$1 + nx + n(n-1)\frac{x^2}{1.2} + n(n-1)(n-2)\frac{x^3}{1.2.3} + \&c.,$$
and of the series for $(1+x)^{n'}$, or
$$1 + n'x + n'(n'-1)\frac{x^2}{1.2} + n'(n'-1)(n'-2)\frac{x^3}{1.2.3} + \&c.$$
will be equivalent to the series for $(1+x)^{n+n'}$, or
$$1 + (n+n')x + (n+n')(n+n'-1)\frac{x^2}{1.2}$$
$$+ (n+n')(n+n'-1)(n+n'-2)\frac{x^3}{1.2.3} + \&c,$$
under the same circumstances.

682. Again, the series for $\frac{1}{(1+x)^n}$, or $(1+x)^{-n}$ (Art. 640,) will be found by replacing n by $-n$ in the series for $(1+x)^n$: we thus get
$$(1+x)^{-n} = 1 - nx + n(n+1)\frac{x^2}{1.2} - n(n+1)(n+2)\frac{x^3}{1.2.3} + \&c :$$
for the product $-n(-n-1)$ may be replaced by $n(n+1)$: the product $-n(-n-1)(-n-2)$ by $-n(n+1)(n+2)$, and similarly for the subsequent terms.

683. The series for $(1-x)^n$ is deducible, in virtue of the same principle (Art. 631), from that of $(1+x)^n$, by changing the signs of those terms which involve the odd powers of x: we thus get, as in Art. 491,
$$(1-x)^n = 1 - nx + n(n-1)\frac{x^2}{1.2} - n(n-1)(n-2)\frac{x^3}{1.2.3} + \ldots :$$
and in a similar manner, the series for $(1-x)^{-n}$ will become
$$1 + nx + n(n+1)\frac{x^2}{1.2} + n(n+1)(n+2)\frac{x^3}{1.2.3} + \ldots,$$
where all the signs and terms are positive.[21]

---

[21] *Ibid.*, pp. 110 – 111. The inconsistency of the punctuation is preserved from the original.

5. PEACOCK'S ATTEMPTED METAPHYSICAL PROOF

We must remember in reading these words that Peacock's Arithmetical Algebra concerns only positive numbers. Symbolical Algebra was formed by allowing unlimited subtraction and assuming all universal identities holding among the positive numbers, along with $b + (a - b) = a$. Had he declared Arithmetical Algebra to concern the integers in general and extended division making sure to include the rule $b \cdot \frac{a}{b} = a$, his Principle of the Permanence of Equivalent Forms would have been refutable:

$$(-1)^{2x} = 1 \text{ for all integers } x, \text{ but } (-1)^{2 \cdot 1/2} = -1 \neq 1. \qquad (94)$$

If one compares the statements of the Principle of the Permanence of Equivalent Forms in the two editions of the *Treatise*, paragraphs 132 (first edition) and 631 (second edition) cited a few pages back, one sees that application of Symbolical Algebra to Arithmetical Algebra is absent from the second edition.[22] Exactly why this is so is unclear. Was it because he feared things like (94), from which the application of the two halves of the Principle of 132 in the passage from integers to rationals would yield

$(-1)^{2x} = 1$ in Arithmetical Algebra $\Rightarrow (-1)^{2x} = 1$ in Symbolical Algebra

$\Rightarrow -1 = 1$ in Symbolical Algebra

$\Rightarrow 1 + 1 = 0$ in Symbolical Algebra

$\Rightarrow 1 + 1 = 0$ in Arithmetical Algebra,

would pop up? Or, was it simply that he wanted to consider the expansion of a function into a power series to be valid even when the series did not converge and thus had no numerical meaning? Robert Woodhouse, a mentor and colleague, as well as a precursor of Peacock's had said as much in his earlier book of 1803 when he stated that the equality of a function to its power series expansion meant only that the series resulted from formal manipulation and laid no claim to numerical significance.[23] Indeed, today we know of examples of functions which do not equal their everywhere convergent power series.

Given the importance of power series and the Binomial Theorem for each other, I should note that this result was found in 1822 by Augustin Louis Cauchy, who showed that the Maclaurin expansion of

$$f(x) = \begin{cases} 0, & x = 0 \\ e^{-1/x^2}, & x \neq 0 \end{cases}$$

was $0 + 0x + 0x^2 + 0x^3 + \ldots$ Numerical equality between $f(x)$ and this series held only for $x = 0$ and nowhere else. Cauchy's result came almost two decades after the publication of Woodhouse's book and had no bearing on his remarks. Cauchy's result was known already by the time of the first edition of Peacock's

---

[22] I confess that I missed the significance of this point until reading Menachem Fisch, "The making of Peacock's *Treatise on Algebra*: a case of creative indecision", *Archive for History of Exact Science* 54 (1999), pp. 137 – 179.

[23] Robert Woodhouse, *The Principles of Analytical Calculation*, Cambridge University Press, Cambridge, 1803, pp. 3, 13.

Treatise. The expansion, however is not worked out algebraically and does not fit the counterexample scheme:

$$f = 0 \text{ in Symbolical Algebra} \Rightarrow f = 0 \text{ in Arithmetical Algebra.}$$

But it too, like (94), gives reason to pause.

For whatever reason, Peacock dropped the requirement that Symbolical Algebra only prove true equivalences. But he wholeheartedly accepted the converse principle that forms equivalent in Arithmetical Algebra would be equivalent in Symbolical Algebra. Hermann Hankel in his later exposition of the principle was more cautious:

> The herein contained seminal principle can be denoted as the Principle of the Permanence of Formal Laws and is this: If two forms expressed in general symbols of arithmetica universalis are identical, then they should also remain identical if the symbols cease to denote simple quantities and thereby the operations acquire also arbitrarily different content.
>
> This principle will guide our steps in the sequel; however it may not be applied thoughtlessly and we will generally allow it to be applied only to the determination of necessary and sufficient rules, insofar as these are independent of one another.[24]

The overall problem dealt with by Peacock and Hankel was the same—justifying the use of ideal elements like negative numbers or fractions or square roots of negative numbers. Peacock's solution was purely formal: just act as if the ideal elements existed and use as rules for dealing with them all identities that held among the genuine entities, be they whole or fractional positive numbers. It was a shotgun fire approach in assuming all identities. Hankel's approach was more refined, assuming only those identities one considers as defining the operations involved. In the case of addition and multiplication, every polynomial can be expanded and simplified using only the associative, commutative, and distributive laws, and an identity,

$$P(x_1, x_2, \ldots, x_n) = Q(x_1, x_2, \ldots, x_n),$$

holds in the integers just in case the simplified expansions are syntactically identical, i.e., these laws suffice to derive all polynomial identities in the positive integers, the integers, the rationals, ... The laws of exponentiation,

$$(ab)^n = a^n b^n, \quad (a^m)^n = a^{mn}, \quad a^{m+n} = a^m \cdot a^n,$$

holding for nonnegative integers, suffice to define exponentiation for rational values of m, n. To Hankel, Peacock's converse half of the (1830) Principle of the Permanence of Equivalent Forms (paragraph 132 copied above) was heuristic, a desideratum for defining the extensions of functions. It was not generally valid, as conflicts like (94) might arise, and did not always work. But when it did, he gave geometric interpretations of the ideal elements as elements of extended domains of numbers, and thus the direct half of the Principle held

---

[24]Hermann Hankel, *Vorlesungen über die complexen Zahlen und ihre Functionen. I. Theil. Theorie der complexen Zahlensysteme*, Leopold Voss, Leipzig, 1867, p. 11.

in general for him.[25] Peacock's assumption of all identities, with the danger of such conflicts as (94) when functions other than $+$ and $\cdot$ are involved and the subsequent collapse of the integers into a collection of only two numbers 0 and 1 ($1+1=0, 1+1+1=1, 1+1+1+1=0$, etc.) ruled out the direct application of the Principle.

By assuming the validity of any identity like the Finite Binomial Theorem, regardless of its method of proof (e.g., induction on the exponent), Peacock readily derived the Binomial Theorem for general exponents, but without the direct half of the Principle he established no connexion between the formal identity and its numerical instances. His proof does not justify the application of the Theorem as in section 3 to the calculation of roots. If, perchance, one can derive $-1 = 1$, i.e., $1+1=0$, in Symbolical Algebra, as with (94), does not one have a symbolical derivation of

$$\sqrt{17} = 4.123\ldots = 1 + 1 + 2.123\ldots = 2.123\ldots?$$

And, if so, why does my calculator yield

$$(2.123\ldots)^2 = 4.507\ldots$$

and not 17? Peacock's demonstration of the Binomial Theorem is not incorrect so much as irrelevant.

6. D'Alembert and The Limit Concept

By 1800 there were several approaches to the Differential Calculus. Newton had envisioned motion as basic, developing a fluxional calculus that some saw as more art than science. He had already spoken vaguely of limits, but the age of precise definitions had not yet arrived and the limits were only spoken about rather loosely. The Calculus, which had emerged from a grab bag of heuristic procedures had no firm foundation and its practitioners had to tread carefully. And while they were doing this, some of them criticised theology for accepting things on faith. This proved to be too much for George Berkeley (1685 – 1753), who was an Anglican bishop as well as an important philosopher of science. His response, *The Analyst, or, a discourse addressed to an infidel mathematician*[26] (1734), is a detailed, fair, and accurate assessment of all that was wrong with the Newtonian approach. Woodhouse praised it as follows:

> The name of Berkeley has occurred more than once in the preceding pages: and I cannot quit this part of my subject without commending the analyst and the subsequent pieces, as forming the most satisfactory controversial discussion of pure science, that ever yet appeared: into what perfection of perspicuity and of logical precision, the doctrine of fluxions may be advanced, is no subject of consideration: But, view the doctrine as Berkeley found it, and its

---

[25]If an identity holds of all elements of an extended domain, it holds in particular for all elements in the original domain.

[26]Portions of it are frequently anthologised. D.E. Smith's and J. Stedall's source books already cited contain extracts, as does Dirk J. Struik's. The full text can be downloaded at Google Books.

defects in metaphysics and logic are clearly made out.

If, for the purpose of habituating the mind to just reasoning, (and mental discipline is all the good the generality of students derive from the mathematics) I were to recommend a book, it should be the *Analyst*. Even those who still regard the doctrine of fluxions as clearly and firmly established by their immortal inventor, may read it, not unprofitably, since, if it does not prove the cure of prejudice, it will be at least the punishment.[27]

Following the devastating critique by Berkeley, Jean le Rond d'Alembert (1717 – 1783) offered an explanation in terms of limits. Both Landen and Joseph Louis Lagrange (1736 – 1813) offered fluxion- and limit-free algebraic approaches. Landen described differentiation as a sort of operator: one simplifies the difference quotient until Δx is no longer in the denominator and then sets $\Delta x = 0$. Lagrange assumed all functions expandable into power series and defined the derivative at a to be the coefficient a_1 of the linear term in the Taylor expansion,

$$f(x) = a_0 + a_1(x - a) + a_2(x - a)^2 + \ldots$$

In his *The Principles of Analytical Calculation* of 1803, Robert Woodhouse found fault with all of these approaches.

Woodhouse's discussion of these matters is well worth reading, especially to those with philosophical/foundational interests. The more strictly mathematically inclined, who have still never forgiven Berkeley, may not care for his praise of Berkeley and may excuse their not taking the time to read Woodhouse on the grounds that we all know how it came out: D'Alembert and the limit concept won. No one speaks of fluxions anymore, Landen is a footnote in histories of mathematics, Lagrange's algebraic approach to Taylor series is of some algebraical interest but of no value in the Calculus, and no one ever heard of Woodhouse's formal approach attempting to overcome the faults of all the rest. I shall assume here the Philistine's cloak and, other than to urge the reader to have a look at what Woodhouse has to say[28], will pass over discussion of the failed attempts and discuss only the limit concept as it applies to series.

Throughout the 18th century the practice had been to deal with infinite series formally, performing one's manipulations on them paying no heed to questions of convergence until the computation was over. Euler distinguished

---

[27]Robert Woodhouse, *The Principles of Analytical Calculation*, University of Cambridge Press, Cambridge, 1803, pp. xvii – xviii. At the end of this passage, Woodhouse adds a footnote that finishes with the words

> The reason why Berkeley's ideas have not obtained a more general reception, seems to be this; unbiassed men, earnest lovers of truth, and moderately skilled in mathematics, read not the Analyst, because they imagined the discussion too deep for them; and professed mathematicians, in judging of an hostile tract, felt a zeal for the honor of their order, and a more than reasonable affection for their favourite study.

[28]The book is available online at Google Books.

between the *value* of a series obtained by formal manipulation, e.g.,

$$1 + 2 + 4 + 8 + \ldots = -1,$$

and the *sum* of the series—the limit of the partial sums when they seemed to converge. Woodhouse accepted this practice, deeming the former analytical calculation. However, the winds of change were in the air and the limit concept would soon begin its rise to power.

Thus Berkeley criticised the fluxional calculus so severely that some response was necessary. D'Alembert's explanation of the operations of the Calculus in terms of limits, which he published in various articles in Denis Diderot's *Encyclopédie*, was one such response. His definition of limit appeared in volume IX (1765):

> LIMIT (*Mathematics*). One says that a magnitude is the *limit* of another magnitude, when the second may approach the first more closely than by a given quantity, as small as one wishes, moreover without the magnitude which approaches being allowed ever to surpass the magnitude that it approaches; so that the difference between such a quantity and its *limit* is absolutely unassignable...
>
> The theory of *limits* is the foundation of the true justification of the differential calculus. *See* DIFFERENTIAL, FLUXION, EXHAUSTION, INFINITE. Strictly speaking, the *limit* never coincides, or never becomes equal to the quantity of which it is the *limit*, but the latter approaches it ever more closely, and may differ from it as little as one wishes.[29]

This definition is not very precise and explicitly rules out the possibility of the magnitude ever equalling the limit. This led one prominent historian of mathematics to postulate a notion of *limit avoidance*, a concept that made no mathematical sense, as was quickly—and forcefully—pointed out by an outspoken mathematician.

The modern notations $\lim_{n \to \infty} a_n$ and $\lim_{x \to a} f(x)$ refer to the limit of the dependent magnitude, but contain within them notations, $n \to \infty$ and $x \to a$, for the limits of the independent variable. There is an assymetry between the two. In discussing the limit as $n \to \infty$ or $x \to a$, we are interested in the behaviour of the sequence $a_1, a_2, a_3, \ldots$ for larger and larger values of n or of the function $f(x)$ for x closer and closer to a; there is no question of what happens when $n = \infty$, and how f behaves *at* a has no bearing on the tendency of f's behaviour around a. So n is not allowed to equal its limit and x is not allowed to equal a. That a_n not be allowed to equal its limit and $f(x)$ never allowed to equal its limit is best explained as an oversight. D'Alembert is defining the notion for a monotone sequence of the limit from the left or right for a function monotone on either side of a. The definition given, though not as quantitatively precise as the modern one, is clear enough in these cases. The examples he cites are of this form. In the article on limits he describes the circumference of the

---

[29]English translation from: Stedall, *op. cit.*, pp. 297 – 298. Stedall includes also excerpts on limits from Wallis, Newton, Maclaurin, and Cauchy.

172 3. THE BINOMIAL THEOREM PROVEN

circle as either the limit of the perimeters of a sequence of inscribed polygons of increasing numbers of sides or as the limit of the perimeters of a sequence of circumscribed polygons. The one is a monotonically increasing sequence which never equals nor surpasses its limit; the other is a decreasing such sequence. Stedall says

> His definition of "limit" in Volume IX was close to Newton's idea of a limit as a bound that could be approached as closely as one chose, and because d'Alembert, like Newton, worked with examples that were primarily geometric, there was still no obvious need to consider quantities that might oscillate from one side of a limit to the other.[30]

With sequences, there would indeed be no need for such a consideration. Either from some point on the elements of the sequence were always on one side of the limit or one could split the sequence into two subsequences which separately approach the limit from above and below.[31]

In an earlier volume of the *Encyclopédie*, in his article on differentials (volume IV, 1754), d'Alembert considers the derivative of the parabola $y^2 = ax$ as the limit of the difference quotient $z/u = \Delta y/\Delta x$ and obtains $a/(2y+z)$:

> So dy/dx is the limit of the ratio of z to u, and this limit is found by making $z = 0$ in the fraction $a/(2y+z)$.
>
> ... This limit is the quantity to which the ratio z/u approaches more and more closely if we suppose z and u to be real and decreasing.[32]

Here again we see the oversight of not imagining a function oscillating infinitely often in a neighbourhood of the point at which the limit is being taken. There had long been a tendency to believe all functions were very smooth and any interval could be broken into a finite number of subintervals on which a given function was monotone and otherwise well behaved. Badly behaved functions were still over half a century away when d'Alembert was proposing limits as the foundation of the Calculus.

Thus far in this book we have not given a careful definition of the notion of limit, but have used the term freely, given a few examples, and generally talked around it. It isn't really necessary for most purposes to know what one is talking about. By the time d'Alembert gave his definition of limit, nearly a century had passed since Newton had made his important discoveries in the Calculus, over a century since Europeans had been calculating limits, and it would still be over another half century before a precise definition would make it into print, so we probably don't need such a definition to read what d'Alembert has to say about the Binomial Theorem, but such a definition will help us see what he did not do.

First we define what it means for a sequence to converge:

---

[30] *Ibid*, p. 297.

[31] Of course, one has still overlooked the mildly complicating possibility of the sequence occasionally equalling the limit.

[32] Struik, *op. cit.*, p. 344. Unless for some strange reason d'Alembert is only interested in one-sided derivatives, by "decreasing" he means "decreasing in absolute value".

3.6.1. Definition. *Let* $a_0, a_1, a_2, \ldots$ *or* $a_1, a_2, a_3, \ldots$ *be a sequence of real numbers. We say that a number* L *is the* limit *of the sequence, written*

$$\lim_{n \to \infty} a_n = L,$$

in case the following holds: for any $\epsilon > 0$ *there is an* n_0 *such that for all* $n > n_0$,

$$|a_n - L| < \epsilon.$$

If such a limit exists, the sequence is said to converge; *otherwise it is said to* diverge.

Sometimes, for emphasis, one phrases the initial two quantifiers as "for any $\epsilon > 0$ *no matter how small* there is an n_0 *so large that*...".

The condition that $|a_n - L| < \epsilon$ simply says that a_n is close to L, and, ϵ being arbitrary, as close as we please. The final two quantifiers assert that, for a given ϵ, elements of the sequence do get that close— and *stay* there: eventually, if one goes far enough out in the sequence, one is always within ϵ of the limit.

It may be helpful to consider a couple of examples.

3.6.2. Example. $\lim_{n \to \infty} \dfrac{1}{1 + \frac{1}{n}} = 1$. *Intuitively, this is clear: as n gets larger, $1/n$ approaches 0, and $1/(1+1/n)$ will tend to $1/(1+0) = 1/1 = 1$. To establish this formally using the definition, let $\epsilon > 0$ and solve the inequality*

$$\left| \frac{1}{1 + \frac{1}{n}} - 1 \right| < \epsilon. \tag{95}$$

Now

$$\frac{1}{1 + \frac{1}{n}} - 1 = \frac{n}{n+1} - 1 = \frac{n}{n+1} - \frac{n+1}{n+1} = \frac{-1}{n+1},$$

whence

$$\left| \frac{1}{1 + \frac{1}{n}} - 1 \right| = \left| \frac{-1}{n+1} \right| = \frac{1}{n+1} < \frac{1}{n}.$$

Thus, provided $\epsilon > 1/n$, i.e., $n > 1/\epsilon$, we will have (95). So choose n_0 to be any integer greater than $1/\epsilon$. Then for any $n > n_0$ we will have $n > 1/\epsilon$, and thus (95).

3.6.3. Example. *Let* $a_n = 1 + (-1)^n$. $a_0, a_1, a_2, \ldots$ *is just the sequence $2, 0, 2, 0, \ldots$ and it has no limit. For, if L were the limit, we could take $\epsilon = 1/2$ and there would be an n_0 such that for all $n > n_0$, $|a_n - L| < 1/2$. Choosing any large enough even integer n yields $|2 - L| < 1/2$, i.e.,*

$$-\frac{1}{2} < 2 - L < \frac{1}{2}, \tag{96}$$

while choosing any large odd integer tells us $|0 - L| < 1/2$, i.e.,

$$-\frac{1}{2} < -L < \frac{1}{2}. \tag{97}$$

Multiplying (97) by -1 reverses the inequalities,

$$-\frac{1}{2} < L < \frac{1}{2},$$

and adding this to (96) yields
$$-1 < 2 - L + L < 1,$$
i.e., $-1 < 2 < 1$, which simply is not true.

Note that our second example explains why we have the second universal quantifier. The sequence $a_0, a_1, a_2, \ldots$ gets as close as we please to 2. In fact, it equals 2 infinitely often. But just as often it veers away. For L to be the limit of the sequence, the elements of the sequence must not only get arbitrarily close to L, but must stay close.

If I seem to be going on and on about this, it is because it really is a complex definition. Mathematicians rarely use such an alternation of quantifiers: $\forall \exists \forall$ ($\forall$ = for all, $\exists$ = there exists). The $\exists \forall$ and $\forall \exists$ combinations are simple enough. $\exists x \forall y$ asserts there to be some x which uniformly satisfies whatever condition follows the quantifiers for all values of y. Ideally an assertion $\exists x \forall y A(x, y)$ would be proven by explicitly producing a value x and proving that $A(x, y)$ is true for all possible values of y. $\forall x \exists y$ asserts that for any choice of x one can find a y satisfying the given condition. The natural way of proving an assertion $\forall x \exists y A(x, y)$ is to exhibit a function $f(x)$ producing the y and establishing $\forall x A(x, f(x))$. In Definition 3.6.1, the function $N_0(\epsilon)$ satisfying the condition,
$$\forall \epsilon > 0 \, \forall n > N_0(\epsilon) \bigl(|a_n - L| < \epsilon \bigr),$$
would be called the *modulus of convergence*.

3.6.4. Definition. *A series,*
$$\sum_{n=0}^{\infty} a_n \quad \text{or} \quad \sum_{n=1}^{\infty} a_n,$$
converges if the sequence $(s_0,)\, s_1, s_2, \ldots$ *of its partial sums,*
$$s_n = \sum_{k=0}^{n} a_k \quad \text{or} \quad s_n = \sum_{k=1}^{n} a_k,$$
respectively, converges. If a series converges, the limit of the sequence of its partial sums is called the limit *of the series.*

We may also refer to the limit of a series as its *sum*.

The natural way of proving that a series converges to a limit is to produce the limit and modulus of convergence. Finding the limit, however, is not so easy. As we saw in Chapter 2, section 8, simple algebraic trickery of the sort that so easily handled the geometric progression will not work in other cases. We have
$$1 - \frac{1}{2} + \frac{1}{3} - \frac{1}{4} + \ldots = \sum_{n=0}^{\infty} \frac{(-1)^n}{n+1} = \ln 2$$
$$1 - \frac{1}{3} + \frac{1}{5} - \frac{1}{7} + \ldots = \sum_{n=0}^{\infty} \frac{(-1)^n}{2n+1} = \frac{\pi}{4},$$

6. D'ALEMBERT AND THE LIMIT CONCEPT

by equations (61) and (59) of that section. And to these examples I cannot resist adding Euler's sum,

$$1 + \frac{1}{4} + \frac{1}{9} + \frac{1}{16} + \ldots = \sum_{n=1}^{\infty} \frac{1}{n^2} = \frac{\pi^2}{6}.$$

These values, $\ln 2, \pi/4, \pi^2/6$, are not algebraic entities coming from the same domain as the series' entries. The first two fall out of the power series expansions of a logarithmic and a trigonometric function, respectively, and Euler's sum, originally established by him through some very complex reasoning about a power series expansion, is nowadays usually presented as a quick application of Fourier series, series that are built up from sines and cosines.

When one cannot determine the limit of a series directly, it may still be possible to determine indirectly that it converges. This is done by appeal to a *convergence test*. One of the earliest such tests is due to Leibniz:

3.6.5. Theorem (Alternating Convergence Test). *Let $a_1, a_2, a_3, \ldots$ be a monotonically decreasing sequence of positive real numbers (i.e., $a_n \geq a_{n+1}$ for all n). The series $\sum_{n=1}^{\infty} (-1)^{n+1} a_n$ converges iff $\lim_{n \to \infty} a_n = 0$.*

One of the first things one learns about series in the standard Calculus course is that, if a series converges, the terms tend to 0. And quick reference to the harmonic series, $\sum \frac{1}{n}$, shows the failure, in general, of the converse. Leibniz claimed, however, that the converse does hold under the conditions that the terms alternate in sign and decrease in size (i.e., in absolute value) to 0.

It may be worth our while to digress a moment to consider the proof of this theorem. Let $a_1 > a_2 > a_3 > \ldots$ be a sequence of positive numbers converging to 0 and consider the series

$$\sum_{n=1}^{\infty} (-1)^{n+1} a_n = a_1 - a_2 + a_3 - a_4 + \ldots$$

and its partial sums

$$s_1 = a_1$$

$$s_{2m} = \sum_{n=1}^{2m} (-1)^{n+1} a_n = (a_1 - a_2) + (a_3 - a_4) + \ldots + (a_{2m-1} - a_{2m})$$

$$s_{2m+1} = \sum_{n=1}^{2m+1} (-1)^{n+1} a_n = a_1 + (-a_2 + a_3) + (-a_4 + a_5)$$

$$+ \ldots + (-a_{2m} + a_{2m+1}).$$

Because $a_n \geq a_{n+1}$ each summand $a_{2k+1} - a_{2k+2}$ is nonnegative and each $-a_{2k} + a_{2k+1}$ is nonpositive. Thus we see quickly that

$$s_{2m+2} = s_{2m} + (a_{2m+1} - a_{2m+2}) \geq s_{2m}$$

$$s_{2m+3} = s_{2m+1} + (-a_{2m+2} + a_{2m+3}) \leq s_{2m+1}.$$

So the subsequence of partial sums with an even number of terms is monotone increasing,
$$s_2 \leq s_4 \leq s_6 \leq \ldots,$$
while the subsequence of partial sums with an odd number of terms is monotone decreasing,
$$\ldots \leq s_5 \leq s_3 \leq s_1.$$
Moreover,
$$s_{2m+2k+1} = s_{2m} + (a_{2m+1} - a_{2m+2})$$
$$+ \ldots + (a_{2m+2k-1} - a_{2m+2k}) + u_{2m+2k+1} \geq s_{2m},$$
whence all the partial sums with odd numbers of terms are greater than all of those with even numbers of terms:
$$s_2 \leq s_4 \leq s_6 \leq \ldots \leq s_5 \leq s_3 \leq s_1. \tag{98}$$

Today we would conclude the existence of the limit in one of three ways. The first is to note that the sequence $s_2, s_4, s_6, \ldots$ of partial sums with even numbers of terms is monotone increasing and bounded above, while $s_1, s_3, s_5, \ldots$ is monotone decreasing and bounded below. We would then appeal to another convergence test:

3.6.6. Theorem (Monotone Convergence Theorem). *Every bounded monotone sequence converges to a limit.*

Thus,
$$\lim_{m \to \infty} s_{2m} = L_e \quad \text{and} \quad \lim_{m \to \infty} s_{2m+1} = L_o$$
exist. But L_e, L_o lie between s_{2m} and s_{2m+1} for all m, whence
$$|L_e - L_o| \leq s_{2m+1} - s_{2m} = a_{2m+1}.$$
As a_{2m+1} can be made smaller than any given $\epsilon > 0$, it follows that $|L_e - L_o|$ is smaller than any $\epsilon > 0$ and must therefore equal 0, i.e., $L_e = L_o$. This common limit of the two subsequences is the limit of the sequence $s_1, s_2, s_3, \ldots$ of partial sums, hence the limit of the series.

The second modern proof is to start with (98) and consider the sequence of closed intervals $I_1, I_2, I_3, \ldots$ given by $I_k = [s_{2k}, s_{2k+1}]$. By (98), one has $I_1 \supseteq I_2 \supseteq I_3 \supseteq \ldots$ One then applies the Nested Interval Property:

3.6.7. Theorem (Nested Interval Property). *If $I_1 \supseteq I_2 \supseteq I_3 \supseteq \ldots$ are closed bounded intervals, $I_k = [a_k, b_k]$, then the intersection $\bigcap_{k=1}^{\infty} I_k$ is nonempty.*

In the present situation this means there is at least one number L that lies in every interval $[s_{2k}, s_{2k+1}]$:
$$s_2 \leq s_4 \leq s_6 \leq \ldots \leq L \leq \ldots \leq s_5 \leq s_3 \leq s_1. \tag{99}$$
Again, the increasing closeness of s_{2k} and s_{2k+1} rules out the existence of a second number satisfying (99) and L will be the limit of the sequence of partial sums, whence the limit of the series.

Leibniz did not appeal to either of these theorems, but treated the expression $\sum(-1)^{n+1}a_n$ as an ideal element S and concluded $s_{2m} \leq S$ and $S \leq s_{2m+1}$ for all m, and thereby establishing the finiteness of S. S now satisfied (99) and, again, was the limit of the series. The third modern proof, available to the mathematical logician, uses what is called Nonstandard Analysis to replace S by S_N for some infinite integer N. The finiteness of S_N guarantees it is infinitely close to some real number that happens to be the sought after limit.

Proofs of Theorems 3.6.6 and 3.6.7 are not given in the standard Calculus course. Proofs, in fact, are beside the point—these results are axioms, commonly called the *completeness of the real number line*. There are several formulations of completeness and in the Real Analysis course one of these is chosen as an axiom and the rest derived from it. Which is an axiom (e.g., the Monotone Convergence Theorem or the Nested Interval Property) and which are to be derived is an arbitrary choice.[33]

These axiomatic considerations emerged in the 19th century and should not be of much concern to us in the present section.

We are now only one simple test away from returning to d'Alembert.

3.6.8. Theorem (Comparison Test). *Let $a_1, a_2, a_3, \ldots$ and $b_1, b_2, b_3, \ldots$ be sequences of nonnegative numbers and suppose*

i. for each n, $a_n \leq b_n$

ii. $\sum_{k=1}^{\infty} b_n$ converges.

Then: $\sum_{k=1}^{\infty} a_n$ converges.

This is an easy corollary to the Monotone Convergence Theorem: Let

$$s_m = \sum_{n=1}^{m} a_n \text{ and } L = \sum_{n=1}^{\infty} b_n.$$

Obviously $s_1 \leq s_2 \leq s_3 \leq \ldots \leq L$, whence Theorem 3.6.6 applies.

And we finally arrive back at d'Alembert with his famous ratio test.

3.6.9. Theorem (D'Alembert's Ratio Test). *Let $a_0, a_1, a_2, \ldots$ be a sequence of positive real numbers and suppose $\lim_{n \to \infty} a_n/a_{n+1} = L$ exists.*

i. if $L > 1$, then $\sum_{n=0}^{\infty} a_n$ converges

ii. if $L < 1$, then $\sum_{n=0}^{\infty} a_n$ diverges.

Proof. i. If $L = 1$ or fails to exist, the test is inconclusive.

---

[33]There is another procedure: one constructs the real numbers from the rationals by an infinitistic construction motivated by one's belief in one of these properties. Unsurprisingly, the property on which the construction was based can be proven to hold for the constructed real numbers.

The ratio test is a simple application of the Comparison Test, using a geometric series to provide the comparison. Suppose, in the important case, $L > 1$, and let $0 < \epsilon < L - 1$. Choose n_0 such that for $n > n_0$

$$\left| \frac{a_n}{a_{n+1}} - L \right| < \epsilon.$$

Then

$$\left| a_{n+1} - \frac{a_n}{L} \right| < \frac{\epsilon a_{n+1}}{L}$$

and

$$a_{n+1} < \frac{a_n}{L} + \frac{\epsilon a_{n+1}}{L}$$

$$a_{n+1}\left(1 - \frac{\epsilon}{L}\right) < \frac{a_n}{L}$$

$$a_{n+1} < a_n \left(\frac{L}{L-\epsilon}\right)\frac{1}{L} = a_n \frac{1}{L-\epsilon}. \tag{100}$$

But from $\epsilon < L - 1$, we conclude $1 < L - \epsilon$ and for $r = 1/(L-\epsilon)$ we have $r < 1$. Thus, for $n > n_0$, $a_{n+1} < a_n r$ and $a_{n_0+k} < a_{n_0} r^k$. Comparing

$$\sum_{k=1}^{\infty} a_{n_0+k} \text{ with } a_{n_0} \sum_{k=1}^{\infty} r^k$$

we have the convergence of

$$\sum_{n=0}^{\infty} a_n = \sum_{n=0}^{n_0} a_n + \sum_{k=1}^{\infty} a_{n_0+k}.$$

ii. If $L < 1$, choose n_0 so that for all $n > n_0$,

$$\left| \frac{a_n}{a_{n+1}} - L \right| < 1 - L.$$

Then

$$|a_n - a_{n+1}L| < (1-L)a_{n+1}$$
$$a_n - a_{n+1}L < a_{n+1} - a_{n+1}L$$
$$a_n < a_{n+1},$$

and, for $n > n_0$, the sequence is growing in value, whence a_n does not tend to 0 and $\sum_{n=0}^{\infty} a_n$ does not converge. □

I have not stated the Ratio Test in full generality. First, and most easily, the test still holds when $L = +\infty$, i.e., when the ratios a_n/a_{n+1} grow without bound. In this case, one simply avoids the initial algebra of the proof, noting that from some n_0 on one has $a_n/a_{n+1} > 2$, whence

$$\sum_{k=1}^{\infty} a_{n_0+k} < \sum_{k=1}^{\infty} \frac{1}{2^k} a_{n_0} = a_{n_0} \sum_{k=1}^{\infty} \frac{1}{2^k} = a_{n_0}.$$

Second, and this is of great importance in the sequel, the Ratio Test also applies when negative terms are allowed and one considers the ratio of absolute

6. D'ALEMBERT AND THE LIMIT CONCEPT

values. This is a deeper result, the exposition of which requires two new notions of convergence and their accompanying convergence tests.

3.6.10. Definition. *A series $\sum_{n=0}^{\infty} a_n$ is absolutely convergent if $\sum_{n=0}^{\infty} |a_n|$ is convergent.*

3.6.11. Theorem. *i. Absolute convergence implies convergence.*
ii. Convergence need not imply absolute convergence.

The second assertion is easy to demonstrate. The alternating harmonic series,
$$\sum_{n=1}^{\infty} \frac{(-1)^{n+1}}{n}$$
converges, but its series of absolute values, the harmonic series $\sum_{n=1}^{\infty} 1/n$, does not.

The first assertion of Theorem 3.6.11 is not as easy to verify. The simplest proof uses another test:

3.6.12. Definition. *A sequence $a_0, a_1, a_2, \ldots$ of real numbers is Cauchy convergent if, for every $\epsilon > 0$ there is an integer n_0 such that for all integers $m, n > n_0$, $|a_n - a_m| < \epsilon$.*

Again, this is not a definition of a type of convergence to a specified limit, but a criterion of convergence:

3.6.13. Theorem (Cauchy Completeness). *A sequence of real numbers is convergent iff it is Cauchy convergent.*

That convergence implies Cauchy convergence is easy to prove: Let L be the limit of the sequence, let $\epsilon > 0$, and choose n_0 so large that for all $n > n_0$ one has $|a_n - L| < \epsilon/2$. Then, for $m, n > n_0$,
$$|a_n - a_m| = |a_n - L + L - a_m| \leq |a_n - L| + |L - a_m| < \frac{\epsilon}{2} + \frac{\epsilon}{2} = \epsilon.$$

The converse implication is more difficult because one must infer the existence of the limit without being able to specify it. It is essentially an axiom, another Completeness Axiom comparable and equivalent to the other completeness axioms mentioned (Monotone Convergence Theorem, Nested Interval Property) and unmentioned (existence of the least upper bound of any bounded set). Its naturalness as an axiom is attested to by the fact that some mathematicians used it without any attempt to justify it—Euler in the 18th century and Cauchy in the 19th. It also seems to have been multiply discovered, apparently receiving its first explicit enunciation by the Portuguese mathematician José Anastácio da Cunha in the 1780s, and its first proof attempt by Bolzano in a monograph of 1817[34], but his proof is not deemed acceptable today. It does, however, seem to fit the view of the nature of the real numbers Bolzano was

---

[34]Bernard Bolzano, *Rein analytischer Beweis des Lehrsatzes, dass zwischen je zwey Werthen, die ein entgegengesetztes Resultat gewähren, wenigstens eine reelle Wurzel der Gleichung liege*, §7; English translation: "Purely analytical proof...", in: Russ, *op. cit.*, pp. 266 – 268.

developing in the 1830s and, I think, constitutes genuine progress. Later mathematicians, regarding Cauchy Completeness as true, based infinitistic constructions of the real numbers on Cauchy convergent sequences of rational numbers and proved the Theorem for the real numbers so constructed.

The justification of Cauchy Completeness is not the issue here[35]; its application is:

Proof of Theorem 3.6.11.i. Let $\sum a_n$ be absolutely convergent and suppose $\epsilon > 0$ is given. There is some n_0 such that for all $m > n > n_0$,

$$\left| \sum_{k=0}^{m} |a_k| - \sum_{k=0}^{n} |a_k| \right| < \epsilon,$$

i.e., $|a_{n+1}| + \ldots + |a_m| < \epsilon$. But

$$\left| \sum_{k=0}^{m} a_k - \sum_{k=0}^{n} a_n \right| = |a_{n+1} + \ldots + a_m| \leq |a_{n+1}| + \ldots + |a_m| < \epsilon,$$

whence $\sum a_n$ is Cauchy convergent, whence convergent. □

With this we can now state d'Alembert's Ratio Test in more general terms:

3.6.14. Corollary. *Let $a_0, a_1, a_2, \ldots$ be a sequence of real numbers and suppose $\lim_{n \to \infty} |a_n/a_{n+1}| = L$ exists.*

i. if $L > 1$, then $\sum_{n=0}^{\infty} a_n$ converges;

ii. if $L < 1$, then $\sum_{n=0}^{\infty} a_n$ diverges.

Proof. i. By the Ratio Test (Theorem 3.6.9) and Theorem 3.6.11.i,

$$L > 1 \Rightarrow \sum |a_n| \text{ converges} \Rightarrow \sum a_n \text{ converges}.$$

ii. This is essentially the same as the proof of Theorem 3.6.11.ii. □

The application of the Ratio Test to a power series is the following.

3.6.15. Theorem. *Let $a_0, a_1, a_2, \ldots$ be a sequence of real numbers and suppose $R = \lim_{n \to \infty} |a_n|/|a_{n+1}|$ exists. Let x be a real number.*

i. if $|x| < R$, then $\sum_{k=0}^{\infty} a_n x^n$ converges

ii. if $|x| > R$, then $\sum_{k=0}^{\infty} a_n x^n$ diverges.

Proof. i. For $a_0, a_1, a_2, \ldots$ all positive and $0 \leq x < R$ this follows from Theorem 3.6.9:

$$\frac{a_n x^n}{a_{n+1} x^{n+1}} = \frac{a_n}{a_{n+1}} \cdot \frac{1}{x} \to \frac{R}{x},$$

---

[35] The interested reader can find proofs discussed in Chapter I, section 5, and Chapter II, sections 4 and 7, of Smoryński, *Adventures in Formalism, op. cit.*

which is > 1 for $x < R$ and < 1 for $x > R$. For $-R < x < 0$, the terms alternate in sign, tend to 0:

$$\left|a_n x^n\right| = a_n |x|^n \to 0 \text{ by the convergence of } \sum a_n |x|^n,$$

and, for large enough n, the absolute values of the terms are decreasing:

$$\left|a_{n+1} x^{n+1}\right| = a_{n+1} |x|^{n+1} \approx \frac{a_n}{R} |x|^{n+1} = |a_n x^n| \cdot \frac{|x|}{R} < |a_n x^n|$$

for $|x| < R$. Hence the Alternating Convergence Test applies.

For $a_0, a_1, a_2, \ldots$ having mixed signs, the proof is identical after absolute value signs are inserted and the more general version of the Ratio Test applied.

ii. Again, the n-th term fails to go to 0 if $|x| > R$. □

The number $R = \lim_{n \to \infty} |a_n|/|a_{n+1}|$ is called the *radius of convergence* of the power series, so called because the theorem holds in the complex plane, the series converging everywhere inside the circle of radius R centred at the origin and diverging everywhere outside the circle. A deep theorem of Complex Analysis tells us that the series diverges at some point on the circle itself, but generally when $|x| = R$ the Theorem offers no information.

For the binomial series,

$$(1+x)^q = \sum_{n=0}^{\infty} \binom{q}{n} x^n,$$

when q is non-integral we have

$$\binom{q}{n} \Big/ \binom{q}{n+1} = \frac{q(q-1) \cdots (q-n+1)}{n(n-1) \cdots 1} \cdot \frac{(n+1)n(n-1) \cdots 1}{q(q-1) \cdots (q-n+1)(q-n)}$$

$$= \frac{n+1}{q-n} = \frac{1+1/n}{q/n - 1} \to \frac{1}{-1} = -1$$

as $n \to \infty$, whence $R = |-1| = 1$ and the series converges for $|x| < 1$.

We have already gone far beyond what is necessary to discuss d'Alembert's application of the Ratio Test to the binomial series, which was to conclude that the series has a radius of convergence of 1. For this, because the coefficients alternate in sign from some point on, the Alternating Convergence Theorem yields the result for $0 < x < 1$ and the simpler version, Theorem 3.6.9, applies for $-1 < x < 0$. I have discussed absolute and Cauchy convergence for the sake of the discussion in the next section, where the following result, a sort of nonconstructive general version of Theorem 3.6.15 will be used.

3.6.16. Lemma. *Let $\sum_{n=0}^{\infty} a_n x_0^n$ converge to a limit. For any x with $|x| < |x_0|$, the series $\sum_{n=0}^{\infty} a_n x^n$ is absolutely convergent.*

Proof. Let $|x| < |x_0|$. By convergence, $|a_n x_0^n| \to 0$ as $n \to \infty$. Thus there is some n_0 such that $|a_n x_0^n| < 1/2$ for $n > n_0$. For such n

$$|a_n x^n| = \left|a_n \left(\frac{x}{x_0}\right)^n x_0^n\right| < |a_n x_0^n| \cdot \left|\frac{x}{x_0}\right|^n < \frac{1}{2}\left|\frac{x}{x_0}\right|^n.$$

Thus

$$\sum_{n=0}^{\infty} |a_n x^n| = \sum_{n=0}^{n_0} |a_n x^n| + \sum_{n=n_0+1}^{\infty} |a_n x^n|$$
$$< \sum_{n=0}^{n_0} |a_n x_0^n| + \frac{1}{2} \sum_{n=n_0+1}^{\infty} \left|\frac{x}{x_0}\right|^n. \tag{101}$$

The first term of the right-hand side of (101) is a finite sum and the second is the sum of a geometric series with ratio $|x/x_0| < 1$, i.e., a convergent series of positive terms. Hence the Comparison Test applies to yield the absolute convergence of the series $\sum_{n=0}^{\infty} a_n x^n$, and thus its convergence. $\square$

This Lemma allows us to generalise the notion of the radius of convergence to all power series which converge at more than the origin. Given a series $\sum a_n x^n$, define the set

$$\mathcal{R} = \left\{ |x_0| \,\Big|\, \sum a_n x_0^n \text{ converges} \right\}.$$

If $\mathcal{R}$ is bounded, we can define the *radius of convergence* R of $\sum a_n x^n$ to be the least upper bound of $\mathcal{R}$; and if $\mathcal{R}$ is unbounded we can take R to be $+\infty$.

Now for the fun fact: D'Alembert neither stated nor proved his Ratio Test. He did apply it implicitly and later mathematicians made the application explicit in extracting the test from his paper. I find the paper most interesting from the perspective of the present book in that it demonstrates the rôle played by the Binomial Theorem in the mathematics of the day. Hence I shall quote from it at length, interrupting occasionally with pertinent annotations.

The paper starts thus:

XXXV-th MEMOIR.[36]
Reflexions on series and on imaginary roots.

§. I.

---

[36] At some point d'Alembert, having run afoul of many of his contemporaries, had difficulty finding an outlet for his mathematical papers. He thus collected many of them together in a series of volumes bearing the title *Opuscules mathématique*. His application of the Ratio Test to the binomial series is memoir 35 in volume 5 published in 1768. The memoir has two parts, the first on the convergence or divergence of series and the second on imaginary roots. The first part covering pages 171 – 183 consists of 32 numbered paragraphs. Paragraphs 1 – 25 are translated here.

6. D'ALEMBERT AND THE LIMIT CONCEPT

Reflexions on divergent or convergent series.

1. If one raises $1 + \mu$ to the power m, the n-th term of the series will be
$$\mu^{n-1} \times \frac{m(m-1)\ldots(m-n+2)}{2.3.4\ldots n-1},$$
& the following, that is to say the $(n+1)$-th, will be
$$\mu^n \times \frac{m(m-1)\ldots(m-n+2)(m-n+1)}{2.3.4\ldots n-1.n},$$
thus the ratio of the $(n+1)$-th term to the n-th term will be $\frac{\mu(m-n+1)}{n}$; but in order that the series be convergent, it is necessary that the ratio (ignoring the plus or minus sign) be $<$ that of unity.

This might be a good place to note the different use of $<$. D'Alembert ignores the sign in making comparison, as exemplified by his reference to a number ν being "negative $\&$ < 1" in paragraph 11, below. The inequality refers to *size*, not *position* in the real number line.

2. Note first that the preceding formula gives the means of forming very quickly the terms of a series: for example, if $m = \frac{1}{2}$, one must multiply the former term first by $\mu \times \frac{1}{2}$ to obtain the second; the second by $-\mu\left(1 - \frac{1}{4}\right)$ to obtain the third; the third by $-\mu\left(1 - \frac{3}{6}\right)$ to obtain the fourth, & so forth, giving
$$1 + \frac{\mu}{2} - \frac{\mu\mu}{2.4} + \frac{\mu^3.3}{2.4.6} - \frac{\mu^4.3.5}{2.4.6.8} + \frac{\mu^5.3.5.7}{2.4.6.8.10}, \&c.$$
If $m = \frac{1}{3}$, one has similarly
$$1 + \frac{\mu}{3} - \frac{2\mu\mu}{3.6} + \frac{\mu^3.2.4}{3.6.8} - \frac{\mu^4.2.4.6}{3.6.8.10} + \frac{\mu^5.2.4.6.8}{3.6.8.10.12}, \&c.^{37}$$

3. And in general, if $m = \frac{p}{q}$, one always multiplies the last term found by $\mu\left(-1 + \frac{p+q}{qn}\right)$. We will now see the conditions which make the series convergent.

The French word "convergente" is translated by my dictionary as "convergent" or "converging". The latter may be a more appropriate translation as he does not use the terms "convergente" and "divergente" in the modern sense that the series as a whole converges or diverges. His usage is best explained by considering a series that behaves nicely for a while and then starts to oscillate wildly or one which oscillates for a bit and then settles down. In the first case we might

---

[37] The factors in the coefficients should increase by 3s, not 2s. Thus the fourth term should be $\frac{\mu^3.2.5}{3.6.9}$ and the fifth $\frac{-\mu^4.2.5.8}{3.6.9.12}$, etc.

consider the series to be converging for a while and then it starts to diverge; in the second the behaviour changes from diverging to converging. Ferraro[38] translates "convergente" as "decreasing" and "divergente" as "increasing". To use the philosophical jargon, this agrees extensionally with d'Alembert's usage; I am not yet convinced about the intension. In any event, what we see here should not be confusing. True, he said in paragraph 1 that the series was convergent when $|x| < 1$ and he now says we will see under which conditions the series will be convergent. One can take the first sentence as announcing the result and the second as announcing the beginning of its derivation. But, he is actually going to locate where the terms of the binomial series begin to decrease at such a rate as to make convergence analysable. *Cf.* also paragraph 17, below.

> 4. Let $\mu = \pm(1 \pm \nu)$, which represents all the possible series; one notes first that in the case where one has $1 - \nu$, we may assume $\nu < 1$; for if ν were > 1, then $+(1 - \nu)$ falls under the case $-1 - \nu$, or $-1 + \nu$, ν being in the latter case < 1; & $-(1 - \nu)$ falls under the case $+1 + \nu$, or $+1 - \nu$, ν being in the latter case < 1.

This lacks detail for an exposition, but meets the standards for a research paper: If $1 < \nu < 2$, then $-1 < 1 - \nu < 0$ (using $<$ in the modern sense) and $1 - \nu = -1 + \nu'$ for some $0 < \nu' < 1$. If $2 \leq \nu$, then $1 - \nu \leq -1$ and $1 - \nu = -1 - \nu'$ for some ν'. And if $\nu \leq -1$, then $\nu = -1 - \nu'$. Similar reasoning applies to $-(1 - \nu)$. The upshot of all this is that when he writes $\mu = 1 - \nu$ or $-1 + \nu$ he can assume $|\nu| < 1$. Because the series will converge only for $|\mu| < 1$, trivially so for $\mu = 0$, he really needs only to consider the cases $-1 < \mu < 0$ and $0 < \mu < 1$. In the first case he can write $\mu = -1 + \nu$ with $0 < \nu < 1$ and in the second case $\mu = 1 - \nu$ with $0 < \nu < 1$. The discussion, with its myriad of variables $\mu, m, n, p, q, \nu, \omega$, is difficult to wade through.

> 5. That said, if $\dfrac{m+1}{n}$ is > 1, the ratio of the $(n+1)$-th term to the n-th found above, will be $\pm(1 \pm \nu)\left(\dfrac{m+1}{n} - 1\right)$, or, ignoring the $\pm$ sign, $(1 \pm \nu)\left(\dfrac{m+1}{n} - 1\right)$; & if $\dfrac{m+1}{n}$ is < 1, the ratio will be $(1 \pm \nu)\left(1 - \dfrac{m+1}{n}\right)$. The second of these is the formula for producing the terms in the series, for it is obvious that the larger n becomes, the more $\dfrac{m+1}{n}$ diminishes.
>
> 6. Now for the first of these ratios to be < 1, it must be, on supposing $n = 1 + \omega$, that $1 \pm \nu < \dfrac{1 + \omega}{m - \omega}$; & for the second ratio, it must be that $1 \pm \nu < \dfrac{1 + \omega}{\omega - m}$.
>
> 7. Now, because ω is always a positive integer, since the integer

---

[38]Giovanni Ferraro, *The Rise and Development of the Theory of Series Up to the Early 1820s*, Springer Science+Business Media, LLC, New York, 2008, p. 303.

$n = 1 + \omega$ is never < 1, it follows from the two preceding formulæ, 1st. that in the case where m is not a positive integer, the only case where the series $(1+\mu)^m$ has an infinite number of terms, the series will be convergent in its later terms if we have $\mu = 1 - \nu$; because by making $\omega = \infty$ in the second formula, we have $\dfrac{1+\omega}{\omega - m} = 1$, & consequently $1 - \nu < \dfrac{1+\omega}{\omega - m}$, at least in the later terms; 2nd. in the contrary case the series will be divergent to the end, if we have $\mu = 1 + \nu$.

The reference to ω equalling ∞, as well as the reference to the infinitely small in the next paragraph, raises the question: Did d'Alembert really allow infinitely large and infinitely small numbers? Leibniz had earlier allowed their use, but considered their mention a *façon de parler*, shorthand for describing what happened in the limit. D'Alembert's contemporary, and occasional mathematical rival, Euler used them as actual objects.

8. From this it follows that the series will prove *false*, for all those μ that are > 1, since the series will be divergent in the end; & on the contrary, when $\mu < 1$ the series can still be used, since it will be convergent to the ends, & that its last term will be infinitely small.

The curious use of the word "false" requires comment. In the present context, it could mean one of two things. First, the series has no limit; it is divergent in the modern sense of the word. But the series is also false in that it does not agree with the root: $(1+\mu)^m$ does not equal $\sum \binom{m}{n}\mu^n$.

9. The formulæ $1 \pm \nu < \dfrac{1+\omega}{m - \omega}$, & $1 \pm \nu < \dfrac{1+\omega}{\omega - m}$ give the two conditions $\pm \nu < \dfrac{1 + 2\omega - m}{m - \omega}$; & $\pm \nu < \dfrac{1 + m}{\omega - m}$; the first formula will be used in the case where ω is $< m$, & the second for the case where ω is $> m$.

10. If m is negative, it will always be the second formula, because then ω (which is always either zero or a positive integer) is $> m$.

11. If μ is < 1, & consequently ν negative & < 1, we will have in the first formula $-\nu < \dfrac{1 + 2\omega - m}{m - \omega}$, 1st. if the numerator is positive, that is to say, on supposing $m = \omega + \rho$, if $1 + \omega$ is $> \rho$; 2nd. if the numerator is negative, & such that $\dfrac{\rho - 1 - \omega}{\rho}$ is a fraction smaller than ν; therefore $1 - \nu < \dfrac{1+\omega}{\rho}$, or $\mu < \dfrac{1+\omega}{\rho}$. So if $1 + \omega$ is $< \rho$, & that μ is $> \dfrac{1+\omega}{\rho}$, the series will be divergent in its first terms.

12. Now as ω can never be < 0, it is necessary, for $1 + \omega$ to be $< \rho$, that ρ be > 1, & hence because $m = \omega + \rho$, that m also be > 1. So if m is positive & smaller than 1, the series always converges on the assumption that ν be negative and < 1.

13. On the same assumption that $\mu < 1$, the second formula gives $-\nu < \dfrac{1+m}{\omega - m}$, 1st. if m is positive, 2nd. if m is negative, & that by making $m = \omega - \rho$, we have $-\nu < \dfrac{1+\omega-\rho}{\rho}$, or because $\rho > \omega$, $\dfrac{\rho-\omega-1}{\rho} < \nu$, or because $\nu = 1 - \mu$, $\mu < \dfrac{1+\omega}{\rho}$.

Paragraphs 14, 15, 17, and 18 interrupt the flow of the demonstration providing counterexamples, cases where an apparently convergent series ultimately diverges and where an apparently divergent series ultimately converges. This is pertinent information, expositionally poorly placed. It should have been done before the demonstration began and provided with a comment that he was now going to determine where the behaviour changed—where the coefficients $\binom{m}{n}$ began to alternate in sign and, most importantly, where the terms $\binom{m}{n}\mu^n$ begin to decrease in absolute value.

14. We do not need to consider the case where m would be > 1; for we have seen that in this case the series would be divergent in its later terms, and therefore would give *false*, although it could appear convergent in it first terms, as would happen, for example, if we had $\mu = 2$, & $n = \frac{1}{3}$ [*sic: it should read* $m = \frac{1}{3}$]; for then the first two terms of the series would be 1, & $+\frac{2}{3}$, & yet the series would give *false*.

15. This convergence of the first terms can even be pushed out further in a series that will eventually also be divergent, and which therefore gives *false*; for example, if one raises $1 + \dfrac{200}{199}$ to the power $\frac{1}{2}$, the series only begins to diverge after the term whose quantity n is such that $\left(1 + \dfrac{1}{199}\right)\left(1 - \dfrac{3}{2n}\right)$ will be > 1, i.e., such that n is > 300.

16. In general, if $\dfrac{1}{\nu} + 1$ is $=$ to a positive integer, which I call k, that gives $\nu = \dfrac{1}{k-1}$ & $1 + \nu = \dfrac{k}{k-1}$, the series converges up to a term whose exponent n is such that we have $n > (1+m)\left(1 + \dfrac{1}{\nu}\right)$ or $(1+m)k$.

17. It is therefore wrong to believe one has real series convergence because it converges even very far into its first terms. It is necessary for this that μ be < 1.

18. For the same reason, it is clear that a series may be divergent in its first terms, though convergent in its last, & therefore good & not false, provided that one pushes it far enough; for example, if $a + b$ is raised to the power -3, & $b > \dfrac{a}{3}$, but $< a$, the series will be in this case we are talking about. In general, for the series to be

always convergent, $-\nu$ must always be $< \dfrac{1+2\omega-m}{m-\omega}$, or $< \dfrac{1+m}{\omega-m}$.

For example, if $\mu = 1 - \dfrac{1}{k}$, m a negative number $= -p$ we have as condition for convergence, $-\dfrac{1}{k} > \dfrac{1-p}{\omega+p}$ & $\dfrac{1}{k} > \dfrac{p-1}{\omega+p}$; and so if ω is $< kp - k - p$, the series will be divergent. For example if $k = 100, p = 2$, the series diverges to the 99th term; since we have $kp - k - p = 200 - 100 - 2 = 98$; therefore, as n or $1+\omega$ will be < 99, we have $\omega < kp - k - p$.

The next two paragraphs finally make the point that from some point on the coefficients of the binomial series alternate in sign and thus the terms are of the same sign if μ is negative and of alternating signs if μ is positive.

19. We note again that since the ratio of the $(n+1)$-th term to the n-th is $\mu\left(\dfrac{m+1}{n} - 1\right)$ or $\mu\left(\dfrac{m-\omega}{1+\omega}\right)$, these two terms are of the same sign, 1st. if μ is positive & $m > \omega$; 2nd. if μ is negative & $m < \omega$; & of different signs, if μ is positive & $m < \omega$, or if μ is negative & $m > \omega$.

20. So if μ is negative, the terms of the series will be of the same sign, starting at the term n, as n or $1+\omega$ is $> 1+m$; in the contrary case, if μ is positive, the terms will have alternating signs starting where n is $> 1+m$.

21. For a series to be as perfect as possible, it is necessary 1st. that all the terms decrease from the first term on; 2nd. that these terms, if possible, have the same sign; since a series whose terms must all be added is more obviously convergent, all things being equal, than a series that should add and subtract the terms alternately. We can always completely satisfy the first condition, splitting the quantity $1 \pm \mu$ in two, one of which is as small as you like compared to the other; for example, if it is proposed to extract the square root of 2, we can put $\sqrt{2}$ in the form $\sqrt{(4-2)}$ or $2\sqrt{(1-\frac{1}{2})}$, or $\sqrt{\left(\dfrac{9-1}{4}\right)} = \dfrac{3}{2} \times \sqrt{(1-\frac{1}{9})}$, or &c. But it is not always possible to make the terms simultaneously of the same sign; since μ is assumed given, one cannot always force μ & $m - \omega$ to have the same sign.

I am not certain what he means by perfection here. If the terms of a series $\sum a_n$ alternate in sign and decrease in absolute value, convergence (in the modern sense) is easy to analyse: the difference between

$$\sum_{n=0}^{k} a_n \text{ and } \sum_{n=0}^{\infty} a_n$$

is less than $|a_{n+1}|$. On the other hand, if all the terms have the same sign, the partial sums are not oscillating, but are heading straight toward their limit. And the next two paragraphs, evidently (to the modern eye) an application of the Comparison Test using a convergent geometric series[39], provide an error estimate.

22. If the terms of the series have the same sign starting after the n-th term, where n is $> 1 + m$, then it is easy to see that the sum of the series, starting at the n-th term, which I call A, is $< A + A\mu + A\mu^2 + A\mu^3$, &c. & conversely $A > A + A\mu\left(\dfrac{\omega - m}{1 + \omega}\right) + A\mu^2\left(\dfrac{\omega - m}{1 + \omega}\right)^2 + A\mu^3\left(\dfrac{\omega - m}{1 + \omega}\right)^3$, &c.

23. Therefore the sum of the terms, beginning with A, will be $< \dfrac{A}{1 - \mu}$ & $> \dfrac{A}{1 - \mu\left(\frac{\omega - m}{1 + \omega}\right)} = \dfrac{A}{1 - \mu + \frac{\mu(m+1)}{1+\omega}}$; which gives a fairly convenient approximation to the sum; the error will be smaller than $\dfrac{A\mu(m+1)}{(1-\mu)(1+\omega-\mu+\mu m)}$.

24. For example, if we want to take the square root of 2, we first put $\sqrt{2}$ in the form $\sqrt{\left(\dfrac{9-1}{4}\right)} = \tfrac{3}{2}\sqrt{(1 - \tfrac{1}{9})}$; then (putting aside the factor $\tfrac{3}{2}$) if we take $n = 10$, or $\omega = 9$, we will have $A = \dfrac{1.3.5.7.9.11.13.15}{9^9.2.4.6.8.10.12.14.16.18}$; & after adding all the first nine terms, the rest of the series will be $< \dfrac{A}{1 - \tfrac{1}{2}}$ & $> \dfrac{A}{1 - \tfrac{1}{2} + \tfrac{1.3}{9.20}}$.

25. If we only had $n = 5$, we would have $A = \dfrac{1.3.5}{9^4.2.4.8}$, & we will have the value of the series by the above formulæ, with an error less than $\dfrac{A.3}{9.10 \times \tfrac{2}{9} \times \left(\tfrac{2}{9} + \tfrac{1}{10}\right)}$.

When confronted with a paper like this, in which there are several cases, numerous variables, some asides, and a statement like "We will now see the conditions which make the series convergent", which may not be a very clear statement of the intention of the paper to one not versed in the mathematical practice of the day, the best strategy to understanding it is not to read it carefully, but to glance over it, determine roughly what is being done, and then make one's own reconstruction of its contents.

One might begin with some of the counterexamples, noting that the initial partial sums might be poor indicators of the convergence or divergence of the series.

3.6.17. Example. *For* $\mu = 2, m = \tfrac{1}{3}$ *we have*

$$(1 + 2)^{1/3} = 1 + \binom{1/3}{1} 2 + \binom{1/3}{2} 2^2 + \ldots,$$

---

[39]And pretty much establishing the Ratio Test in this case.

the partial sums of which are

$$1$$
$$1 + \frac{1}{3} \cdot 2 = 1 + \frac{2}{3} = 1.\overline{6}$$
$$1 + \frac{1}{3} \cdot 2 + \frac{1}{3}\left(\frac{-2}{3}\right) \cdot 4 = 1 + \frac{2}{3} - \frac{8}{9} = .\overline{7}$$
$$1 + \frac{1}{3} \cdot 2 + \frac{1}{3}\left(\frac{-2}{3}\right) \cdot 4 + \frac{1}{3}\left(\frac{-2}{3}\right)\left(\frac{-5}{3}\right) \cdot 8 = 3.74\ldots$$

Note that the first two terms are converging on $\sqrt[3]{3} \approx 1.44\ldots$.

3.6.18. Exercise. *On the TI-83 Plus store* $200/199$ *in* M, $1/2$ *in* Q, *and enter the following program:*

PROGRAM:DALEM
:1→S
:1→T
:For(K,1,N)
:((Q−(K−1))/K)∗T∗M→T
:S+T→S
:End
:S

This will calculate the binomial series for $(1 + 200/199)^{1/2}$ *out to the term of degree n stored in* N. *Run the program for* $n = 50, 100, 150, \ldots$ *to verify that it seems to be converging very slowly to* $\sqrt{1 + 200/199} \approx 1.415989098$.

3.6.19. Exercise. *Write a program* DALEM2 *by inserting the line*

:S→L₁(K)

before the End *command of* DALEM *and replacing the final* S *by* L₁. *Store* $1/2$ *in* M, -3 *in* Q, *and* 25 *in* N. *Next store* $99/100$ *in* M *and* -2 *in* Q *for* $N = 25, 50, 75$. *Lists cannot have more than* 999 *entries on the TI-83 Plus, but one can run* DALEM *for several values of* N *greater than* 100 *to verify d'Alembert's final remarks in his paragraph* 18.

Having given such examples, the problem addressed by d'Alembert becomes clear—to determine when the binomial series converges and how close an approximation to the sum that the partial sums are. The answer to the convergence question had long been known: The series

$$(1 + \mu)^m = \sum_{n=0}^{\infty} \binom{m}{n} \mu^n$$

converges for $|\mu| < 1$. And, as we saw before reading d'Alembert's paper, this follows from the Alternating Convergence and Ratio Tests. A modern version of d'Alembert's paper would then mention this, either citing the two tests as known or establishing them from scratch. D'Alembert did not do this. He did not prove or cite either test, nor prove in any way that the series was convergent or that it converged to $(1 + \mu)^m$. He *assumed* these facts.

190   3. THE BINOMIAL THEOREM PROVEN

The two things d'Alembert had to establish were i. where the binomial coefficients begin to alternate in sign, and ii. where the terms start to decrease in absolute value. The first of these tasks is easily disposed of. For m not a whole number,
$$\binom{m}{n+1} = \frac{m-n}{n+1}\binom{m}{n}$$
and the terms will be of opposite signs so long as $m - n$ is negative, i.e., $n > m$. For the second task, we want the ratio of two successive terms,
$$\left|\frac{\binom{m}{n}\mu^n}{\binom{m}{n+1}\mu^{n+1}}\right| = \frac{|n+1|}{|m-n|} \cdot \frac{1}{|\mu|}$$
to be greater than 1. For $n > m$ we have $|m - n| = n - m$, so we want
$$\frac{n+1}{n-m} \cdot \frac{1}{|\mu|} > 1,$$
i.e.,
$$n + 1 > (n-m)|\mu|$$
$$n(1 - |\mu|) > -1 - m|\mu|$$
$$n > \frac{-1 - m|\mu|}{1 - |\mu|}. \tag{102}$$

Thus, if we take n larger than m and $\dfrac{-1 - m|\mu|}{1 - |\mu|}$, the terms of the series
$$\sum_{k=n}^{\infty} \binom{m}{k}\mu^k$$
decrease in absolute value and either alternate in sign (μ positive) or have the same sign (μ negative) and the size of this sum as the error can be estimated.

The bound (102) is negative for m positive, whence it will be satisfied in that case for $n > 0$ and one need only consider the condition $n > m$. For negative m, however, the right-hand-side of (102) can be a bit large. For the example cited at the end of paragraph 18, $m = -2, \mu = 99/100$, one has
$$\frac{-1 - m|\mu|}{1 - |\mu|} = \frac{-1 - (-2) \cdot \frac{99}{100}}{1 - \frac{99}{100}} = \frac{-100 + 2 \cdot 99}{100 - 99} = \frac{-100 + 198}{1} = 98.$$
Thus the convergence of the series for $(1+99/100)^{-2}$, in the sense of d'Alembert, begins when the exponent reaches 99. Ferraro informs us:

> Some remarks are appropriate. First d'Alembert did not depart from the basic tenets of the 18th century conception: a series was not an autonomous object but the result of a transformation of a given closed analytical expression. Indeed, d'Alembert did not determine the sum of
> $$\sum_{n=0}^{\infty} \binom{m}{n} x^n :$$

He assumed the development of the function $(1 + x)^m$ is $\sum_{n=0}^{\infty} \binom{m}{n} x^n$ (it is to be imagined that, according to d'Alembert, this relation was derived by [the] usual formal methods).

Second, it is true that d'Alembert used the technique of inequalities, but this technique is a tool for numerical evaluation of a function. In no case did he use the technique to prove the existence of a limit. D'Alembert's [sic] *did not know the ratio test*, if by this term we intend a convergence criterion by which we establish if the series has a finite sum. For him the condition

$$\frac{a_{n+1}}{a_n} < 1$$

served to establish where the series approximated its known sum, and $(209)^{40}$ is not used to prove the existence of the sum but was only a procedure to determine the bounds of errors.[41]

Ferraro's introduction to d'Alembert's paper in his history of infinite series begins with the words

In 1768, d'Alembert published an innovative paper, *Réflexions sur les suites et sur les racines imaginaires*; its novel feature was the fact that the problem of approximation was associated with the determination of an explicit error estimate.[42]

All this is true, and the fact that d'Alembert chose the binomial series as his example to illustrate the problem speaks to the importance of the Binomial Theorem and its series, but it doesn't seem to get us any closer to a proof of the Binomial Theorem, and my inclusion of it here may begin to raise doubts about my ability to organise material coherently. To this I say that d'Alembert is relevant to the present chapter for two reasons. First, his emphasis on the notion of limit as the bedrock on which the Calculus is founded, even if he did not apply it directly to series, prepared the way for the eventual proof. Second, even if he did not enunciate "his" Ratio Test, and his bounds in paragraph 22 assume the sum exists, it is hard for a modern mathematician to read this paper and not see the Ratio Test lurking behind that paragraph. D'Alembert came very near to proving the convergence of the binomial series for $(1+x)^q$ for $|x| < 1$ and its divergence for $|x| > 1$. All that was needed was the recognition of the problem. After that, of course, would come the problem of showing that, where $|x| < 1$, the limit of the series was indeed $(1 + x)^q$. This would have to wait another half century and the appearance on the scene of Bernard Bolzano.

7. The New Maths of the 19th Century

The turn of the century, from the 18th to the 19th, marks the beginning of a new era of mathematical rigour. Hermann Hankel, mathematician and

---

[40] An equation in Ferraro's text giving the upper bound of paragraph 22 above.
[41] Ferraro, *op. cit.*, pp. 304 – 305.
[42] *Ibid.*, p. 303.

historian of mathematics, described the transition as it pertained to infinite series in an encyclopædia article on limits:

§19. C r i t i c a l   P e r i o d. The time of the—if I may call it thus—naïve faith in the benign nature of series had celebrated its last hurrah in Lagrange. The critical period, prepared by Lagrange himself, began to be slowly expressed in doubts over the applicability of divergent series, in the stress on various paradoxes from the existing theory. The new scientific consciousness strove for a complete comprehension of the essence of series. In addition, all the properties of infinite series which one had hitherto derived out of the special nature of convergent p o w e r series would be brought into question suddenly by those marvellous series, which to be sure Euler and Lagrange already introduced, but really had first been grasped in their essence by F o u r i e r. One knew that it required a renewed, exact investigation of the fundamental concepts in order to escape from this labyrinth. Fortunately, C a u c h y undertook this task and in 1821 in his famous *Cours d'analyse algébrique* founded the theory of series on the above given simple concept of convergence, by which the sum of a series is the l i m i t of the sums of n members and divergent series appear as insubstantial shadow pictures.[43]

This work was not the first, however, which contained an exact investigation of convergence. Long before the appearance of his epoch-making work G a u s s already had a complete knowledge of the true nature of the subject, about which the famous investigation of the convergence of the hypergeometric series and scattered remarks bear sufficient witness. That he, however, nowhere gave a further development of the principles, so that proof would, like all of his works at the time, be marvelled at but very little studied, and it remained momentarily without influence on the development of the science.

It went even worse for another contemporary who then and today has remained almost entirely unknown among mathematicians: We have to lay down the claim for the p r i o r i t y   o f   t h e   f i r s t   r i g o r o u s   d e v e l o p m e n t   o f   t h e   s e r i e s for the common transcendental functions on behalf of the excellent B e r n h a r d   B o l z a n o who in an unassuming work[44] had given them with unsurpassable sharpness already in 1816. Bolzano's conceptions of the convergence of series are clear and correct throughout, his operations with infinite series all rigorously proven, and there is nothing

---

[43]Here Hankel inserts a footnote: "It is striking that with the influence which the consideration of Fourier series had on the development of the correct basic concepts, C a u c h y could falsely assert in his *Cours d'anal.*, p. 131, the statement that the sum of a convergent power series is continuous, so long as its individual members are—a theorem that would first be expressly rejected by Abel."

[44]A footnote identifies *Der binomische Lehrsatz* (*op. cit.*) as this work.

to find fault with in the development of those series for real arguments, to which he restricted himself. In the preface he gives an apt critique of previous works on the Binomial Theorem and the then customary unrestrained use of infinite series. In short he possessed everything which places him in the same height as Cauchy in this respect, only not the Frenchman's characteristic art to dress up his thoughts and express them in a most agreeable manner. So Bolzano remained unknown and would soon be forgotten; Cauchy was the lucky one whom one lauded as reformer of science and whose elegant writings in short time found general circulation.

The French mathematicians of that time grasped immediately the ideas developed by Cauchy, and the newer German school, as they found their focal point since 1826 particularly in C r e l l e ' s commendable mathematical journal, accepted this rigour in the use of infinite series as an unpleasant necessity. In the first volume of this journal one finds a discourse by A b e l, in which the binomial series is in this way derived for complex values of the exponent and base[45], and through this, as well as through some splendid observations on the convergence of power series, that which Cauchy had already taught would be completed.

What he expressed already in 1826, "The divergent series are in general fatal things, and it i s a d i s g r a c e to base any proof on them" [46], is what has since become the guiding principle of all mathematicians.[47]

Power series are particularly well-behaved and, provided one stays within the radius of convergence, one can treat them as oversized polynomials, differentiating and integrating them term-by-term, though no one had ever actually proved this. In the mid-18th century, however, a new type of series arose. D'Alembert had proposed the problem of analysing the shape of a vibrating string and Euler introduced trigonometric series as possible solutions, with Lagrange furthering their study. In the opening years of the 19th century, Joseph Fourier found them useful in the theory of heat, so much so that these series became known as *Fourier series* in his honour. They were not as well-behaved as power series and their study would be a central concern of analysis well into the 20th century, the resulting theory, Harmonic Analysis, being one of the

---

[45]Hankel adds a footnote: "The derivation of Abel appears still not generally appreciated; for one finds it in abbreviated form in the textbooks, which [form] however in no way possesses sufficient rigour and generality."

[46]Umberto Bottazzini, *The Higher Calculus: A History of Real and Complex Analysis from Euler to Weierstrass*, Springer-Verlag New York, Inc., New York, 1986, p. 87, cites an even stronger formulation of this by Abel: "Divergent series are in their entirety an invention of the devil and it is a disgrace to base the slightest demonstration on them". The third chapter of Bottazzini's book gives an excellent survey of the broader aspects of the history introduced by Hankel.

[47]Hermann Hankel, "Grenze", in: J.S. Ersch and J.G. Gruber (eds.), *Allgemeine Encyklopädie der Wissenschaften und Künste*, vol. 90, F.A. Brockhaus, Leipzig, 1871, pp. 209 – 210.

most beautiful mathematics has to offer. But that is not our concern in the present book. They are mentioned because the difficulties they offered brought into question everything one knew about series. The time was ripe for providing a firm foundation for infinite series. The founders were Gauss, Bolzano, Cauchy, and Abel.

Although D'Alembert had considered limits of functions and had essentially proven the convergence of the binomial series, he had not considered series as functions in their own right but as expansions. Gauss viewed series as functions defined only for those arguments at which they converged. In 1812 he wrote a paper published the following year, *Disquisitiones generales circa seriem infinitum* $1 + \frac{\alpha\beta}{1\cdot\gamma}x + \frac{\alpha(\alpha+1)\beta(\beta+1)}{1\cdot 2\cdot\gamma(\gamma+1)}xx + \frac{\alpha(\alpha+1)(\alpha+2)\beta(\beta+1)(\beta+2)}{1\cdot 2\cdot 3\cdot\gamma(\gamma+1)(\gamma+2)}x^3 +$ etc.[48], on the *hypergeometric function*:

$$1 + \frac{\alpha\beta}{1\cdot\gamma}x + \frac{\alpha(\alpha+1)\beta(\beta+1)}{1\cdot 2\cdot\gamma(\gamma+1)}x^2 + \frac{\alpha(\alpha+1)(\alpha+2)\beta(\beta+1)(\beta+2)}{1\cdot 2\cdot 3\cdot\gamma(\gamma+1)(\gamma+2)}x^3 + \ldots$$

Gauss accepted as known the Ratio Test and noted that the series converged for $|x| < 1$ and diverged for $|x| > 1$. He also gave a rigorous study of the convergence of the series in the case $|x| = 1$, developing a stronger convergence test for this purpose. As Hankel noted, Gauss did not pursue the ideas presented in this paper further and it was left to Bolzano, Cauchy, and Abel to develop the theory of series rigorously. The paper merits mention here, however, because, aside from being a milestone in the discussion of infinite series, the paper relates to the Binomial Theorem in that the hypergeometric series is a family of series that represents many different functions of x by varying the parameters α, β, γ. The binomial series falls out readily enough:

$$F(-n, \beta, \beta, -x) =$$
$$= 1 + \frac{(-n)\beta}{1\cdot\beta}(-x) + \frac{(-n)(-n+1)\beta(\beta+1)}{1\cdot 2\cdot\beta(\beta+1)}x^2$$
$$+ \frac{(-n)(-n+1)(-n+2)\beta(\beta+1)(\beta+2)}{1\cdot 2\cdot 3\cdot\beta(\beta+1)(\beta+2)}x^3 + \ldots$$
$$= 1 + \frac{n}{1}x + \frac{n(n-1)}{2\cdot 1}x^2 + \frac{n(n-1)(n-2)}{3\cdot 2\cdot 1}x^3 + \ldots$$
$$= 1 + \binom{n}{1}x + \binom{n}{2}x^2 + \binom{n}{3}x^3 + \ldots$$

Gauss also gave expressions for $\ln x, \sin x, \cos x, e^x$ and other functions, in each case, however, proving only that the representation equalled the power series

---

[48] I do not know if there is an English translation, but there is a German translation: Carl Friedrich Gauss (Heinrich Simon, translator) *Allgemeine Untersuchungen über die unendliche Reihe* $1 + \frac{\alpha\beta}{1\cdot\gamma}x + \frac{\alpha(\alpha+1)\beta(\beta+1)}{1\cdot 2\cdot\gamma(\gamma+1)}xx + \frac{\alpha(\alpha+1)(\alpha+2)\beta(\beta+1)(\beta+2)}{1\cdot 2\cdot 3\cdot\gamma(\gamma+1)(\gamma+2)}x^3 +$ u.s.w., Verlag von Julius Springer, Berlin, 1888. The translation covers the original paper published in 1813, an announcement published in 1812, and an unpublished fourth section. Bottazzini *op. cit.*, and Ferraro, *op. cit.*, provide summaries in English of Gauss's work, the latter providing greater detail.

expansion and not that the expansion equalled the function in question. Thus, in the binomial case, he showed that the binomial series for $(1+x)^n$ could be expressed as shown, but he did not prove it equalled $(1+x)^n$, a fact he accepted as given.

Of particular relevance here are the works of Bolzano, Cauchy, and Abel cited by Hankel. Bolzano did indeed give the first almost rigorous proof of the Binomial Theorem and, as Hankel notes, hardly anyone noticed. Indeed, it was Hankel's announcement of Bolzano's work in his encyclopædia article that began the slow recognition of Bolzano's stature as a mathematician of the first rank. Bolzano did not consider results in isolation. He was a systematist and his proof of the Binomial Theorem and his rigorous approach to series were part of a general foundational programme the true extent of which would only be revealed over a century later.

Whereas Bolzano was philosophically motivated, Cauchy's motivation appears to have been pædagogical: his important work here was in his textbooks. That Cauchy had a completely rigorous proof of the Binomial Theorem would be denied by later mathematicians because the proof depended on two "false lemmas" and "illegitimate" interchanges of limits. In recent decades, however, it came to be realised that Cauchy's false lemmas are true if one reads his ambiguous definitions of continuity and convergence in the right way: his notions of continuity and convergence are stricter than the modern ones and pretty much coincide with what we call uniform continuity and uniform convergence (both concepts to be defined in the sequel). Thus it was Abel who was generally credited with having given the first correct proof of Newton's Binomial Theorem in a remarkable paper of 1826. We will take a look at the proofs of these three men, Bolzano, Cauchy, and Abel, in the next several sections.

8. Bolzano's Proof

We opened this chapter with a quotation from Bolzano's exposition of the Binomial Theorem. The chronology of the Binomial Theorem prior to this publication of Bolzano in 1816, as discussed here, is this:

1665 Newton discovers the Theorem

1676 Newton describes the Theorem and its genesis in two letters to Leibniz

1744 Euler attempts a proof

1758 Landen attempts a proof

1768 D'Alembert proves in essence that the series converges for $|x| < 1$

1812 Hutton attempts another limit-free proof.

D'Alembert's careful discussion of the closeness of the approximation of partial sums to their limit aside, this has been a history of ignoring the key issue and trying to prove an analytic result through algebraic manipulation. From

Bolzano on, with the exception of Peacock whose attempted proof is best regarded as an anachronistic aberration, this would change. The Binomial Theorem would be seen for what it is—a problem about convergence to a limit—and attacked head on. Simultaneously, the laying of the foundations of the Calculus would begin. The Binomial Theorem would be conquered by Bolzano, Cauchy, and Abel; and the foundations of the Calculus laid by Bolzano, Cauchy, and Karl Weierstrass, with a bit of help from Abel.

In the preface of his monograph on the Binomial Theorem, Bolzano began his criticism of earlier proof attempts with the words

> First of all I think that the *meaning* of the theorem itself has been, I will not say incorrectly understood, but at least not very clearly presented.[49]

In this work Bolzano eschews infinitesimally small numbers and infinite sums, taking the binomial equation

$$(1+x)^n = 1 + \binom{n}{1}x + \binom{n}{2}x^2 + \ldots,$$

not as an assertion of equality, but as an abbreviation for the assertion that the limit of the partial sums of the series on the right equals the value of the expression on the left for any choice of x and n.

Bolzano's monograph is divided into a preface and 75 numbered sections. The preface, from which I have taken the above quote, discusses the problem, including criticism of previous attempts to prove the Theorem. Following the preface, the rest of the work is carefully organised into discussions of a succession of topics. For the most part the succession proceeds thus: proof of the Finite Binomial Theorem, proof that the only possible power series expansion is the binomial (i.e., uniqueness), proof of the Binomial Theorem for $|x| < 1$ (i.e., existence), and miscellaneous topics (Multinomial Theorem, expansions of exponential and logarithmic functions). Not all of this needs to concern us here. Indeed, Bolzano advises the reader not interested in the uniqueness result or the extras to start reading in §38 and stop after §48. Here, we can skip §§1 – 10 which are devoted to a rigorous proof of the Finite Binomial Theorem.

In §11 he begins by considering the binomial equation for $(1+x)^n$ for $n = -1$, i.e., the geometric series. He notes that, since

$$(1+x)^{-1} = \frac{1}{1+x} = 1 - x + x^2 - x^3 + \ldots \pm \frac{x^r}{1+x}, \tag{103}$$

the binomial equation

$$(1+x)^{-1} = \frac{1}{1+x} = 1 - x + x^2 - x^3 + \ldots \pm x^r$$

can only hold when

$$x^r = \frac{x^r}{1+x},$$

i.e., when $x = 0$:

---

[49]Russ, *op. cit.*, p. 157.

8. BOLZANO'S PROOF

We therefore see from this example that the binomial equation certainly does not hold for every value of n and x.

§12

Corollary 2. But if x is a *proper fraction*, then the remarkable situation occurs that the binomial series $1 - x + x^2 - x^3 + \cdots \pm x^r$ can be brought as close to the value $(1+x)^{-1}$ as desired, merely by sufficiently increasing its [number of] terms.[50]

There follows a marvellously rigorous proof for $0 < x < 1$ that
$$\lim_{r \to \infty} \left(1 - x + x^2 - x^3 + \ldots \pm x^r\right) = \frac{1}{1+x}.$$

By (103),
$$\left|\frac{1}{1+x} - \left(1 - x + x^2 - x^3 + \ldots \pm x^r\right)\right| = \left|x^r - \frac{x^r}{1+x}\right| = \frac{x^{r+1}}{1+x}.$$

To make this less than D (his notation—we used ϵ in Definition 3.6.1 on page 173, above), he shows it suffices to take
$$r > \frac{\frac{1}{D(1+x)} - 1}{u} - 1,$$

where $u = 1/x - 1$. He then notes that almost the same proof works when $-1 < x < 0$. Thus he has proven, in modern terms and with modern rigour, that the sum of the geometric series $1 - x + x^2 - \ldots$ is $1/(1+x)$ when $|x| < 1$.

I find this proof marvellous because I had previously read Bolzano's initially unfathomable proof of the same result in his posthumously published *Paradoxien des Unendlichen*[51] [*Paradoxes of the Infinite*]. When the third edition was published by Alois Höfler (ed.) with annotations by analyst Hans Hahn around 1920[52], Hahn criticised the proof given there in no uncertain terms:

> The definition given here of the concept of s u m is so abstract, and so lacking in clarity, that it is difficult to fix its meaning precisely.[53]

> One will also have objected to B.'s execution of his proof, that the sums trotted out in it are tied to no precise concept, and that the computations carried out with them... are grounded on nothing.[54]

Hahn then felt compelled to present the accepted definition of the sum as a limit along with a proof that the limit of the geometric series was as declared.

Bolzano's baffling proof from the *Paradoxien* actually is grounded on something, namely his unfinished work on the foundations of the Calculus. Bolzano

---

[50] Russ, *op. cit.*, p. 171. Here, by "proper fraction", Bolzano means that $|x| < 1$; he is not assuming x is rational.

[51] Fr. Přihonský (ed.), *Dr. Bernard Bolzanos Paradoxien des Unendlichen*, C.H. Reclam, Leipzig, 1851.

[52] Bernard Bolzano, *Paradoxien des Unendlichen*, Verlag von Felix Meiner, Leipzig. The dates given on the title page (1920) and cover (1921) disagree slightly.

[53] *Ibid.*, p. 134.

[54] *Ibid.*, p. 137.

carried out this work in the 1830s, but he ran into technical difficulties he never resolved, and it was not published until the 1960s (in more definitive form in the 1970s) and his genius has only begun to be appreciated in this past half century.

But we are with Bolzano in 1816 and he tried to avoid infinite sums altogether by referring directly to the partial sums as approximations.

In §§13 – 22 Bolzano formally introduces the notion of a quantity "which can become smaller than any given quantity" or "which can become as small as desired". Such quantities serve as a sort of shorthand for discussing limits of functions. To understand these quantities, we should first look at the formal definitions of limit and continuity in their modern formulations.

3.8.1. Definition. *Let f be a function defined everywhere near a point a, but not necessarily at a, i.e., f is defined for all $x \neq a$ in some interval (α, β) with $a \in (\alpha, \beta)$. The number L is the limit of $f(x)$ as x approaches a, written*

$$\lim_{x \to a} f(x) = L,$$

if, for every $\epsilon > 0$ there is a $\delta > 0$ such that for all $x \in (\alpha, \beta)$,

$$0 < |x - a| < \delta \implies |f(x) - L| < \epsilon. \tag{104}$$

This is analogous to the definition of the limit of a sequence. The difference is that in sequences n approached ∞ by getting larger than some determined n_0 and here x approaches a by being no farther distant from a than δ. The new feature is the inequality $0 < |x - a|$, i.e., the assumption $x \neq a$. This corresponds to the fact that n can never equal ∞. The important thing here is the behaviour of $f(x)$ as x approaches a, not its behaviour at a itself. This is best explained by considering a simple example.

3.8.2. Example. *Let $f(x) = \dfrac{x^2 - 1}{x - 1}$.* This function is undefined at $x = 1$, but it has a limit as x approaches 1. In fact, if one graphs it on the calculator using the ZDecimal setting, one will see a straight line $y = x + 1$. Hitting the Trace button and moving to the right one sees successively at the bottom of the screen

    X=0   Y=1

    X=.1   Y=1.1

    ⋮

    X=.8   Y=1.8

    X=.9   Y=1.9

    X=1   Y=

    X=1.1 Y=2.1

The lack of any value for Y at X=1 reflects the fact that $f(1)$ is undefined. However, one can see that as x tends to 1, the value of $f(x)$ is tending steadily to 2.

To verify that $\lim_{x \to 1} f(x) = 2$, let ϵ be given and solve the inequality

$$\left| \frac{x^2 - 1}{x - 1} - 2 \right| < \epsilon.$$

To this end, note that for $x \neq 1$,

$$\frac{x^2 - 1}{x - 1} - 2 = x + 1 - 2 = x - 1,$$

and one will have $|f(x) - 2| < \epsilon$ for $x \neq 1$ so long as $|x - 1| < \epsilon$. I.e., we can choose $\delta = \epsilon$.

Requiring the implication (104) to hold also when $x = a$ would imply that $|f(a) - L| < \epsilon$ for all $\epsilon > 0$, i.e., $f(a) = L$. This is the definition of continuity:

3.8.3. Definition. *Let f be a function defined for all x in an interval (α, β). f is continuous at a if $\lim_{x \to a} f(x) = f(a)$, i.e., if, for all $\epsilon > 0$ there is a $\delta > 0$ such that for all $x \in (\alpha, \beta)$,*

$$|x - a| < \delta \Rightarrow |f(x) - f(a)| < \epsilon. \tag{105}$$

In proving that a function is continuous at a point or that the limit is some given number, one has, given $\epsilon > 0$, to find a $\delta > 0$ making (105) or (104) true. This is not always as trivial as in Example 3.8.2.

3.8.4. Example. For $a > 0$, $\lim_{x \to a} \dfrac{1}{x} = \dfrac{1}{a}$. To see this, let $\epsilon > 0$ and consider

$$\left| \frac{1}{x} - \frac{1}{a} \right| = \left| \frac{a - x}{ax} \right| = \frac{|a - x|}{|ax|} = \frac{|x - a|}{ax}$$

for $x > 0$. If we suppose, in fact, that $x > a/2$, we will have

$$\frac{|x - a|}{ax} < \frac{|x - a|}{a} \cdot \frac{2}{a} < \epsilon$$

provided $|x - a| < \dfrac{a^2}{2}\epsilon$. Thus choose $\delta = \min\{a^2\epsilon/2, a/2\}$ (the latter to make $x > a/2$).

The construction of ϵ-δ-proofs, often jokingly called *epsilontics*, can be a bit intricate and algebraically demanding. In my experience, most students in a freshman level Calculus course can master proofs for functions like $f(x) = \dfrac{a}{x + b}$ or $f(x) = a(x+b)^2$ for specific values of a, b, but only a few can go beyond this. Students who have not been forced to do such proofs in the Calculus course are often incapable of mastering even these on their introduction to rigour in an upper division Analysis course.

Such precise, strict definitions of limit and continuity were missing in the early decades of the Calculus. We read d'Alembert's less formal definition of limit in the preceding section that came about a century after Newton took up the subject. Bolzano and around 1820 Cauchy would come close to the precise definition, which would finally be perfected by Weierstrass in lecturing on the subject to his students. In place of such definitions, mathematicians resorted to the use of *infinitesimals*—quantities smaller in absolute value than any positive real number. Although such quantities were not real numbers themselves, one would calculate with them the way one would with ordinary real numbers. And there were additional helpful rules: the sum of any finite number of infinitesimals is infinitesimal (or 0), the product of an infinitesimal and a finite (non-zero) real number was again an infinitesimal. In terms of infinitesimals, Definition 3.8.1 would read:

$$\lim_{x \to a} f(x) = L \text{ iff } f(x + \omega) \text{ is infinitesimally close to } L$$

$$\text{for any infinitesimal } \omega$$

And one would prove theorems like:

3.8.5. Example. *Let, for* $i = 1, 2$, $\lim_{x \to a} f_i(x) = L_i$. *Then* $\lim_{x \to a} (f_1(x) + f_2(x)) = L_1 + L_2$.

Proof. Let ω be infinitesimal, find infinitesimals Ω_1 and Ω_2 such that $f_i(x + \omega) = L_i + \Omega_i$, and observe

$$f_1(x + \omega) + f_2(x + \omega) = (L_1 + \Omega_1) + (L_2 + \Omega_2)$$
$$= (L_1 + L_2) + (\Omega_1 + \Omega_2) = (L_1 + L_2) + \Omega,$$

where $\Omega = \Omega_1 + \Omega_2$, *being the sum of two infinitesimals, is again an infinitesimal.* □

Early in his mathematical development Bolzano eschewed the infinite, both the infinitely large and infinitely small (infinitesimals). His definition of limit, however, referred to the arbitrarily small and, instead of discovering epsilontics as, with our modern training, we would consider the most natural way to deal with such, he replaced infinitesimals by certain *variable quantities* which can become as small as we please. He denoted them by ω and Ω, with and without indices[55]. Bolzano uses the lower case ω to denote those quantities one assumes can be made as small as one chooses, and the upper case Ω for those quantities one concludes can be made as small as desired by making the ω's sufficiently small (or n's sufficiently large).

Following their introduction, Bolzano lays down some simple rules for calculating with ω's and Ω's. For example §15 states the

> *Lemma* If each of the quantities $\omega, \omega_1, \omega_2, \ldots, \omega_m$ can become as small as desired while the (finite) number of them does not alter, then their algebraic *sum* or *difference* is also a quantity which can become as small as desired, i.e.
>
> $$\omega \pm \omega_1 \pm \omega_2 \pm \ldots \pm \omega_m = \Omega.$$
>
> *Proof.* For if the *sum* of these quantities is to be $< D$, where D designates some finite quantity, then if there is a constant number n of them, each of them may be taken $< \frac{D}{n}$, which is possible as a

---

[55] His indices are numbers placed above ω or Ω. Here I shall use ordinary subscripts to save space and some small amount of labour typesetting.

8. BOLZANO'S PROOF

consequence of the assumption. Then certainly $\omega \pm \omega_1 \pm \omega_2 \pm \ldots \pm \omega_m < D$, even if the terms of this sum should all be positive, and all the more so in any other case.[56]

This and other familiar properties of infinitesimals are established for his arbitrarily small variable quantities in §§15 – 22, paving the way for their application.

In §23 Bolzano determines the derivative of x^n:

Lemma. *The quantity*
$$\frac{(x+\omega)^n - x^n}{\omega}$$
can be brought as close to the value nx^{n-1} as desired, if ω is taken small enough: n and x may denote whatever desired, provided x is not $= 0$. That is,
$$\frac{(x+\omega)^n - x^n}{\omega} = nx^{n-1} + \Omega.\ [57]$$

The restriction that $x \neq 0$ is only necessary for $n < 1$ where x^{n-1} would fail to be defined at $x = 0$. Since Bolzano is interested in real numbers, he should also disallow negative values of x when n is rational with odd numerator and even denominator. Or he could add the clause, "provided x^{n-1} is defined".

Bolzano's proof is fairly rigorous when rational exponents are allowed. Unlike Landen, Euler, and Hutton, he also considers irrational exponents and here his proof, or at least his exposition, is lacking.

Proof of the Lemma. (The reader might wish to replace the proofs of cases 2 and 3 on a first reading by the simpler proof given on pages 207 – 208, below.) Let x, ω, n be given as above.
1. If $n = 0$, the result is trivial.
2. Let $n = p/q$, where p, q are positive integers. Write
$$(x+\omega)^n = (x+\omega)^{p/q} = x^{p/q}\left(1 + \frac{\omega}{x}\right)^{p/q} \quad (106)$$
and set $\omega_1 = \omega/x$ to simplify notation. ω_1, being the product of a quantity that can be made as small as we please and a finite number, is itself a quantity that can be made as small as we please[58].

3.8.6. Sublemma. *For any positive integer k and any real number y,*
$$(1+y)^k = 1 + u, \quad \text{where} \quad \begin{cases} u > 0, & \text{if } y > 0 \\ -1 < u < 0, & \text{if } -1 < y < 0. \end{cases}$$

Proof. By induction on $k \geq 1$.
For $k = 1$, $(1+y)^k = 1 + y$, whence $u = y$.

---

[56] Russ, *op. cit.*, p. 173.
[57] *Ibid.*, p. 176.
[58] I really wish Bolzano had accepted infinitesimals: "infinitesimal" is so much simpler to state than "a quantity that can be made as small as we please".

For the induction step, note that
$$(1+y)^{k+1} = (1+y)^k(1+y) = (1+u)(1+y) = 1+u+y+uy.$$
If $y > 0$, then $u' = u + y + uy$ is the sum of three positive numbers and is positive itself.

If $-1 < y < 0$, $u' = y + (1+y)u$ is negative since $y, u < 0$ and $1+y > 0$. It cannot be less than -1 since then $(1+y)^{k+1}$ would be negative, and thus a negative product of two positive numbers. □

Returning to the second stage of the proof of the Lemma, assume $\omega_1 > 0$, i.e., ω, x are both positive or both negative.

3.8.7. Sublemma. $(1+\omega_1)^{p/q} = 1 + \Omega$, i.e., $(1+\omega_1)^{p/q} = 1 + u$ where $u > 0$ can be made as small as desired.

Proof. Note that if $(1+\omega_1)^{p/q} \leq 1$, then
$$(1+\omega_1)^{p/q} = \frac{1}{1+u} \quad \text{for some } u \geq 0$$
$$(1+\omega_1)^p = \frac{1}{(1+u)^q}.$$

By Sublemma 3.8.6, $(1+\omega_1)^p > 1$ and $(1+u)^q > 1$, whence $1 > 1/(1+u)^q$. But then
$$(1+\omega_1)^p > 1 > \frac{1}{(1+u)^q} = (1+\omega_1)^p,$$
a contradiction. Thus $(1+\omega_1)^{p/q} > 1$ and we can write $(1+\omega_1)^{p/q} = 1+u$, for some $u > 0$.

To see that u can be made as small as we please, let $\epsilon > 0$ be given[59]. To make $u < \epsilon$, it suffices to make $(1+\omega_1)^{p/q} < 1+\epsilon$, i.e.,
$$(1+\omega_1)^p < (1+\epsilon)^q. \tag{107}$$

Expand the left power of (107) using the Finite Binomial Theorem:
$$1 + p\omega_1 + \binom{p}{2}\omega_1^2 + \ldots + \binom{p}{p}\omega_1^p < (1+\epsilon)^q$$
$$\omega_1\left(p + \binom{p}{2}\omega_1 + \ldots + \binom{p}{p}\omega_1^{p-1}\right) < (1+\epsilon)^q - 1.$$

Now choose $\omega_2 < \omega_1$ satisfying
$$\omega_2 < \frac{(1+\epsilon)^q - 1}{p + \binom{p}{2}\omega_1 + \ldots + \binom{p}{p}\omega_1^{p-1}}.$$
Then
$$\omega_2\left(p + \binom{p}{2}\omega_2 + \ldots + \binom{p}{p}\omega_2^{p-1}\right)$$

---

[59]Bolzano uses D, but ϵ is so familiar in this context to a modern mathematician that I feel it conveys the meaning better. I am following the spirit of Bolzano's exposition, but not the letter.

$$< \omega_2 \left(p + \binom{p}{2}\omega_1 + \ldots + \binom{p}{p}\omega_1^{p-1} \right) < (1+\epsilon)^q - 1,$$

and
$$(1+\omega_2)^p < (1+\epsilon)^q.$$

This yields
$$(1+\omega_2)^{p/q} = 1 + \Omega_1, \tag{108}$$

as claimed, where we choose ω_2 to replace ω_1 (and thus we choose $\omega' = x\omega_2$ to replace $\omega = x\omega_1$) □

Back to the Lemma, we have (108), where we have replaced u by Ω_1 to remind us notationally that it is a variable number that can be made as small as desired. Thus
$$(1+\omega_2)^p = (1+\Omega_1)^q$$

and, expanding both sides, we have

$$1 + p\omega_2 + \binom{p}{2}\omega_2^2 + \ldots + \binom{p}{p}\omega_2^p = 1 + q\Omega_1 + \binom{q}{2}\Omega_1^2 + \ldots + \binom{q}{q}\Omega_1^q$$

$$\omega_2 \left(p + \binom{p}{2}\omega_2 + \ldots + \binom{p}{p}\omega_2^{p-1} \right) = \Omega_1 \left(q + \binom{q}{2}\Omega_1 + \ldots + \binom{q}{q}\Omega_1^{q-1} \right)$$

$$\frac{p + \binom{p}{2}\omega_2 + \ldots + \binom{p}{p}\omega_2^{p-1}}{q + \binom{q}{2}\Omega_1 + \ldots + \binom{q}{q}\Omega_1^{q-1}} = \frac{\Omega_1}{\omega_2}. \tag{109}$$

Applying the rules of §§15 – 22 for dealing with variable quantities that..., Bolzano concludes the left-hand side of (109) to be $p/q + \Omega_2$, i.e.,

$$\frac{\Omega_1}{\omega_2} = \frac{p}{q} + \Omega_2.$$

We are almost finished with the present case. Recalling $\omega' = x\omega_2$,

$$\frac{(x+\omega')^{p/q} - x^{p/q}}{\omega'} = x^{p/q}\frac{(1+\omega_2)^{p/q} - 1}{x\omega_2} = x^{p/q-1} \cdot \frac{\Omega_1}{\omega_2}$$

$$= x^{p/q-1}\left(\frac{p}{q} + \Omega_2\right)$$

$$= \frac{p}{q}x^{p/q-1} + x^{p/q-1}\Omega_2 = \frac{p}{q}x^{p/q-1} + \Omega,$$

since the product of the finite $x^{p/q-1}$ with the arbitrarily small Ω_2 is itself arbitrarily small.

This completes the proof in the case where $\omega_1 = \omega/x$ is positive. Bolzano does not treat the case in which ω_1 is negative, other than to say that the proof is similar, which indeed it is.

204   3. THE BINOMIAL THEOREM PROVEN

3. Let n be a negative rational number, $n = -p/q$ for some positive integers p, q. Then[60]
$$(x+w)^n = (x+w)^{-p/q} = x^{-p/q}(1+w/x)^{-p/q} = x^{-p/q}(1+\Omega_1)^{-1},$$
using the same notation as before, prior to replacing w by w' and w_1 by w_2. Therefore
$$\frac{(x+w)^{-p/q} - x^{-p/q}}{w} = \frac{x^{-p/q}(1+\Omega_1)^{-1} - x^{-p/q}}{w}$$
$$= x^{-p/q}\frac{1-(1+\Omega_1)}{w} \cdot \frac{1}{1+\Omega_1}$$
$$= x^{-p/q}\frac{-\Omega_1}{w} \cdot \frac{1}{1+\Omega_1} = x^{-p/q}\frac{-\Omega_1}{w_1 x} \cdot \frac{1}{1+\Omega_1}$$
$$= -x^{-p/q}\frac{p + \binom{p}{2}w_1 + \ldots + \binom{p}{p}w_1^{p-1}}{x\left(q + \binom{q}{2}\Omega_1 + \ldots + \binom{q}{q}\Omega_1^{q-1}\right)} \cdot \frac{1}{1+\Omega_1}$$
$$= -x^{-p/q-1} \cdot \frac{p}{q} + \Omega_2 = nx^{n-1} + \Omega,$$
again using the obvious arithmetic rules involving w's and Ω's.

4. The irrational case is not well-handled by Bolzano. His approach is indicated by the opening lines:

> Finally, if n is *irrational* then there is always a fraction $\frac{p}{q}$ (positive or negative) which comes as close as required to n. But then it follows from the definition of the *concept* of an irrational power that the quantity $a^{p/q}$ gives a value as close to that value of a^n as desired, if $\frac{p}{q}$ comes as close to the value n as desired.[61]

The idea is that for each x the function $f(m) = x^m$ is continuous on the rationals and it can be extended by continuity to the real numbers. Thus, if n is irrational, one would define
$$f(m) = \lim_{p/q \to m} f(p/q), \ p/q \text{ rational.}$$
He would then use this fact to show
$$\frac{(x+w)^n - x^n}{w} = \frac{(x+w)^{p/q} - x^{p/q}}{w} + \Omega_1 = \frac{p}{q}x^{p/q-1} + \Omega_2 = nx^{n-1} + \Omega_3.$$
The definition of f and the justification of this last string of equations are matters of greater subtlety than Bolzano imagined. His proof in this case is easily corrected with our modern understanding of this subtlety, so Bolzano deserves a good deal of partial credit for his incomplete proof.

The subtle point concerns the notion of *uniform continuity*. We already know what it means for a function on an interval to be continuous at a point a. The obvious definition of continuity of a function on the interval is that it

---

[60] Bolzano (*op. cit.*, p. 23) neglects the x in the denominator of the third term. Russ (*op. cit.*, p. 178) reproduces Bolzano's oversight without comment.

[61] Russ, *op. cit.*, p. 178.

be continuous at all points thereof. We can generalise this to allow functions not defined throughout the interval, e.g., defined only for rational arguments:

3.8.8. Definition. *Let f be a function defined on a set I of real numbers. f is continuous on I if, for all $a \in I$ and all $\epsilon > 0$ there is a $\delta > 0$ such that for all $x \in I$,*
$$|x - a| < \delta \implies |f(x) - f(a)| < \epsilon.$$

A function can be continuous on the rational numbers and fail to have a continuous extension to the reals.

3.8.9. Example. *The function $f(x) = \dfrac{1}{x^2 - 2}$ is continuous at all rational numbers, but has no limit at $x = \sqrt{2}$.*

Guaranteeing a function from the rational numbers can be extended to a continuous function on the reals by taking limits requires the stronger condition of uniform continuity:

3.8.10. Definition. *Let f be a function defined on a set I of real numbers. f is uniformly continuous on I if, for all $\epsilon > 0$ there is a $\delta > 0$ such that for all $a, x \in I$,*
$$|x - a| < \delta \implies |f(x) - f(a)| < \epsilon.$$

The density of the rationals in the reals guarantees that every irrational number n is the limit of rationals p/q and the uniform continuity of f guarantees the limit of $f(p/q)$ exists. Now, I do not wish to go into detail on the subject here. The details will be given in the next section. I wish only to explain that i. the "*concept* of an irrational power" relies for its definition on the uniform continuity in an interval of exponentiation involving rational powers, and ii. Bolzano's proof of his Lemma on differentiation of powers makes an unannounced use of uniformity.

I might make one last remark on the subject before returning to the proof of Bolzano's Lemma. If we look at the positioning of the quantifiers in the two definitions,

continuity: $\forall a \, \forall \epsilon \, \exists \delta \, \forall x$
uniform continuity: $\forall \epsilon \, \exists \delta \, \forall a \, \forall x$,

we see that the choice of δ in the latter case is uniform for all a: it doesn't depend on a. Stated differently, the function producing δ, called the *modulus of continuity*, takes both a and ϵ as arguments in the continuous case—thus $\delta(a, \epsilon)$—, while it has only the argument ϵ in the uniformly continuous case—thus $\delta(\epsilon)$.

The necessity of assuming uniformity went unnoticed by mathematicians of the day and minor controversies arose among later mathematical historians as they debated among themselves whether certain proofs were incorrect because they used uniformity without mention or they were indeed correct because the authors of the proofs intended uniformity even though they didn't use the word. Definitions of limit and continuity were often written in natural language, which is ambiguous on this point. Bolzano's use of variable numbers and Cauchy's later use of infinitesimals do not dispel this ambiguity. It has been claimed that Bolzano later understood the distinction between continuity and uniform continuity and that Cauchy only considered uniform continuity. Be that as it may, Bolzano's exposition fails to make the distinction and rests on uniform continuity.

Let us now, finally, return to Bolzano's proof. He immediately follows our last citation with the words:

> Therefore $(x + \omega)^n = (x + \omega)^{p/q} + \Omega$ and $x^n = x^{p/q} + \Omega_1$ where Ω and Ω_1 can become as small as desired for the same x and ω (merely by changing p/q). Therefore also in
> $$\frac{(x+\omega)^n - x^n}{\omega} = \frac{(x+\omega)^{p/q} - x^{p/q}}{\omega} + \frac{\Omega - \Omega_1}{\omega}$$
> the term $\dfrac{\Omega - \Omega_1}{\omega}$ will, by §15 and §17, be able to become as small as desired. But by what has just been proved,
> $$\frac{(x+\omega)^{p/q} - x^{p/q}}{\omega}$$
> comes as close to the value $\frac{p}{q}.x^{p/q-1}$ as desired.[62]

This completes the proof. □

This last part of the proof is an expositional morass. Why can $(\Omega - \Omega_1)/\omega$ become as small as we please? $\Omega - \Omega_1$ can become small, but so can ω, making the ratio something like $0/0$. One must order the limit operations more carefully than Bolzano has. First, choose δ_1 such that, for ϵ_1 to be determined,
$$\left|\frac{p}{q} - n\right| < \delta_1 \Rightarrow |y^n - y^{p/q}| < \epsilon_1$$
for all y in some neighbourhood of x (uniform continuity is used here), say for $|y - x| < \delta_2$. We can also choose δ_3 so that
$$0 < |y - x| < \delta_3 \Rightarrow \left|\frac{y^{p/q} - x^{p/q}}{y - x} - \frac{p}{q}x^{p/q-1}\right| < \epsilon_2.$$
Now let $|\omega| < \min\{\delta_2, \delta_3\}$ and $y = x + \omega$, and choose δ_1 for some $\epsilon_1 < \omega^2$:
$$\left|\frac{\Omega - \Omega_1}{\omega}\right| < \frac{2\omega^2}{|\omega|} < 2|\omega| = \Omega_2.$$

And
$$\frac{(x+\omega)^n - x^n}{\omega} = \frac{(x+\omega)^{p/q} - x^{p/q}}{\omega} + \Omega_3, \text{ with } |\Omega_3| < \Omega_2$$
$$= \left(\frac{p}{q}x^{p/q-1} + \Omega_4\right) + \Omega_3$$
$$= (nx^{n-1} + \Omega_5) + \Omega_4 + \Omega_3$$
$$= nx^{n-1} + \Omega_6.$$

---

[62]Russ, op. cit., p. 178.

8. BOLZANO'S PROOF

So we can say that Bolzano proved the differentiation rule for x^n for rational values of n, and had most of the ingredients for putting together a proof for irrational values of n, but did not completely succeed in this last task.

We can see in §23 what Hankel meant (page 193, above) about Bolzano's inability to express his thoughts in the most agreeable manner. Instead of offering a direct statement of the Lemma,

$$\frac{d}{dx}x^n = nx^{n-1},$$

he avoided reference to the infinitary procedure of differentiation through his translation into the language of variable quantities,

$$\frac{(x+\omega)^n - x^n}{\omega} = nx^{n-1} + \Omega.$$

His use, however, of variable quantities that can be made as small as desired was not that different from the way in which his contemporaries used infinitesimals, so this may not have been so great an expository weakness then as it would be now. The proof itself, however, could have been laid out more intelligibly:

1. The case $n = 0$ is trivial and one would today copy Bolzano's statement of this fact.

2. The case for a positive rational exponent would be split.

2a. First one would assume n to be a positive integer p and repeat the calculation given in Chapter 2 on page 84 to establish directly that $dx^p/dx = px^{p-1}$.

2b. For $n = p/q$, with p, q both positive integers, one would rewrite the difference quotient,

$$\frac{(x+\omega)^{p/q} - x^{p/q}}{\omega} = \frac{((x+\omega)^p)^{1/q} - (x^p)^{1/q}}{\omega},$$

and use the identity

$$u^q - v^q = (u-v)(u^{q-1} + u^{q-2}v + \ldots + uv^{q-2} + v^{q-1})$$

to obtain

$$\frac{(x+\omega)^{p/q} - x^{p/q}}{\omega} = \frac{(x+\omega)^p - x^p}{\omega} \cdot$$

$$\frac{1}{(((x+\omega)^p)^{1/q})^{q-1} + (((x+\omega)^p)^{1/q})^{q-2}(x^p)^{1/q} + \ldots + ((x^p)^{1/q})^{q-1}}.$$

One would now take the limit as $\omega \to 0$:

$$\frac{d}{dx}x^{p/q} = \frac{d}{dx}x^p \cdot \frac{1}{(x^{p/q})^{q-1} + (x^{p/q})^{q-2}x^{p/q} + \ldots + (x^{p/q})^{q-1}}$$

$$= px^{p-1} \cdot \frac{1}{q(x^{p/q})^{q-1}} = \frac{p}{q}x^{p-1-(p/q)(q-1)}$$

$$= \frac{p}{q}x^{p-1-p+p/q} = \frac{p}{q}x^{p/q-1}.$$

We could, of course, have appealed to the Chain Rule, as we did in Chapter 2, but, insofar as the point here is to work from first principles, this exposition incorporates the proof of that Rule in this special case. This happens again in the next case.

3. For $n = -p/q$, write $n = -m$ for m positive and observe

$$\frac{(x+w)^{-m} - x^{-m}}{w} = \frac{1}{w}\left(\frac{1}{(x+w)^m} - \frac{1}{x^m}\right)$$

$$= \frac{1}{w} \cdot \frac{1}{(x+w)^m x^m}(x^m - (x+w)^m)$$

$$= \frac{-1}{(x+w)^m x^m} \cdot \frac{(x+w)^m - x^m}{w}.$$

Taking the limit as $w \to 0$ yields

$$\frac{d}{dx}x^{-m} = \frac{-1}{x^m x^m} \cdot \frac{dx^m}{dx} = -\frac{1}{x^{2m}}mx^{m-1} = -mx^{-m-1} = nx^{n-1}.$$

4. For n irrational, one would proceed as explained above following the end of Bolzano's garbled treatment of this case.

The rest of Bolzano's treatment of the determination of the coefficients of the binomial expansion assuming $(1+x)^n$ can be expanded into a power series is harder to read. Essentially he wants to show that a power series can be differentiated term-by-term, but he fails to do so. The key step in this is the following false Lemma of §29:

> *Lemma.* Suppose a function of x, $F_r x$ of arbitrarily many terms, formed according to a particular rule, has the property that either for all x or all x within certain limits a and b, it can become as small as desired, merely by increasing its number of terms r. Suppose furthermore that $f_r x$ denotes a second function of the same arbitrary number of terms, which depends on the former in such a way that for every value of x within a and b, the equation
>
> $$\frac{F_r(x+w) - F_r x}{w} = f_r x + \Omega$$
>
> holds, in which Ω can become as small as desired if the same holds for w. Then I claim that the function f_r also has the property, that it can become as small as desired for the same values of x as for $F_r x$, if its number of terms r is taken large enough.[63]

In plain English, Bolzano claims that, if

i. $F_r(x) = \sum_{k=0}^{r} G_k(x),$

ii. $\lim_{r \to \infty} F_r(x) = 0$, i.e., $\sum_{k=0}^{\infty} G_k(x) = 0$

iii. for each k, $G'_k(x) = g_k(x)$, and

iv. $f_r(x) = \sum_{k=0}^{r} g_k(x),$

---

[63] *Ibid.*, p. 181.

then $\lim_{r\to\infty} f_r(x) = \lim_{r\to\infty} \sum_{k=0}^{r} g_k(x) = 0$, i.e., $\sum_{k=0}^{\infty} g_k(x) = 0$.

This is false in general.

3.8.11. Example. *We first find a sequence $F_0, F_1, F_2, \ldots$ of differentiable functions with derivatives $f_0, f_1, f_2, \ldots$ such that for all x*

$$\lim_{r\to\infty} F_r(x) = 0, \text{ but } \lim_{r\to\infty} f_r(0) \neq 0.$$

The definition is very simple:

$$F_r(x) = \frac{x}{1+rx^2}.$$

Note that $F_r(0) = 0$ and, for $x \neq 0$,

$$F_r(x) = \frac{x}{1+rx^2} = \frac{\frac{x}{r}}{\frac{1}{r}+x^2} \to \frac{0}{0+x^2} = 0, \text{ as } r \to \infty.$$

Thus $\lim_{r\to\infty} F_r(x) = 0$ for all x. Now

$$f_r(x) = \frac{d}{dx} F_r(x) = \frac{1 \cdot (1+rx^2) - x(r \cdot 2x)}{(1+rx^2)^2} = \frac{1 - rx^2}{(1+rx^2)^2},$$

whence $f_r(0) = \frac{1-0}{(1+0)^2} = 1 \neq 0$.

To obtain a series out of this, define

$$G_0(x) = F_0(x)$$
$$G_{k+1}(x) = F_{k+1}(x) - F_k(x),$$

and observe

$$\sum_{k=0}^{r} G_k(x) = F_0(x) + (F_1(x) - F_0(x)) + \ldots + (F_{r-1}(x) - F_{r-2}(x))$$
$$+ (F_r(x) - F_{r-1}(x))$$
$$= (F_0(x) - F_0(x)) + (F_1(x) - F_1(x)) + \ldots + (F_{r-1}(x) - F_{r-1}(x))$$
$$+ F_r(x)$$
$$= F_r(x).$$

Thus, if we define $g_k(x) = G'_k(x)$, we have

$$\sum_{k=0}^{r} g_k(x) = \sum_{k=0}^{r} \frac{dG_k(x)}{dx} = \frac{d}{dx} \sum_{k=0}^{r} G_k(x) = \frac{d}{dx} F_r(x) = f_r(x),$$

and

$$\sum_{k=0}^{r} g_k(0) = f_r(0) = 1 \neq 0.$$

3.8.12. Exercise. *Show that the sequence*
$$F_r(x) = \frac{\sin rx}{\sqrt{r}}$$
provides another counterexample.

Bolzano wanted to apply this Lemma to an assumed power series representation of the binomial function,
$$(1+x)^n = \sum_{k=0}^{\infty} a_k x^k, \qquad (110)$$
thus to
$$F_r(x) = (1+x)^n - \sum_{k=0}^{r} a_k x^k$$
$$f_r(x) = n(1+x)^{n-1} - \sum_{k=0}^{r} k a_k x^{k-1}.$$
Assuming (110), i.e.,
$$\lim_{r \to \infty} F_r(x) = 0,$$
he would thus conclude
$$\lim_{r \to \infty} f_r(x) = 0,$$
i.e.,
$$n(1+x)^{n-1} = \sum_{k=0}^{\infty} k a_k x^{k-1},$$
i.e., one can differentiate the right-hand side of (110) term-by-term. Thus, what Bolzano intended by his Lemma of §29 was the following:

Lemma. Suppose $G(x) = \sum_{k=0}^{\infty} G_k(x)$ for all x between a and b, and G and each G_k are differentiable for such x. Then
$$G'(x) = \sum_{k=0}^{\infty} G'_k(x).$$

It is true that one can differentiate power series term-by-term, but it is not generally true that the derivative of a differentiable function that can be expressed as the sum of an infinite series of differentiable functions is given by differentiating the series term-by-term. Since the result is false, Bolzano's proof must be incorrect. The correct result, as stated in most textbooks is as follows:

3.8.13. Theorem. *Let $G_0, G_1, G_2, \ldots$ be a sequence of functions differentiable on an interval $[a, b]$. Suppose the series $\sum_{k=0}^{\infty} G'_k(x)$ converges uniformly on $[a, b]$ to some function and $\sum_{k=0}^{\infty} G_k(x)$ converges at at least one point $x_0 \in [a, b]$.*

Then: $\sum_{k=0}^{\infty} G_k(x)$ converges uniformly on $[a,b]$ to a function $G(x)$ and $G'(x) = \sum_{k=0}^{\infty} G'_k(x)$.

Bolzano lacks the concept of uniform convergence (which I will define shortly) and the assumption that the series of derivatives converges. A careful study of his attempted proof should show whether or not these are hidden assumptions he overlooked. If so his proof would become correct when the assumptions are made explicit. If this is the case, it will take a better mind than mine to verify the fact. The changing values of ω and subscripts n are confusing and make it hard to determine—as was the case with his proof of the Lemma of §23—the interdependencies of the limits. (Do the choices of ω and n depend on each other?) What I can say is that if one sorts through the proof, the elements are there for a correct proof of a weaker result, one yet sufficiently strong for his purposes:

3.8.14. Theorem. *Let $G_0, G_1, G_2, \ldots$ be a sequence of functions continuously differentiable on an interval $[a,b]$. Suppose both series $\sum_{k=0}^{\infty} G_k(x)$ and $\sum_{k=0}^{\infty} G'_k(x)$ converge uniformly on $[a,b]$. If $G(x) = \sum_{k=0}^{\infty} G_k(x)$, then $G'(x) = \sum_{k=0}^{\infty} G'_k(x)$.*

The phrase "continuously differentiable" expresses not only that a derivative exists, but that it is continuous.

It is convenient to define uniform convergence for *sequences* first. To this end, so long as we are digressing from Bolzano's monograph, I will revert to the usual notation of $f_0, f_1, f_2, \ldots$ for functions.

3.8.15. Definition. *A sequence $f_0, f_1, f_2, \ldots$ of functions defined on a set X converges pointwise to a function f on X if, for each $x \in X$,*
$$\lim_{n \to \infty} f_n(x) = f(x),$$
i.e., if
$$\forall x \in X \forall \epsilon > 0 \exists n_0 \forall n > n_0 \big(|f_n(x) - f(x)| < \epsilon \big)$$
(where, here, n and n_0 denote positive integers). The sequence converges uniformly to f on X if n_0 depends only on ϵ and not on x:
$$\forall \epsilon > 0 \exists n_0 \forall n > n_0 \forall x \in X \big(|f_n(x) - f(x)| < \epsilon \big).$$

3.8.16. Definition. *A series $\sum_{k=0}^{\infty} g_k$ of functions defined on a set X converges pointwise (uniformly) to a function f on X if the sequence of partial sums,*
$$f_n(x) = \sum_{k=0}^{n} g_k(x),$$
converges pointwise (respectively, uniformly) to f on X.

3.8.17. Exercise. *Examine the following series for pointwise and uniform convergence on the specified sets:*

i. $f_n(x) = \dfrac{x}{1+nx^2}$ *on* $(-\infty, \infty)$ *for* $n = 0, 1, 2, \ldots$

ii. $f_n(x) = \dfrac{x}{1+nx^2}$ *on* $[0, \infty)$ *for* $n = 0, 1, 2, \ldots$

iii. $f_n(x) = \dfrac{\sin nx}{\sqrt{n}}$ *on* $(-\infty, \infty)$ *for* $n = 0, 1, 2, \ldots$

iv. $f_n(x) = \sum_{k=0}^{n} x^k$ *on* $(-1, 1)$ *for* $n = 0, 1, 2, \ldots$

v. $f_n(x) = \sum_{k=0}^{n} x^k$ *on* $(-1/2, 1/2)$ *for* $n = 0, 1, 2, \ldots$

The notion of uniform convergence has its own history, albeit not so epic a tale as that of the Binomial Theorem.[64] In abbreviated form, in the early years of the age of rigour mathematicians were still feeling their way and their definitions were not as pedantically precise as they are today. Definitions of continuity and convergence were ambiguous and could be read as defining the pointwise or the uniform notions. It took some decades for the distinction to be realised and the distinct notions separated. Bolzano would later recognise the distinction between continuity and uniform continuity, but apparently not that between convergence and uniform convergence. Cauchy's celebrated definitions of continuity and convergence are ambiguous, but he seems not to have been interested in the pointwise concepts, interpreting the requirements in a manner consistent with the uniform concepts. The result was that some of Cauchy's results would at first be deemed by Abel as admitting exceptions and later by most mathematicians as simply being false. Further rumination on Cauchy's proof of Theorem 3.8.18, below, would result in the isolation of the notion of uniform convergence and the definitive determination of Cauchy's supposed error.

Among those results Cauchy proved that are deemed false is the continuity of the limit of a convergent sequence of continuous functions. This is false if by "convergent" one means "pointwise convergent", as illustrated by the sequence,

$$f_n(x) = \frac{1 - nx^2}{(1+nx^2)^2},$$

each element of which is continuous at all real numbers, which converges pointwise everywhere, and the limit of which is discontinuous at 0. Cauchy's result is true, however, if one takes "convergent" to mean "uniformly convergent":

3.8.18. Theorem. *Let $f_0, f_1, f_2, \ldots$ be a sequence of continuous functions converging uniformly on $[a, b]$ to a function f. Then: f is continuous.*

Proof. Let $\epsilon > 0$ and let $x_0 \in [a, b]$. Choose n_0 so large that for all $n > n_0$ and all $u \in [a, b]$ one has $|f_n(u) - f(u)| < \epsilon/3$. Choose any $n > n_0$. By the

---

[64] *Cf.*, e.g., Bottazzini, *op. cit.*, Chapter 5, section 4 (pp. 202 – 208) for a short account.

continuity of f_n there is a $\delta > 0$ such that for all $x \in [a,b]$,
$$|x - x_0| < \delta \Rightarrow |f_n(x) - f_n(x_0)| < \frac{\epsilon}{3}.$$
Let $|x - x_0| < \delta$ and observe
$$\begin{aligned}|f(x) - f(x_0)| &= |f(x) - f_n(x) + f_n(x) - f_n(x_0) + f_n(x_0) - f(x_0)| \\ &\leq |f(x) - f_n(x)| + |f_n(x) - f_n(x_0)| + |f_n(x_0) - f(x_0)| \\ &< \frac{\epsilon}{3} + \frac{\epsilon}{3} + \frac{\epsilon}{3} = \epsilon.\end{aligned}$$ □

With this Theorem as a lemma, we can similarly derive the sequential version of Theorem 3.8.14:

3.8.19. Theorem. *Let $f_0, f_1, f_2, \ldots$ be a sequence of functions converging uniformly to f on $[a,b]$ and suppose each f_n is continuously differentiable and $f_0', f_1', f_2', \ldots$ converges uniformly to some function g on $[a,b]$. Then, for all $x \in [a,b]$, $f'(x) = g(x)$.*

Proof. Let $\epsilon > 0$ be given.
Note that for $x, x+h \in [a,b]$ and any n,
$$\left| \frac{f(x+h) - f(x)}{h} - \frac{f_n(x+h) - f_n(x)}{h} \right|$$
$$\leq \frac{|f(x+h) - f_n(x+h))| + |f_n(x) - f(x)|}{|h|}.$$

By the uniform convergence of the sequence $f_0, f_1, f_2, \ldots$, for any choice of h, we can find n_1 so large that for all $n > n_1$ and all $u \in [a,b]$,
$$|f(u) - f_n(u)| < \frac{|h|\epsilon}{6}.$$
For such n,
$$\left| \frac{f(x+h) - f(x)}{h} - \frac{f_n(x+h) - f_n(x)}{h} \right| < \frac{\frac{|h|\epsilon}{6} + \frac{|h|\epsilon}{6}}{|h|} = \frac{\epsilon}{3}. \tag{111}$$

By the Mean Value Theorem, there is some ξ between x and $x+h$ such that
$$f_n(x+h) - f_n(x) = f_n'(\xi) \cdot h,$$
i.e.,
$$\frac{f_n(x+h) - f_n(x)}{h} = f_n'(\xi).$$

Thus (111) yields
$$\left| \frac{f(x+h) - f(x)}{h} - f_n'(\xi) \right| < \frac{\epsilon}{3}. \tag{112}$$

Now
$$\left| \frac{f(x+h) - f(x)}{h} - g(x) \right| =$$

$$= \left| \frac{f(x+h) - f(x)}{h} - f'_n(\xi) + f'_n(\xi) - g(\xi) + g(\xi) - g(x) \right|$$
$$\leq \left| \frac{f(x+h) - f(x)}{h} - f'_n(\xi) \right| + \left| f'_n(\xi) - g(\xi) \right| + \left| g(\xi) - g(x) \right|$$
$$< \frac{\epsilon}{3} + \left| f'_n(\xi) - g(\xi) \right| + \left| g(\xi) - g(x) \right| \tag{113}$$

By the uniform convergence of $f'_0, f'_1, f'_2, \ldots$ we can find n_2 so large that for all $n > n_2$ and all $u \in [a, b]$,
$$\left| f'_n(u) - g(u) \right| < \frac{\epsilon}{3}.$$

Thus, for any $n > n_0 = \max\{n_1, n_2\}$, (113) yields
$$\left| \frac{f(x+h) - f(x)}{h} - g(x) \right| < \frac{\epsilon}{3} + \frac{\epsilon}{3} + \left| g(\xi) - g(x) \right|. \tag{114}$$

Now we have shown for any h such that $x, x + h \in [a, b]$, there is a $\xi \in [a, b]$ between x and $x + h$ so that (114) holds. By Theorem 3.8.18, g is continuous and there is a $\delta > 0$ such that for any $u \in [a, b]$,
$$|u - x| < \delta \Rightarrow |g(u) - g(x)| < \frac{\epsilon}{3}.$$

But $|\xi - x| < |h|$, so choosing $|h| < \delta$ yields $|g(\xi) - g(x)| < \epsilon/3$ and (114) becomes
$$\left| \frac{f(x+h) - f(x)}{h} - g(x) \right| < \epsilon,$$

i.e.,
$$\lim_{h \to 0} \frac{f(x+h) - f(x)}{h} = g(x),$$

i.e., $f'(x) = g(x)$. □

Theorem 3.8.14 follows immediately from Theorem 3.8.19 by identifying G_k with f_k, G'_k with f'_k, and G with f.

The hypotheses of Theorem 3.8.14 are much stronger than those Bolzano assumed, but it doesn't matter: they hold for power series in any closed interval within the radius of convergence. This is actually a pair of assertions, one for the power series itself and one for the series obtained by termwise differentiation.

These assertions are the following two lemmas:

3.8.20. Lemma. *Let the power series $\sum_{n=0}^{\infty} a_n x^n$ have a radius of convergence $R > 0$. Then, for any r satisfying $0 < r < R$, the series $\sum_{n=0}^{\infty} a_n x^n$ converges uniformly on $[-r, r]$.*

3.8.21. Lemma. *Let the power series $\sum_{n=0}^{\infty} a_n x^n$ have a radius of convergence $R > 0$. Then the series $\sum_{n=1}^{\infty} n a_n x^{n-1}$ also has radius of convergence R.*

Both of these lemmas require their own lemmas in order to be proven, but before we discuss these, let us first finish off the determination of the coefficients of any possible power series expansion of the binomial function $(1 + x)^n$, n now representing, following Bolzano, any real exponent.

By Lemmas 3.8.20 and 3.8.21, if the power series $\sum a_k x^k$ has radius R of convergence, and $0 < r < R$, then

$$\sum a_k x^k \text{ and } \sum k a_k x^{k-1}$$

both converge uniformly to functions, say, f and g on $[-r, r]$. By Theorem 3.8.14, $f' = g$. Thus, if

$$f(x) = \sum_{k=0}^{\infty} a_k x^k$$

then

$$f'(x) = \sum_{k=1}^{\infty} k a_k x^{k-1}, \quad f''(x) = \sum_{k=2}^{\infty} k(k-1) a_k x^{k-2}, \ldots$$

and

$$f(0) = a_0, \quad f'(0) = 1 \cdot a_1, \quad f''(0) = 2 \cdot 1 \cdot a_2, \ldots$$

whence

$$a_0 = f(0), \quad a_1 = \frac{f'(0)}{1}, \quad a_2 = \frac{f''(0)}{2!}, \ldots$$

à la Maclaurin. And, modulo the proofs of Lemmas 3.8.20 and 3.8.21 we have proven the uniqueness of the coefficients of the power series expansion of any function that admits such an expansion.

As for the determination of the form, Bolzano compares the coefficients of $(1 + x)^n$ with those of $\frac{1}{n}(1 + x) \sum k a_k x^{k-1}$ as we did in sections 2 and 4, above, in discussing Landen's and Hutton's attempted proofs of the Binomial Theorem.

The proof of Lemma 3.8.20 requires us to define the notions of pointwise and uniform Cauchy convergence of sequences of functions, and to prove the equivalence of these notions with pointwise and uniform convergence, respectively. The formulation of the definitions I leave as an exercise for the reader. And the equivalence of pointwise Cauchy convergence with pointwise convergence is an immediate consequence of the equivalence (Theorem 3.6.13) for numerical sequences. Likewise, that uniform Cauchy convergence of a sequence of functions follows from its uniform convergence is a trivial analogue of the proof given in the last section (page 179, above) that convergent numerical sequences are Cauchy convergent. The converse implication, however, would seem not to be an axiom as it was in the numerical case and to require a proof:

3.8.22. Lemma. *Let $f_0, f_1, f_2, \ldots$ be a uniformly Cauchy convergent sequence of functions defined on a set X. There is a function f such that $f_0, f_1, f_2, \ldots$ converges uniformly to f.*

Proof. By uniform Cauchy convergence, for any $\epsilon > 0$ there is an integer n_0 such that for all integers $m, n > n_0$ and all $x \in X$,

$$|f_m(x) - f_n(x)| < \epsilon.$$

Thus, for each $x_0 \in X$, the sequence $f_0(x_0), f_1(x_0), f_2(x_0), \ldots$ is a Cauchy convergent numerical sequence and thus has a limit L. Define $f(x_0) = L$.

To prove that the sequence converges uniformly to f on X, let $\epsilon > 0$ be given. Choose n_0 so large that for all $x \in X$ one has
$$|f_m(x) - f_n(x)| < \frac{\epsilon}{2}$$
whenever $m, n > n_0$.

Let $n > n_0$.

For each $x \in X$, $\lim_{n \to \infty} f_n(x) = f(x)$ whence there is a number n_1 such that for all $m > n_1$,
$$|f_m(x) - f(x)| < \frac{\epsilon}{2}.$$
Choose $m > \max\{n_0, n_1\}$ and observe
$$\begin{aligned}|f_n(x) - f(x)| &= |f_n(x) - f_m(x) + f_m(x) - f(x)| \\ &\leq |f_n(x) - f_m(x)| + |f_m(x) - f(x)| \\ &< \frac{\epsilon}{2} + \frac{\epsilon}{2} = \epsilon.\end{aligned}$$
Thus, given ϵ we found n_0 so large that for all $n > n_0$ and all $x \in X$, $|f_n(x) - f(x)| < \epsilon$. Thus the sequence $f_0, f_1, f_2, \ldots$ converges uniformly to f on X. □

Proof of Lemma 3.8.20. Let R be the radius of convergence of $\sum a_n x^n$, and $0 < r < r' < R$. By the definition of the radius of convergence, $\sum a_n x_0^n$ converges for any x_0 with $|x_0| = r'$ and the proof of Lemma 3.6.16 applies. In particular, for $|x| = r$, formula (101) reads
$$\sum_{n=0}^{\infty} |a_n x^n| \leq \sum_{n=0}^{n_0} |a_n x^n| + \frac{1}{2} \sum_{n=n_0+1}^{\infty} \left(\frac{r}{r'}\right)^n,$$
where n_0 is independent of x. Uniform Cauchy convergence immediately follows. □

The proof of Lemma 3.8.21 depends on the calculation of the exact value of R. In the case where $\lim_{n \to \infty} |a_n|/|a_{n+1}|$ exists, one considers
$$\lim_{n \to \infty} \left|\frac{n a_n}{(n+1) a_{n+1}}\right| = \lim_{n \to \infty} \left(\frac{n}{n+1} \cdot \frac{|a_n|}{|a_{n+1}|}\right)$$
$$= \left(\lim_{n \to \infty} \frac{n}{n+1}\right) \left(\lim_{n \to \infty} \frac{|a_n|}{|a_{n+1}|}\right) = 1 \cdot R = R.$$
In the general case it can be shown that
$$\frac{1}{R} = \overline{\lim_{n \to \infty}} \sqrt[n]{|a_n|},$$
where $\overline{\lim}$, sometimes written lim sup, is the *limit supremum* of the sequence, the limit of the sequence of *suprema*, or upper bounds, of the sequence of sets
$$X_n = \left\{\sqrt[k]{|a_k|} \,\Big|\, k \geq n\right\}.$$
Because
$$\lim_{n \to \infty} \sqrt[n]{n} = 1,$$

it will follow that
$$\varlimsup_{n\to\infty} \sqrt[n]{n|a_n|} = \varlimsup_{n\to\infty} \sqrt[n]{|a_n|} = \frac{1}{R},$$
and $\sum_{n=0}^{\infty} na_n x^{n-1}$ has the same radius of convergence as $\sum_{n=0}^{\infty} a_n x^n$. I leave it to the reader to look up the proofs of these assertions.

With this last remark we have finished our exposition of the completion of Bolzano's proof that, if $(1+x)^n$ has a power series expansion, then that expansion is $\sum \binom{n}{k} x^k$, where once again we rejoin Bolzano in using n for an arbitrary real exponent. His proof was incomplete in that he missed the concepts of uniform continuity and uniform convergence. Possibly this is due to his reliance on his ω, Ω notation, which makes it easy to overlook the dependencies of one variable on another. And he misstated the important Theorem 3.8.13, mistakenly believing he had proven the convergence of the series of derivatives as well as the continuity of these functions. What matters most, however, is that, unlike his predecessors, Bolzano realised that termwise differentiability required a proof and he had many of the ingredients for such a proof, including the crucial application of the Mean Value Theorem made above, but simply did not put these ingredients together well enough to have noticed what was lacking. And, of course, he recognised, as Landen and Hutton hadn't, that proving that the power series for $(1+x)^n$, *if* it converged to $(1+x)^n$, had to be the binomial series did not prove that the binomial series does in fact converge to $(1+x)^n$. This he set out to do beginning in §38.

Today, given what has already been proven, that a power series can be differentiated termwise, we can use the popular modern proof cited following our exposition of Hutton's attempted proof (pp. 159 – 160, above) to conclude the Binomial Theorem without further ado. Bolzano didn't think of this, but instead presented a carefully thought out version of Euler's proof.

He begins in §38 with the following

Theorem. If two binomial series,
$$1 + px + p\frac{p-1}{2}x^2 + \ldots + p\frac{p-1}{2}\cdots\frac{p-(r-1)}{r}x^r$$
and
$$1 + qx + q\frac{q-1}{2}x^2 + \ldots + q\frac{q-1}{2}\cdots\frac{q-(s-1)}{s}x^s$$
in which p and q denote any kind of quantity, are multiplied together and the product is arranged by powers of x, then all terms of the product starting from the first up to the term x^r or x^s according to whether r or s is the smaller number, are identical with the equally many terms of the binomial series belonging to $(1+x)^{p+q}$.[65]

As he is going to be interested in the truth of this for r and s as large as one pleases, nothing is lost in increasing the smaller number to the larger, i.e., in considering only the case $r = s$. The result, in modern terms, is the

---

[65]Russ, *op. cit.*, p. 196.

3.8.23. Lemma. *For any* x,
$$\left(\sum_{k=0}^{r}\binom{p}{k}x^k\right) \cdot \left(\sum_{k=0}^{r}\binom{q}{k}x^k\right) = \sum_{k=0}^{r}\binom{p+q}{k}x^k + \text{ terms of higher degree}.$$

The proof of this Lemma consists of multiplying the factors on the left and gathering like terms. For $n \leq r$, the term of degree n is
$$\binom{p}{0}x^0\binom{q}{n}x^n + \binom{p}{1}x^1\binom{q}{n-1}x^{n-1} + \ldots + \binom{p}{n}x^n\binom{q}{0}x^0,$$
with coefficient
$$\binom{p}{0}\binom{q}{n} + \binom{p}{1}\binom{q}{n-1} + \ldots + \binom{p}{n}\binom{q}{0} = \binom{p+q}{n} \tag{115}$$

by Vandermonde's Theorem cited in our discussion of Euler's attempted proof. Bolzano proves Vandermonde's Theorem first for the case where p and q are positive integers and then makes a Peacockian-sounding statement to the effect that the equations will also hold for arbitrary real numbers p, q—negative, fractional, and irrational—as well as for positive integers. This is true in the present case because, for each n, the two sides of (115) are polynomials in the variables p, q and if they are identically equal in the integers, they are identically equal when expanded and simplified and thus have the same coefficients—and thus are identical in the reals.

For $r < n \leq 2r$, there are also terms of degree n in the product. Thus the product will indeed have the form given by the Lemma. In §39, Bolzano reminds the reader that when $|x| < 1$, the terms of higher degree will be very small. Determining how small will be the main task of the next section.

§40 offers a formal statement of the result just mentioned:

Theorem. *The value of the product of the two series*
$$1 + px + p\frac{p-1}{2}x^2 + \ldots + p\frac{p-1}{2}\cdots\frac{p-(r-1)}{r}x^r$$
and $1 + qx + q\frac{q-1}{2}x^2 + \ldots + q\frac{q-1}{2}\cdots\frac{q-(s-1)}{s}x^s$

differs from the value of the series
$$1 + (p+q)x + (p+q)\left(\frac{p+q-1}{2}\right)x^2 + \ldots$$
$$+ (p+q)\left(\frac{p+q-1}{2}\right)\cdots\left(\frac{p+q-t+1}{t}\right)x^t$$

by a quantity which can be made smaller than any given quantity if r, s, t *are taken large enough and* x *is a proper fraction.*[66]

In essence, he is asserting the lemma:

---

[66] *Ibid.*, p. 200. I remind the reader that by "proper fraction", Bolzano means that $|x| < 1$; he is not assuming x is rational.

3.8.24. Lemma. *Let $|x| < 1$.*

$$\left(\sum_{k=0}^{\infty} \binom{p}{k} x^k\right) \cdot \left(\sum_{k=0}^{\infty} \binom{q}{k} x^k\right) = \sum_{k=0}^{\infty} \binom{p+q}{k} x^k. \tag{116}$$

The proof he gives contains a minor error and omits some details, but is easily completed.

The key is, of course to consider the terms in the product

$$\left(\sum_{k=0}^{r} \binom{p}{k} x^k\right) \cdot \left(\sum_{k=0}^{r} \binom{q}{k} x^k\right)$$

that do not occur in

$$\sum_{k=0}^{r} \binom{p+q}{k} x^k.$$

These are

$$\binom{q}{r} x^r \binom{p}{1} x$$

$$+ \left(\binom{q}{r-1} x^{r-1} + \binom{q}{r} x^r\right) \binom{p}{2} x^2$$

$$+ \left(\binom{q}{r-2} x^{r-2} + \binom{q}{r-1} x^{r-1} + \binom{q}{r} x^r\right) \binom{p}{3} x^3$$

$$\vdots$$

$$+ \left(\binom{q}{1} x + \binom{q}{2} x^2 + \ldots + \binom{q}{r} x^r\right) \binom{p}{r} x^r$$

$$= \binom{q}{r} \binom{p}{1} x^{r+1}$$

$$+ \left(\binom{q}{r-1} + \binom{q}{r} x\right) \binom{p}{2} x^{r+1}$$

$$+ \left(\binom{q}{r-2} + \binom{q}{r-1} x + \binom{q}{r} x^2\right) \binom{p}{3} x^{r+1}$$

$$\vdots$$

$$+ \left(\binom{q}{1} + \binom{q}{2} x + \ldots + \binom{q}{r} x^r\right) \binom{p}{r} x^{r+1}.$$

If P, Q are the largest absolute values of the binomial coefficients

$$\binom{p}{k} \text{ and } \binom{q}{k}, \text{ respectively, for } k = 1, 2, \ldots, r,$$

he notes that this sum is bounded above by

$$PQ|x|^{r+1} \left(1 + (1 + |x|) + (1 + |x| + |x|^2) + \ldots + (1 + |x| + \ldots + |x|^r)\right). \tag{117}$$

He sets out to determine P, Q noting correctly that

$$\left|\binom{p}{k}\right| = \left|\frac{p(p-1)\cdots(p-k+1)}{k!}\right| = \frac{|p|\cdot|p-1|\cdots|p-k+1|}{k!}$$

$$\leq \frac{|p|\cdot(|p|+1)\cdots(|p|+k-1)}{k!} = \left|\binom{-|p|}{k}\right|,$$

and incorrectly that, if p is negative the sequence

$$\binom{p}{1}, \binom{p}{2}, \ldots, \binom{p}{r}$$

increases in absolute value to a maximum at $\binom{p}{r}$. This is only true if $|p| > 1$; for $|p| < 1$ the sequence is decreasing. Thus, for $|p| > 1$ one has

$$P = \frac{|p|\cdot(|p|+1)\cdots(|p|+k-1)}{k!},$$

and for $|p| \leq 1$ one has

$$P = |p|.$$

And, of course, similar results hold for Q. This complicates matters slightly, forcing us to consider two cases.

1. One of $|p|, |q|$ is greater than 1. Letting $m = \max\{|p|, |q|\}$, we have

$$P, Q \leq \frac{m(m+1)\cdots(m-r+1)}{r!}$$

and PQ can be replaced in (117) by

$$M^2 = \left(\frac{m(m+1)\cdots(m-r+1)}{r!}\right)^2.$$

Moreover, the individual sums occurring on the right in (117) can be calculated:

$$1 + |x| + \ldots + |x|^k = \frac{1 - |x|^{k+1}}{1 - |x|}.$$

Thus (117) becomes

$$M^2|x|^{r+1}\left(\frac{1-|x|}{1-|x|} + \frac{1-|x|^2}{1-|x|} + \ldots + \frac{1-|x|^{r+1}}{1-|x|}\right)$$

$$= M^2|x|^{r+1}\frac{1}{1-|x|}\left(r+1 - |x|(1+|x|+\ldots+|x|^r)\right)$$

$$= M^2|x|^{r+1}\frac{1}{1-|x|}\left(r+1 - |x|\frac{1-|x|^{r+1}}{1-|x|}\right)$$

$$= \frac{M^2}{1-|x|}(r+1)|x|^{r+1} - \frac{M^2}{(1-|x|)^2}|x|^{r+2} + \frac{M^2}{(1-|x|)^2}|x|^{2r+2}.$$

(118)

Bolzano's task now is to show that

$$M^2(r+1)|x|^{r+1}, \quad M^2|x|^{r+2}, \quad M^2|x|^{2r+2}$$

8. BOLZANO'S PROOF 221

all tend to 0 as r tends to $+\infty$. For $|x| < 1$, the largest of these is clearly the first, whence it suffices to show

$$\lim_{r\to\infty} M^2(r+1)|x|^{r+1} = 0.$$

Bolzano's justification for this is a hint, not a proof, but if one follows it and fills in the details one gets the following: Let

$$a_r = M^2(r+1) = \left(\frac{m(m+1)\cdots(m-r+1)}{r!}\right)^2 \cdot (r+1).$$

Then

$$\frac{a_r}{a_{r+1}} = \left(\frac{m(m+1)\cdots(m-r+1)}{m(m+1)\cdots(m-r+1)(m+r)}\right)^2 \cdot \left(\frac{(r+1)!}{r!}\right)^2 \cdot \frac{r+1}{r+2}$$

$$= \frac{1}{(m+r)^2} \cdot \frac{(r+1)^2}{1} \cdot \frac{r+1}{r+2} = \left(\frac{r+1}{m+r}\right)^2 \cdot \frac{r+1}{r+2}$$

$$\to 1, \text{ as } r \to \infty.$$

Thus by Theorem 3.6.15 the radius of convergence of the power series

$$\sum_{r=0}^{\infty} a_r x^r$$

is 1, whence $a_r x^r \to 0$ as $r \to \infty$ for $|x| < 1$. I.e.,

$$M^2(r+1)x^r \to 0 \text{ as } r \to \infty.$$

Multiplying by an extra factor of x merely speeds up convergence slightly.

2. If both $|p|, |q| \leq 1$, one can replace M by a constant and the proof given in case 1 simplifies. I leave the details to the reader.

With this we have completed the proof of Lemma 3.8.24. I am not that pleased with the proof, disliking the wastefully huge bound (117) and the necessity of considering two cases.

Simpler proof of Lemma 3.8.24. Note that, for $r < n \leq 2r$, the coefficient of x^n in the product

$$\left(\sum_{k=0}^{r}\binom{p}{k}x^k\right) \cdot \left(\sum_{k=0}^{r}\binom{q}{k}x^k\right)$$

is

$$a_n = \binom{p}{r}\binom{q}{n-r} + \binom{p}{r+1}\binom{q}{n-r+1} + \ldots + \binom{p}{n-r}\binom{q}{r}.$$

Note that

$$|a_n| \leq \sum_{k=r}^{n-r}\left|\binom{p}{k}\binom{q}{n-k}\right| \leq \sum_{k=0}^{n}\left|\binom{p}{k}\binom{q}{n-k}\right|$$

$$\leq \sum_{k=0}^{n}\left|\binom{-|p|}{k}\binom{-|q|}{n-k}\right| = \binom{-|p|-|q|}{n}$$

by Vandermonde's Theorem. Thus

$$\left|\sum_{n=r+1}^{2r} a_n x^n\right| \le \sum_{n=r+1}^{2r} |a_n x^n|$$

$$\le \sum_{n=r+1}^{2r} \binom{-|p|-|q|}{n}|x|^n \le \sum_{n=r+1}^{\infty} \binom{-|p|-|q|}{n}|x|^n,$$

this last sum tending to 0 as r tends to ∞ by the convergence of the binomial series

$$\sum_{k=0}^{\infty} \binom{|p|-|q|}{n} x^n$$

for $|x| < 1$. □

Whatever proof one prefers, once one has Lemma 3.8.24, Euler's proof of the Binomial Theorem for rational exponents carries over. The result then extends to the irrational case by continuity. Aside from handling the irrational case, there is one small wrinkle we neglected to mention in discussing the earlier proof. Granted that the series for the exponent $1/n$ converges to an n-th root of $1+x$, when n is even there are two real roots, one positive and one negative. Which root does the series converge to? Bolzano's resolution of this issue does not satisfy. Yet today it is a simple matter. Define

$$f(x) = \sum_{k=0}^{\infty} \binom{1/n}{k} x^k,$$

n a positive even integer. Then $f(0) = 1$ is positive. By Lemma 3.8.20 and Theorem 3.8.18, f is continuous. By the Intermediate Value Theorem, which Bolzano announces in *Der binomische Lehrsatz* as being prepared for publication and which appeared in another monograph[67] the following year, if $f(x)$ took on any negative value for some $x \in (-1, 1)$, it would follow that $f(x_0) = 0$ for some $x_0 \in (-1, 1)$. But $f(x)^n = 1+x$ and one would have $1 + x_0 = 0$, i.e., $x_0 = -1$, contrary to the location of x_0 in the open interval $(-1, 1)$. Thus the series must always converge to the positive root.

With this we end our discussion of Bolzano's proof of the Binomial Theorem. His proof was not flawless and his exposition less than pleasing, but the errors were minor and easily corrected and the gaps easy to fill in. All in all I accept Hankel's description of Bolzano's proof of the Binomial Theorem as the first rigorous one ever given. It, however, went unnoticed and a new rigorous proof would be given half a decade later by Augustin Louis Cauchy, but he too would not receive credit, albeit for an entirely different reason.

9. Cauchy's Proof

Much of Cauchy's foundational work is startlingly similar to Bolzano's, so much so that the historian Ivor Grattan-Guinness postulated a dependence of Cauchy on the work of Bolzano. This so infuriated the mathematician Hans

---

[67]Bolzano, *Rein analytischer Beweis*, op. cit.

Freudenthal that he came to the defence of Cauchy in a paper titled "Did Cauchy plagiarize Bolzano?"[68] Judith V. Grabiner later made a careful study[69] of Cauchy and his predecessors, citing common influences as the probable cause of similarities.

> The familiarity of Cauchy and Bolzano with the work of their common predecessors can be documented. Once this is done, the similarities between the work of Cauchy and Bolzano will point out strongly how important these predecessors were; in the period 1815–1825, any genius of sufficient magnitude, seeking rigor in analysis, could have done these things. There were two such geniuses.[70]

The real difference was that between the effects the two men had on posterity.

> Bolzano's work also could have served as a starting point for the rigorization of analysis. Cauchy was the man who taught rigorous analysis to all of Europe, however, while Bolzano's works went almost unread until the 1860s. This is not only because of the magnificent clarity of exposition in Cauchy's books; the reasons are partly social and institutional. The Ecole Polytechnique in Paris, where Cauchy delivered his lectures, was the first and foremost scientific school in Europe. Most of the leading French mathematicians and mathematical physicists of the age went there; many leading mathematicians read the courses of lectures written there as soon as they were published as books. Paris was the center of the mathematical world, and many mathematicians not lucky enough to be French came there to study. In comparison, Bolzano worked in relative isolation in Prague, did not hold an important teaching position, published many of his papers as separate pamphlets because he had no ready access to a prestigious journal, and, finally, was known as a philosopher and theologian, rather than as a mathematician.[71]

Both men were originators of the new rigour and neither could complete the task:

> No mathematical subject is ever perfected overnight. Neither Cauchy nor Bolzano had solved by 1825 all the outstanding problems of analysis. There were two major lacunae in Cauchy's work at this time. First, he—and also Bolzano—did not yet appreciate the distinction between convergence and uniform convergence or that between continuity and uniform continuity.[72]

The confusion between pointwise and uniform properties led to

---

[68] *Archive for History of Exact Sciences* 7 (1971), pp. 375 – 392.

[69] Judith Grabiner, *The Origins of Cauchy's Rigorous Calculus*, The MIT Press, Cambridge (Mass.), 1981. A reprint was published by Dover in 2005.

[70] *Ibid.*, pp. 11 – 12.

[71] *Ibid.*, pp. 14 – 15.

[72] *Ibid.*, p. 12. In the last decade or so it has been revealed that later Bolzano had in fact understood the latter distinction. *Cf.* Russ, *op. cit.*, pp. 575 – 577.

> Cauchy's famous "proof," in 1821, of the false theorem that an infinite series of continuous functions is continuous; in 1816, Bolzano too seems to have believed that an infinite series of continuous functions was continuous. In 1826, Abel, in his study of the continuity of the sum of a power series, published a counterexample to Cauchy's false theorem, but Abel did not identify the error in the proof. The elucidation of the difference between convergence and uniform convergence by men like Stokes, Weierstrass, and Cauchy himself was still more than a decade away.[73]

The quotation marks around the word "proof" in this last quote point to the reason Cauchy is not universally credited with having given a valid proof of the Binomial Theorem. His proof relies on a pair of his notorious "false lemmas". The first pair of quotation marks indicates the opinion of the user that the proof in question is not valid. Grabiner's opinion has been the consensus among mathematical historians for some time.[74] My use of quotation marks on the phrase "false lemmas" indicates only that this phrase has occasionally been applied to describe results of Cauchy that have generally been deemed false by the mathematical community. But they could also be used here in the same sarcastic sense in which the first pair is applied: not everyone agrees that the lemmas are false. Put more succinctly: read "proof" and "false lemmas" as "so-called proof" and "so-called false lemmas", respectively.

The two "false lemmas" in question are the following two theorems, both from *Cours d'analyse algébrique*[75]. From Chapter II (p. 47),

> *Theorem 1.*—If the variables $x, y, z, \ldots$ have given quantities $X, Y, Z, \ldots$ as their limits, and if the function $f(x, y, z, \ldots)$ is continuous in each of the variables $x, y, z, \ldots$ in the neighbourhood of this system of particular values
> $$x = X, y = Y, z = Z, \ldots,$$
> then $f(x, y, z, \ldots)$ has $f(X, Y, Z, \ldots)$ for its limit.

And, from Chapter VI (p. 120):

> *Theorem 1.*—When the various terms of the series (1) [i.e., $u_0 + u_1 + u_2 + \ldots$] are functions of the same variable x, continuous with respect to this variable in the neighbourhood of a particular value for which the series is convergent, the sum s of the series is also, in the neighbourhood of this particular value, a continuous function of x.[76]

---

[73] Grabiner, *op. cit.*, p. 12.

[74] *Cf.*, e.g., footnote 43 on page 192, above.

[75] Augustin Louis Cauchy, *Cours d'analyse algébrique de l'École Royale Polytechnique*, Paris, 1821.

[76] All else aside, the statement of this theorem is incorrect. As Detlef Laugwitz ("Infinitely small quantities in Cauchy's textbooks", *Historia Mathematica* 14 (1987), pp. 258–274; here: p. 265) points out, "for which" should read "in which"—Cauchy assumes convergence at all points in the neighbourhood.

The first of these states that a function continuous in each variable separately is continuous as a function of several variables. The second asserts that the limit of a convergent series of continuous functions is continuous. Both results, as stated, are false. We have already seen the latter in our discussion of Bolzano. As to the former, the standard counterexample is the following.

3.9.1. Example. *Let*

$$f(x,y) = \begin{cases} \dfrac{xy}{x^2+y^2}, & \langle x,y \rangle \neq \langle 0,0 \rangle, \\ 0, & \langle x,y \rangle = \langle 0,0 \rangle. \end{cases}$$

For each fixed x_0 (y_0), the function

$$g(y) = f(x_0, y) \ \ (h(x) = f(x, y_0), \ respectively)$$

is continuous for all y (x, respectively). But f is not continuous at $\langle 0,0 \rangle$.

3.9.2. Exercise. *Prove the assertions of Example 3.9.1.*

What is one to make of these Theorems? Detlef Laugwitz (1932 – 2000), one of the leaders of the re-evaluation of Cauchy, sums it up nicely:

> Both theorems are incorrect when interpreted in the by now common conceptual framework of analysis (which obviously cannot have been Cauchy's framework). Both theorems become correct as soon as one adds assumptions on uniformity (which, at least in the form by now common, were never used by Cauchy). The theorems are correct in any of the modern theories of infinitesimals...

> The three attitudes mentioned (Cauchy erred; Cauchy forgot about essential assumptions; Cauchy was correct, but only when put against a modern background) are unsatisfactory from the point of view of a historian. The first one, shared by a majority, is inadequate even psychologically: Is it believable that Cauchy, the exponent of rigor, should make mistakes at the lowest level of his calculus? Nevertheless: "For instance, it is well known that he asserted the continuity of the sum of a convergent series of continuous functions; Abel gave a counterexample, and *it is clear that Cauchy himself knew scores of them*".[77]

The "three attitudes" referred to by Laugwitz are examples of what is known among historians of science as "Whig history", the natural tendency to interpret the past through modern eyes.

> The 'Whig' interpretation of history has had a powerful influence within the history of science. This is because science has appeared to historians to be particularly progressive. Some historians of science have, therefore, seen the present state of scientific knowledge as an

---

[77]Laugwitz, *op. cit.*, pp. 259 – 260. The final quotation is from Freudenthal's article on Cauchy in the *Dictionary of Scientific Biography* (Charles Scribner's Sons, New York, 1971). The italics were added by Laugwitz, who did not repeat the next sentence from Freudenthal: "It is less known that later Cauchy correctly formulated and applied the uniform convergence that is needed here."

absolute against which earlier attempts to understand Nature could be evaluated. This was particularly true before the 1950s.[78]

The emergence of Nonstandard Analysis around 1960 prompted a second look at Cauchy. Until then, historians had dismissed Cauchy's reference to the infinitely large and infinitely small. Now that one knew how to handle such concepts safely it was possible to reconsider Cauchy's use of them and it was suggested that Cauchy's definition of continuity was not to be restricted to actual real numbers and likewise his definition of the convergence of a sequence or series of functions was to apply to functions at what are now called *hyperreal numbers*—elements in an extension of the real numbers differing from real numbers by infinitesimal amounts. When one accepts this, it becomes clear that Cauchy's definitions of continuity and convergence yield uniform continuity and uniform convergence for the real restrictions of the functions in question and Cauchy's proof of the Binomial Theorem is indeed correct. A detailed discussion of this is too great a digression for presentation here[79]. What is important here is that Cauchy's proof is completely rigorous and acceptable to the modern mathematician as soon as one reads enough uniformity into the terms "continuous" and "convergent".

Whereas Bolzano's monograph had the proof of the Binomial Theorem as its central concern and Abel's paper bearing the binomial series in its title would also devote itself largely to the Binomial Theorem, Cauchy's textbook proves the result in passing: it is a straightforward application of powerful results established in his rigorisation of analysis. There are details missing having to do with the establishment of background results. For example, on defining his notion of continuity, he states without proof that it is easy to prove the continuity of the "simple" functions, including

$f(x) = x^a$, for $x > 0$ and a a real constant
$g(x) = A^x$, A a positive constant and x real.

Was this omission a gap in rigour or an expository decision?

A more serious omission is a proof that x^a exists for a irrational (or, A^x exists for x irrational). He does outline the definition of exponentiation in an appendix, citing the extension of the definition to irrational exponents by appeal to "continuity", i.e., uniform continuity. He had the tools to prove that this could be done, but at least in the *Cours d'analyse* did not carry out the necessary construction.

---

[78]C.B. Wilde, "Whig history", in: W.F. Bynum, E.J. Browne, and Roy Porter (eds.), *Dictionary of the History of Science*, Princeton University Press, Princeton, 1981.

[79]I can, however, offer a few references. Laugwitz, *op. cit.*, has already been cited. Detlef Spalt, *Die Vernunft im Cauchy-Mythos* (Verlag Harri Deutsch, Thun and Frankfurt am Main, 1996) offers a book-length re-interpretation of Cauchy's foundational work. Chapter 5 discusses the theorem on convergent series and the Binomial Theorem; and Chapter 12 reviews the interpretations of Cauchy from 1966 up till 1996. Also, I offer a simplified presentation of Laugwitz's discussion of Cauchy in terms of Nonstandard Analysis in Chapter III, section 6.4, of *Adventures in Formalism*, College Publications, 2012.

9. CAUCHY'S PROOF

Advanced textbooks today often postpone a formal definition of exponentiation x^y until after the exponential function,
$$\exp(x) = e^x,$$
has been defined as the limit
$$\exp(x) = \lim_{n \to \infty} \left(1 + \frac{x}{n}\right)^n,$$
the sum of the infinite series
$$\exp(x) = \sum_{k=0}^{\infty} \frac{x^k}{k!},$$
or the inverse to the natural logarithm defined by
$$\ln(x) = \int_1^x \frac{dt}{t}.$$
Using such a definition one can with varying amounts of effort derive the usual laws of exponentiation and show $\exp(x)$ and $h(x,y) = \left(e^{\ln x}\right)^y$ to be continuous functions. And, one can show that $h(x,q)$ agrees with x^q for all rational values of q, whence h is the continuous extension x^y to all real arguments y.

Here I shall give the more direct proof from the basic definitions. It is a slight digression from Cauchy's exposition of a proof of the Binomial Theorem, but not so great a digression as the modern textbook treatments cited above, and is part of the proof when fully worked out in the modern manner.

Let us begin with Cauchy's definition of continuity. It is a bit long and is not what one would expect:

> Let $f(x)$ be a function of the variable x, and let us suppose that, for every value of x between two given limits, this function always has a unique and finite value. If, beginning from one value of x lying between these limits, we assign to the variable x an infinitely small increment α, the function itself increases by the difference
> $$f(x+\alpha) - f(x),$$
> which depends simultaneously on the new variable α and on the value of x. Given this, the function $f(x)$ will be a *continuous* function of this variable within the two limits assigned to the variable x if, for every value of x between these limits, the numerical value[80] of the difference
> $$f(x+\alpha) - f(x)$$
> decreases indefinitely with that of α. In other words, *the function $f(x)$ will remain continuous with respect to x within the given limits if, within these limits, an infinitely small increase of the variable always produces an infinitely small increase of the function itself.*[81]

---

[80] "numerical value" = "absolute value".
[81] Cauchy, *op. cit.*, p. 43. I have copied the English translation from Bottazzini, *op. cit.*, pp. 104 – 105. The passage is often translated: C.H. Edwards, Jr., *The Historical Development of the Calculus* (Springer-Verlag New York, Inc., New York, 1979), pp. 310 – 311; Laugwitz, *op. cit.*, p. 261 (French original plus Edwards's translation); Stedall, *op. cit.*, pp. 312 –

The thing to notice about this definition is that continuity, as Cauchy defines it, is not continuity at a point, but continuity at all points in an interval, i.e., continuity in an interval. The limits referred to are the endpoints of the interval in which the continuity is being considered. As to the matter of linguistics, Laugwitz and Grabiner say that Cauchy uses the words "included" [*comprise*] when the closed interval is intended and "lying between" [*renfermé*] when the open interval is meant. Bearing this in mind, Cauchy's definition of continuity can be rephrased as follows:

3.9.3. Definition. *A function f having a finite value for every element x in an interval (a,b) is continuous on the interval if for all $x \in (a,b)$ and all infinitesimal α, $f(x+\alpha) - f(x)$ is infinitesimal.*

This is somewhat stronger than pointwise continuity for all *real* $x \in (a,b)$ because Cauchy assumes the statement true of *all* $x \in (a,b)$, including such hyperreals as $a+\alpha$ for positive infinitesimals α. If $f(a+\alpha)$ has a finite value, it will differ from a real number L by an infinitesimal amount and $\lim_{x \to a+} f(x)$ will equal L. The boundedness of f on (a,b) follows, as, in fact, does the uniform continuity of the restriction of f to real arguments of (a,b).

3.9.4. Example. $f(x) = \dfrac{1}{x}$ *is not continuous on $(0,1)$ in Cauchy's sense.* For, if α is a positive infinitesimal, $f(\alpha)$ cannot have a finite value:

$$f(\alpha) = \frac{1}{\alpha} > n,$$

for all positive integers n. However, f is continuous on every interval (a,b) for which a,b are real and $0 < a < b$. For, if $x \in (a,b)$ and α is infinitesimal,

$$|f(x+\alpha) - f(x)| = \left|\frac{1}{x+\alpha} - \frac{1}{x}\right| = \left|\frac{x - (x+\alpha)}{x(x+\alpha)}\right|$$

$$= \frac{|\alpha|}{x(x+\alpha)} < \frac{|\alpha|}{a^2}.$$

But $|\alpha|/a^2$ is the product of an infinitesimal and a finite number, whence an infinitesimal.

If we define f to be continuous at a real number x_0 if it is continuous in such an interval (a,b) containing x_0, we see that f is continuous at every real number in $(0,1)$, but it is not continuous on $(0,1)$ itself.

Notice that, in terms of ϵ and δ, the proof of continuity within (a,b) translates to: To make $|f(x+\alpha) - f(x)| < \epsilon$, it suffices to make $|\alpha|/a^2 < \epsilon$, i.e., $|\alpha| <$

---

315 (photoreproduction of the original and an English translation). The *Cours d'analyse* is available online in the original French, as well as two of the three German translations (those by C.L.B. Huzler (1828) and Carl Itzigsohn (1885)). There is finally an English translation of the full textbook: Robert E. Bradley and C. Edward Sandifer (editors and translators), *Cauchy's Cours d'analyse: An Annotated Translation* (Springer Science+Business Media, LLC, New York, 2009). The passage occurs on p. 26 of this book. The differences in wording among the various translations being minor, I did not bother retyping the quotation to bring it into uniformity with Bradley and Sandifer upon acquisition of the book, which I assume will be the standard among English readers.

$\delta = a^2 \epsilon$. Uniform continuity is thus established on $(a, +\infty)$. Cauchy nowhere makes a combined use of ϵ's and δ's in his *Cours d'analyse*. He does use them, however, in his proof of the Mean Value Theorem in his continuation, the *Calcul infinitésimal*[82].

> Unlike most of his predecessors, Cauchy did much more than simply define the derivative; he used his definition to prove theorems about derivatives, and thus created the first rigorous theory of derivatives. The crucial theorem of that theory concerns bounds on the difference quotient: If $f(x)$ is continuous between $x = x_0, x = X$, and if A is the minimum of $f'(x)$ on that interval while B is the maximum, then $A \leqslant [f(X) - f(x_0)]/(X - x_0) \leqslant B$. [Cauchy expressed $\leqslant$ verbally.] In this proof, Cauchy translated his definition of derivative into the language of delta-epsilon inequalities: "Designate by δ and ε two very small numbers; the first being chosen in such a way that, for numerical values of i less than δ, and any value of x between x_0 and X, the ratio $f(x+i) - f(x)/i$ always remains greater than $f'(x) - \varepsilon$ and less than $f'(x) + \varepsilon$."
>
> In this delta-epsilon inequality—the first appearance in history, incidentally, of the delta-epsilon notation—Cauchy expressed the crucial property of the derivative in terms any modern mathematician would recognise. The only shortcoming of this translation of Cauchy's verbal definition is that he assumed his δ would work for all x on the given interval, an assumption equivalent to that of the uniform convergence of the differential quotient. Nevertheless, his use of the inequality to translate the definition is a major achievement. Cauchy knew exactly what he meant by "the derivative is the limit of the quotient of infinitesimal differences," and he was really the first person in history to know this.[83]

Not all maths historians agree with Grabiner that Cauchy's ϵ-δ definition of differentiation as a notion of uniform differentiability was a shortcoming. In light of his use of uniformity in the proofs of his "false lemmas" and in his proof of the Fundamental Theorem of the Calculus in the *Calcul infinitésimal*, as well as the equivalence of his other definitions with the uniform concepts when interpreted in Nonstandard Analysis, it strikes some, the present author included, as more reasonable to assume Cauchy intended some uniformity[84]. I note too that the natural way to prove $f(x + \alpha) - f(x)$ is infinitesimal is to prove that $|f(x + \alpha) - f(x)|$ is less than any positive real ϵ [85], and that forcing

---

[82] Augustin Louis Cauchy, *Résumé des leçons données a l'école royale polytechnique sur le calcul infinitésimal*, Paris, 1823.

[83] Grabiner, *op. cit.*, p. 115. The bracketed comment on Cauchy's expressing $\leqslant$ verbally is Grabiner's; I have only altered her passage in correcting one small typographical error. Incidentally, Grabiner includes a translation of the full passage using ε and δ in an appendix (pp. 168 – 170). While this may be the first occurrence of ε and δ together, Cauchy uses ε in this context in the *Cours d'analyse*—cf. Bradley and Sandifer, *op. cit.*, p. 35.

[84] Or would have, had the distinction already been raised.

[85] Or, D, to use Bolzano's notation.

such an inequality is done by making α small, say, less than δ in absolute value for some explicitly determined δ. And, in carrying out such a procedure, establishing the bounds is usually done uniformly in some neighbourhood—as was the case in Example 3.9.4.

I have three goals for the present section. The first is to prove Cauchy's "false lemmas", phrasing them in currently acceptable terms. The second is to fill in the details Cauchy omitted—define and prove the continuity of exponentiation. Only after these tasks have been completed can we, as a sort of anticlimax, present Cauchy's derivation of the Binomial Theorem.

Cauchy's proofs of his so-called false lemmas are correct and acceptable in Nonstandard Analysis provided one interprets his definitions of "continuity" and "convergence" in a manner consistent with his usage. This statement is either a deep insight or a tautology; I cannot decide which. Consider his proof that a function continuous in each variable separately is continuous as a function of several variables. He assumes, for, say, two variables, f is defined on $(a,b) \times (c,d)$, continuous in each variable separately. He lets $\langle x_0, y_0 \rangle \in (a,b) \times (c,d)$, chooses α, β infinitesimal, and then observes

$$f(x_0 + \alpha, y_0 + \beta) - f(x_0, y_0)$$
$$= f(x_0 + \alpha, y_0 + \beta) - f(x_0 + \alpha, y_0) + f(x_0 + \alpha, y_0) - f(x_0, y_0).$$

Now
$$f(x_0 + \alpha, y_0) - f(x_0, y_0) \text{ is infinitesimal}$$
because $h(x) = f(x, y_0)$ is continuous. "Similarly",
$$f(x_0 + \alpha, y_0 + \beta) - f(x_0 + \alpha, y_0) \text{ is infinitesimal}$$
because $g(y) = f(x_0 + \alpha, y)$ is continuous. But g is defined via a *hyperreal* parameter $x_0 + \alpha$. Cauchy's application of the familiar nonstandard definition of the continuity of a standard function has been applied to a nonstandard function. This involves strong hidden assumptions and has strong uniformity consequences.

Making these hidden assumptions explicit using the familiar ϵ-δ terminology is more subtle than Laugwitz's remark that the result will "become correct as soon as one adds assumptions on uniformity" might lead one to believe.

3.9.5. Example. *Let f be as in Example 3.9.1, and consider the behaviour of f on the closed square $[-1,1] \times [-1,1]$. For any $\langle x_0, y_0 \rangle \in [-1,1] \times [-1,1]$,*
i. the function $g(y) = f(x_0, y)$ is uniformly continuous on $[-1,1]$, and
ii. the function $h(x) = f(x, y_0)$ is uniformly continuous on $[-1,1]$.
But: f is not continuous at $\langle 0,0 \rangle$.

3.9.6. Exercise. *Prove the assertions of Example 3.9.5.*

The proper assumption appears to be the *equicontinuity* of the families of functions obtained by fixing the values of the respective variables.

3.9.7. Definition. *Let $\mathcal{F}$ be a family of functions defined on an interval I. $\mathcal{F}$ is equicontinuous if, for all $\epsilon > 0$ there is a $\delta > 0$ such that for each $f \in \mathcal{F}$*

and all $x, y \in I$ one has
$$|x-y| < \delta \Rightarrow |f(x) - f(y)| < \epsilon.$$

Note that every element f of an equicontinuous family $\mathcal{F}$ is uniformly continuous.

In terms of equicontinuity, the ϵ-δ analogue to Cauchy's lemma is the following:

3.9.8. Theorem. *Let f be a function defined on $(a,b) \times (c,d)$. Define, for $\langle x, y \rangle \in (a,b) \times (c,d)$,*
$$g_x(y) = f(x,y), \quad h_y(x) = f(x,y).$$
$$\mathcal{G} = \{g_x \,|\, x \in (a,b)\}, \quad \mathcal{H} = \{h_y \,|\, y \in (c,d)\}.$$
If the sets $\mathcal{G}$ and $\mathcal{H}$ are equicontinuous, then f is uniformly continuous on $(a,b) \times (c,d)$.

Proof. Let $\epsilon > 0$ be given. Choose δ_1, δ_2 such that for all $x \in (a,b)$,
$$\forall y_1 y_2 \in (c,d) \left[|y_1 - y_2| < \delta_1 \Rightarrow |g_x(y_1) - g_x(y_2)| < \frac{\epsilon}{2} \right],$$
and for all $y \in (c,d)$,
$$\forall x_1 x_2 \in (a,b) \left[|x_1 - x_2| < \delta_2 \Rightarrow |h_y(x_1) - h_y(x_2)| < \frac{\epsilon}{2} \right].$$
Let $\delta = \min\{\delta_1, \delta_2\}$, choose $\langle x_1, y_1 \rangle, \langle x_2, y_2 \rangle \in (a,b) \times (c,d)$, and observe: For $|x_1 - x_2| < \delta, |y_1 - y_2| < \delta$,
$$\begin{aligned}|f(x_1, y_1) - f(x_2, y_2)| &= |f(x_1, y_1) - f(x_1, y_2) + f(x_1, y_2) - f(x_2, y_2)| \\ &\leq |f(x_1, y_1) - f(x_1, y_2)| + |f(x_1, y_2) - f(x_2, y_2)| \\ &\leq |g_{x_1}(y_1) - g_{x_1}(y_2)| + |h_{y_2}(x_1) - h_{y_2}(x_2)| \\ &< \frac{\epsilon}{2} + \frac{\epsilon}{2} = \epsilon.\end{aligned}$$
$\square$

To conclude simple continuity, one can weaken the conditions slightly.[86]

3.9.9. Theorem. *Let f be a function defined on $(a,b) \times (c,d)$. Define, for $\langle x, y \rangle \in (a,b) \times (c,d)$,*
$$g_x(y) = f(x,y), \quad h_y(x) = f(x,y).$$
$$\mathcal{G} = \{g_x \,|\, x \in (a,b)\}, \quad \mathcal{H} = \{h_y \,|\, y \in (c,d)\}.$$
If every function g_x and h_y is uniformly continuous and one of the sets $\mathcal{G}$ and $\mathcal{H}$ is equicontinuous, then f is continuous.

3.9.10. Exercise. *Prove Theorem 3.9.9.*

The proper ϵ-δ formulation of Cauchy's second so-called false lemma is simply the following pair of theorems.

---

[86] In fact, Theorem 3.9.9 is a more faithful interpretation of Cauchy's Theorem 1 as stated and its proof than is Theorem 3.9.8.

3.9.11. Theorem. *Let $f_0, f_1, f_2, \ldots$ be a sequence of functions continuous on an interval I. If the series*
$$\sum_{n=0}^{\infty} f_n(x)$$
converges uniformly on I to f, then f is continuous on I.

3.9.12. Theorem. *Let $f_0, f_1, f_2, \ldots$ be a sequence of functions uniformly continuous on an interval I. If the series*
$$\sum_{n=0}^{\infty} f_n(x)$$
converges uniformly on I to f, then f is uniformly continuous on I.

Given that "continuity" to Cauchy is equivalent, not to continuity but to uniform continuity, the second of these is the more faithful of the two translations. If I is a closed bounded interval $[a, b]$, a theorem of the theory of the Calculus tells us that f is continuous on I iff it is uniformly so and there is thus no difference between the two theorems cited in that case. If one pays close attention to the precise assumptions of results, one notices that theorems of the Calculus are usually stated for functions continuous on closed intervals and differentiable on open ones. Continuous functions on closed bounded intervals are better behaved than those on open intervals. And differentiability usually expresses itself through the application of the Mean Value Theorem, which only requires the derivative to exist in the interior. Cauchy was a pioneer in this theory, working before this realisation and often cites results, like Theorems 3.9.11 and 3.9.12 (or even Theorems 3.9.8 and 3.9.9) for open intervals where we would now state them for closed bounded intervals.

In any event, we have already discussed uniform convergence in the preceding section, proving in Theorem 3.8.18 that the uniform limit of a sequence of continuous functions is continuous. The modifications of the proof given there to yield proofs of Theorems 3.9.11 and 3.9.12 are minor and I leave them to the reader.

The difficulty in obtaining "the correct" conversions of the "false lemmas" into the language of ϵ's and δ's was one of subtlety. With our next task—that of establishing the continuity of exponentiation—the difficulty is technical. "Difficulty" refers here to the problem of giving a short yet rigorous and intelligible exposition, not to the mathematical side of the problem, which is routine. But it is more involved than one (perhaps Cauchy?) might have expected.[87]

Cauchy defines infinitesimals to be variable quantities that tend to 0. In this he sounds like Bolzano, and we can identify Cauchy's α with Bolzano's ω, and, in the definition of continuity, identify the statement that $f(x + \alpha) - f(x)$ is infinitesimal with Bolzano's equation $f(x + \alpha) - f(x) = \Omega$. That said, we have basically already proven the following lemma.

---

[87]"Although he had been the first to define continuity, it seems Cauchy never proved the continuity of any particular function". (Freudenthal, *op. cit.*, p. 137)

9. CAUCHY'S PROOF

3.9.13. Lemma. *Let $0 < a < b$ and p/q be rational. The function $f(x) = x^{p/q}$ is continuous on (a, b) in Cauchy's sense.*

Proof. Let p/q be positive and let α be infinitesimal. For all $x \in (a, b)$,

$$|f(x+\alpha) - f(x)| = \left|(x+\alpha)^{p/q} - x^{p/q}\right|$$

$$= x^{p/q}\left|\left(1 + \frac{\alpha}{x}\right)^{p/q} - 1\right|$$

$$< b^{p/q}\left|\left(1 + \frac{\alpha}{x}\right)^{p/q} - 1\right|. \quad (119)$$

Because $x > 0$, $1/x$ is finite and $\alpha \cdot (1/x)$, being the product of an infinitesimal and a finite number, is infinitesimal, say ω_1. By Sublemma 3.8.7 of the preceding section (page 202, above),

$$\left(1 + \frac{\alpha}{x}\right)^{p/q} = (1 + \omega_1)^{p/q} = 1 + \Omega_1.$$

Thus (119) reads

$$|f(x+\alpha) - f(x)| < b^{p/q}|1 + \Omega_1 - 1| = b^{p/q}\Omega_1 = \Omega,$$

again since the product of an infinitesimal and a finite number is infinitesimal.

If p/q is negative, the proof is similar, but with $a^{p/q}$ serving as an upper bound on $x^{p/q}$. □

Our discussion of the continuity of exponentiation will be carried out in stages. First we will treat fixed positive integral powers, then fixed roots, then fixed positive rational powers and fixed negative powers, and finally fixed irrational powers. When that is finished we can consider exponentiation with a fixed base and variable exponent. Finally, we consider exponentiation as a function of two variables. Along the way, we establish monotonicity and existence, where necessary, as well as continuity.

I should remind the reader that we are filling in some major gaps in Cauchy's presentation. There is a lot of work involved here and the reader may prefer, on first reading, to skip past this discussion to page 243, below.

We begin with a simple lemma:

3.9.14. Lemma. *Let m be a positive integer and let $f : [0, \infty) \to [0, \infty)$ be defined by $f(x) = x^m$. Then: f is strictly increasing, i.e., for $x, y \in [0, \infty)$*

$$x < y \Rightarrow f(x) < f(y).$$

In particular, f is one-to-one.

Proof. Let $0 \leq x < y$ and write $y = x + h$ with $h > 0$. By Sublemma 3.8.6 of page 201, above,

$$y^m = (x+h)^m = x^m\left(1 + \frac{h}{m}\right)^m = x^m(1+u)$$

for some $u > 0$. Thus

$$y^m = x^m + x^m u > x^m.$$ □

Alternate Proof. Let x, y, h be as before and apply the Finite Binomial Theorem:
$$y^m = (x+h)^m = x^m + mx^{m-1}h + \text{ other nonnegative terms}$$
$$> x^m. \qquad \square$$

3.9.15. Lemma. *Let m be a positive integer. The function $f(x) = x^m$ is uniformly continuous on any closed bounded interval $[a,b]$.*

Proof. Let $\epsilon > 0$, $M = \max\{|a|, |b|\}$, and $x, y \in [a, b]$. Observe
$$|x^m - y^m| = |(x-y)(x^{m-1} + x^{m-2}y + \ldots + xy^{m-2} + y^{m-1})|$$
$$\leq |x-y|\left(|x^{m-1}| + |x^{m-2}y| + \ldots + |xy^{m-2}| + |y^{m-1}|\right)$$
$$\leq |x-y| \cdot \left(M^{m-1} + \ldots + M^{m-1}\right) = |x-y| \cdot mM^{m-1}.$$
To make $|x^m - y^m| < \epsilon$, it thus suffices to take $|x-y| < \delta = \dfrac{\epsilon}{mM^{m-1}}$. $\square$

The second stage, proving the uniform continuity of $g(x) = x^{1/m}$, requires us to explain definitely what is meant by $x^{1/m}$. If m is odd, every real number x has a unique m-th root, but if m is even, only nonnegative reals have such a root—two in fact for positive values of x. To this end, we think of $f(x) = x^m$ as a function

$$f : (-\infty, \infty) \to (-\infty, \infty) \text{ for odd } m$$
$$f : [0, \infty) \to [0, \infty) \text{ for even } m.$$

In either case, f is one-to-one and onto. g is thus taken to be the inverse function $g(x) = f^{-1}(x)$,

$$g : (-\infty, \infty) \to (-\infty, \infty) \text{ for odd } m$$
$$g : [0, \infty) \to [0, \infty) \text{ for even } m.$$

So long as we are being rigorous, we will verify the "obvious" facts that f is one-to-one and onto. In doing so, I will ignore negative arguments in the odd case. Because the function f is an *odd function* for m odd, i.e., $f(-x) = -f(x)$, the handling of the negative case is a routine algebraic reduction to the positive case. Thus, from here on, we consider $f(x) = x^m$ to map $[0, \infty)$ to $[0, \infty)$ regardless of whether m is odd or even.

That $f(x) = x^m$ is one-to-one for positive integral m is a trivial consequence of Lemma 3.9.14. The complementary assertion, that it maps $[0, \infty)$ onto $[0, \infty)$, however, is a deep result. It depends on a result generally stated but not proven in the standard Calculus course: the Intermediate Value Theorem.

3.9.16. Theorem (Intermediate Value Theorem). *Let f be a continuous function defined on $[a, b]$ and suppose $f(a)$ and $f(b)$ are of opposite signs. Then there is a $c \in (a, b)$ such that $f(c) = 0$.*

This Theorem was first proven by Bolzano in 1817[88] and again by Cauchy in the third appendicial note to his *Cours d'analyse*. And both men applied

---
[88]Bolzano, *Rein analytischer Beweis, op. cit.*

9. CAUCHY'S PROOF

the Theorem to polynomials in general, which means both possessed proofs of the existence of m-th roots.

The proofs given by Bolzano and Cauchy of the Intermediate Value Theorem necessarily depend on a completeness axiom. Unsurprisingly, Cauchy uses the convergence of a Cauchy convergent sequence. Bolzano, in outlining his proof uses the Least Upper Bound Principle, but he proved this Principle by appeal to the convergence of Cauchy sequences, which he justified by what I would describe as an appeal to his unfinished theory of real numbers, a theory unpublished before the second half of the twentieth century. If, however, one ignores this extra argument and accepts either the Least Upper Bound Principle or the convergence of Cauchy sequences as an axiom, Bolzano's proof becomes perfectly rigorous.

Bolzano's proof of Theorem 3.9.16. Let f be continuous on $[a, b]$ and assume for the sake of definiteness that $f(a)$ is negative and $f(b)$ positive. Let

$$X = \{x \in [a, b] \mid f(x) < 0\}.$$

b is an upper bound on X. Let c be the least upper bound of X. The claim is that $f(c) = 0$.

Suppose $f(c) > 0$. By the continuity of f at c, for x near c, $f(x)$ is also positive: Choose $\delta > 0$ so that for $x \in [a, b]$, if $|x - c| < \delta$, then

$$|f(x) - f(c)| < \frac{f(c)}{2}.$$

Thus

$$-\frac{f(c)}{2} < f(x) - f(c) < \frac{f(c)}{2},$$

whence

$$0 < \frac{f(c)}{2} = f(c) - \frac{f(c)}{2} < f(x).$$

Let c_0 be any element of $[a, b]$ in $(c - \delta, c)$. For $x > c_0$ either $x \in (c - \delta, c + \delta)$ and $f(x)$ is positive, or $x \geq c + \delta$ and is greater than some upper bound of X, whence $f(x)$ is positive. Hence, for all $x \in [a, b]$,

$$f(x) < 0 \Rightarrow x < c_0,$$

i.e., c_0 is an upper bound on X. But $c_0 < c$ and c is the least upper bound. Thus we have a contradiction and cannot have $f(c) > 0$.

Similarly, $f(c)$ cannot be < 0. Thus $f(c) = 0$. □

For strictly increasing functions f, like $f(x) = x^m$, the last part of the argument is perhaps a bit clearer.

Cauchy's proof is slightly more complicated. One partitions $[a, b]$ into k subintervals of equal length with endpoints $a = x_0 < x_1 < \ldots < x_k = b$. Either $f(x_i) = 0$ for some i and one chooses $c = x_i$ or there is an i for which $f(x_i) < 0 < f(x_{i+1})$. Choose x_i to be c_1 and partition $[x_i, x_{i+1}]\ldots$ One ends up with a monotonically increasing Cauchy sequence $c_1, c_2, \ldots$, the limit of which will be the zero of f.

3.9.17. Corollary. Let m be a positive integer and let $f : [0, \infty) \to [0, \infty)$ be defined by $f(x) = x^m$. Then f is onto: for any $d \geq 0$ there is a $c \in [0, \infty)$ such that $f(c) = d$, i.e., $\sqrt[m]{d}$ exists.

Proof. Apply Theorem 3.9.16 to $g(x) = f(x) - d$ on the interval $[0, d+1]$. □

Thus we have seen that the function $f(x) = x^{1/m}$ is defined for all nonnegative real numbers x. The continuity of f comes for free.

3.9.18. Lemma. Let $f : [a, b] \to [c, d]$ be strictly increasing and onto. Then f is continuous.

Proof. Let $\epsilon > 0$ be given and let $x \in [a, b]$. Suppose, to simplify the argument, x is not a or b. Let

$$\alpha = \max\{f(x) - \epsilon, c\}, \quad \beta = \min\{f(x) + \epsilon, d\},$$

so that $(\alpha, \beta) \subseteq (f(x) - \epsilon, f(x) + \epsilon)$. Because f is onto, there are $\alpha', \beta' \in [a, b]$ such that $f(\alpha') = \alpha, f(\beta') = \beta$. Because f is strictly increasing, we have $\alpha' < x < \beta'$. If we now let $\delta = \min\{x - \alpha', \beta' - x\}$, we have $(x - \delta, x + \delta) \subseteq (\alpha', \beta')$ and

$$|y - x| < \delta \Rightarrow y \in (x - \delta, x + \delta) \Rightarrow y \in (\alpha', \beta')$$
$$\Rightarrow f(y) \in (\alpha, \beta) \Rightarrow f(y) \in (f(x) - \epsilon, f(x) + \epsilon)$$
$$\Rightarrow |f(y) - f(x)| < \epsilon.$$

The argument for $x = a$ or $x = b$ is similar. □

3.9.19. Corollary. Let m be a positive integer. Then $f(x) = x^{1/m}$ is continuous on $[0, \infty)$.

For, being the inverse of a strictly increasing function, $f(x) = x^{1/m}$ is also strictly increasing. In fact, we can now prove the general monotonicity result:

3.9.20. Lemma. Let p, q be integers with $q > 0$ and $f(x) = x^{p/q}$.
i. if $p/q > 0$, f is strictly increasing
ii. if $p/q < 0$, f is strictly decreasing.

Proof. i. $f(x) = (x^p)^{1/q}$ is the composition of two strictly increasing functions and is thus strictly increasing:

$$x < y \Rightarrow x^p < y^p \Rightarrow (x^p)^{1/q} < (y^p)^{1/q}.$$

ii. $f(x) = x^{p/q} = 1/x^{|p/q|}$ is the composition of a strictly decreasing and a strictly increasing function and is thus strictly decreasing:

$$x < y \Rightarrow x^{|p/q|} < y^{|p/q|} \Rightarrow \frac{1}{y^{|p/q|}} < \frac{1}{x^{|p/q|}}. \quad □$$

We can now conclude the continuity of $f(x) = x^{p/q} = (x^p)^{1/q}$ and that of $f(x) = x^{-p/q} = 1/x^{p/q}$ by appeal to the following Lemma.

3.9.21. Lemma. Let $f(x) = g(h(x))$ be the result of composing two functions. If g, h are (uniformly) continuous, then f is (uniformly) continuous.

9. CAUCHY'S PROOF

Proof. We handle the uniform case. Let $\epsilon > 0$ be given and choose δ_1, δ so that

$$|u - v| < \delta_1 \Rightarrow |g(u) - g(v)| < \epsilon$$
$$|x - y| < \delta \Rightarrow |h(x) - h(y)| < \delta_1.$$

Then

$$|x - y| < \delta \Rightarrow |h(x) - h(y)| < \delta_1$$
$$\Rightarrow |g(h(x)) - g(h(y))| < \epsilon$$
$$\Rightarrow |f(x) - f(y)| < \epsilon. \qquad \square$$

A concern may now arise. The proof of Lemma 3.9.18 only established continuity, not uniform continuity. Bolzano would eventually prove that continuity on a closed bounded interval entails uniform continuity, so we have pretty much established the uniform continuity of $f(x) = x^\mu$ on any interval $[a, b]$ for all rational values of μ. It will be convenient, however, to have a better proof.

The obvious proof of the uniform continuity of $f(x) = x^{p/q}$ in appropriate neighbourhoods $[a, b]$ is given by a careful reworking of the proof of Sublemma 3.8.7 of the preceding section. I am going to cheat, however, and call upon Bolzano's calculation of the derivative of f and the Mean Value Theorem. Both Bolzano and Cauchy gave proofs of the Mean Value Theorem, albeit in later works than those reported on here, so I am not straying too far from the path of faithfulness to history.

3.9.22. Theorem. *Let R be a positive rational number and, for p/q rational numbers for which $|p/q| \leq R$, define*

$$f_{p/q}(x) = x^{p/q}$$

and

$$\mathcal{F} = \{f_{p/q} \mid |p/q| \leq R\}.$$

The set $\mathcal{F}$ is equicontinuous on $[a, b]$ for $0 < a < b$. In particular, each $f_{p/q}$ is uniformly continuous on $[a, b]$.

Proof. By the Mean Value Theorem,

$$\left| x^{p/q} - y^{p/q} \right| = |x - y| \cdot \left| \frac{p}{q} \right| \cdot \theta^{p/q - 1} \qquad (120)$$

for some θ between x and y. The factor $|p/q|$ is clearly bounded above by R. We can give explicit bounds for $\theta^{p/q-1}$ for the cases $p/q > 1, p/q = 1$, and $p/q < 1$.

If $p/q > 1$, the function $f(\theta) = \theta^{p/q-1}$ is strictly increasing, whence it is bounded above by $b^{p/q-1}$, but $b^{p/q-1} \leq b^{R-1}$.

If $p/q = 1$, then $f(\theta) = \theta^0 = 1$.

If $p/q < 1$, then $p/q - 1 < 0$ and $f(\theta) = \theta^{p/q-1}$ is strictly decreasing, whence it is bounded above by $a^{p/q-1} \leq a^{-R-1}$.

If we set $M = \max\{b^{R-1}, 1, a^{-R-1}\}$, we see that for all $x, y \in [a, b]$ formula (120) yields

$$\left| x^{p/q} - y^{p/q} \right| \leq |x - y| \cdot RM$$

which is less than ϵ whenever $|x - y| < \delta = \dfrac{\epsilon}{RM}$. □

As I mentioned earlier, both Bolzano and Cauchy had proofs of the Mean Value Theorem, albeit given in later works than those under consideration, so the proof given here is not too much of a cheat, but it does seem to violate the spirit of Cauchy's presentation as well as a strict adherence to *Methodenreinheit*, the call for *purity of methods*: it seems a cheat to call on differentiability to prove a theorem about continuity. Sometimes, however, the savings afforded by appeal to "inappropriate" advanced methods are too great to resist.

Still, I am feeling a bit guilty about cheating here and thus I ought to give a careful ϵ-δ proof of this Theorem along the lines of the proof of Sublemma 3.8.7. However, I just don't feel like sorting through the details and have, after much soul searching, come up with an alternative way to soothe my conscience:

3.9.23. Exercise. *Provide a detailed ϵ-δ proof of Theorem 3.9.22 that makes no appeal to the Mean Value Theorem.*

Even with all of this we have still not finished discussing the continuity of the function $f(x, \mu) = x^\mu$. There are still the problems of defining x^μ for irrational exponents, proving the uniform continuity of the individual functions $f_x(\mu) = x^\mu$ for irrational μ on intervals $[-R, R]$, and discussing the uniform continuity of f itself on $[a, b] \times [-R, R]$ for $[a, b] \subseteq (0, \infty)$ and $R > 0$.

I quoted what Bolzano had to say about the problem in *Der binomische Lehrsatz* on page 204, above. For convenience I repeat the quote here:

> Finally, if n is *irrational* then there is always a fraction $\frac{p}{q}$ (positive or negative) which comes as close as required to n. But then it follows from the definition of the *concept* of an irrational power that the quantity $a^{p/q}$ gives a value as close to that value of a^n as desired, if $\frac{p}{q}$ comes as close to the value n as desired.

This is obviously an implicit appeal to continuity and the pre-supposed existence of a quantity a^n for irrational n. Cauchy states the definition of A^B for irrational B more explicitly:

> When B is an irrational number, we can then obtain rational numbers with values approaching it more and more closely. We easily prove that under the same hypothesis, powers of A indicated by the rational numbers in question approach more and more closely towards a certain limit. This limit is the power of A of degree B.[89]

Neither Bolzano nor Cauchy proves the existence of x^μ for irrational exponents μ, but Cauchy explicitly defines the value and promises us that there is an easy proof that it exists.

3.9.24. Definition. *Let μ be an irrational number and $x > 0$.*

$$x^\mu = \lim_{q \to \mu} x^q, \ q \ rational,$$

provided this limit exists.

---

[89] Bradley and Sandifer, *op. cit.*, p. 274.

9. CAUCHY'S PROOF

The proof that this limit exists is not so much a matter of difficulty as of subtlety.

3.9.25. Example. *Let f be defined on $\mathbb{Q} = \{q \mid q \text{ is rational}\}$ by*

$$f(q) = \begin{cases} q, & q < \sqrt{2} \\ 1+q, & q > \sqrt{2}. \end{cases}$$

f is strictly increasing and continuous on $\mathbb{Q}$, but $\lim_{q \to \sqrt{2}} f(q)$ does not exist.

It follows from this example that not all monotone or continuous functions on $\mathbb{Q}$ have continuous extensions to all the reals. What is required is uniform continuity.

3.9.26. Theorem. *Let f be defined and uniformly continuous for all rationals $q \in [a,b]$. There is a unique uniformly continuous extension $\overline{f}$ of f to all reals $\mu \in [a,b]$.*

Proof. There are two parts to the proof. First, we find a candidate L for $\overline{f}(\mu)$. Then we show that $\overline{f}$, so defined, is uniformly continuous on $[a,b]$.

For the first part of the proof, let $\mu \in [a,b]$ be irrational. Bolzano would eventually develop a theory of real numbers as "infinite number expressions" which could be arbitrarily well-approximated by rationals. Cauchy regarded real numbers as measurements, explaining rationals as resulting from the repetition of a unit a fixed number of times, subdivision into a fixed number of equal parts, or a combination of these two operations. After intuitively introducing the notion of limit, he remarks on irrationals,

> Thus, for example, an irrational number is the limit of the various fractions that give better and better approximations to it.[90]

He repeats this assertion in defining the product of a real number A with an irrational number B[91] in a phrasing paralleling that that for exponentiation which we cited on the immediately preceding page.[92]

The upshot of this is that there is a sequence $q_1, q_2, q_3, \ldots$ of rational numbers in $[a,b]$ converging to μ. By convergence, the sequence $q_1, q_2, q_3, \ldots$ is Cauchy convergent and, for any $\delta > 0$ there is an n_0 so large that for all $m, n > n_0$,

$$|q_m - q_n| < \delta.$$

Let $\epsilon > 0$ be given and choose $\delta > 0$ such that, for all rational $p, q \in [a,b]$,

$$|p - q| < \delta \Rightarrow |f(p) - f(q)| < \epsilon.$$

Thus, for all $m, n > n_0$, we have $|f(q_m) - f(q_n)| < \epsilon$. Thus, the sequence $f(q_1), f(q_2), f(q_3), \ldots$ is Cauchy convergent and has a limit L. Define $\overline{f}(\mu)$ to be L.

---

[90] *Ibid.*, p. 6.
[91] *Ibid.*, p. 271.
[92] Cauchy does not do the same for addition, presumably because it has an intuitive meaning for measurements. The first full discussion of the real numbers would be given in the 1840s by Karl Weierstrass, but not published until the 1870s, at which time several solid foundations of the real number system were being published.

For each rational value of $\mu \in [a,b]$, we may also choose a sequence $q_1, q_2, q_3, \ldots$ of elements of $[a,b]$ convergent to μ, e.g., the constant sequence $q_n = \mu$. Thus for each $\mu \in [a,b]$ we can choose a sequence $q_n \to \mu$ and define
$$\overline{f}(\mu) = \lim_{n\to\infty} f(q_n), \text{ where } q_n \to \mu.$$

For μ rational, $\overline{f}(\mu) = \lim_{n\to\infty} f(q_n) = f(\mu)$ by the continutiy of f at μ or, if we chose the constant sequence $q_n = \mu$, by the fact that $f(q_n)$ then equals $f(\mu)$. For μ irrational, the value of $\overline{f}(\mu)$ could very well depend on the exact choice of sequence $q_n \to \mu$. That it does not, i.e., that the assigned value of $\overline{f}(\mu)$ is independent of the choice of sequence can be proven directly, but it also follows from the continuity of $\overline{f}$ and we need not prove it directly.

To establish the uniform continuity of $\overline{f}$, let $\epsilon > 0$ and choose $\delta_1 > 0$ so that, for all rational $p, q \in [a,b]$,
$$|p - q| < \delta_1 \Rightarrow |f(p) - f(q)| < \frac{\epsilon}{3}.$$

Let $\mu, \nu \in [a,b]$, let $p_1, p_2, p_3, \ldots$ and $q_1, q_2, q_3, \ldots$ be the sequences of rationals in $[a,b]$ converging to μ, ν, respectively, from which $\overline{f}(\mu)$ and $\overline{f}(\nu)$ have been defined:
$$\overline{f}(\mu) = \lim_{n\to\infty} f(p_n), \quad \overline{f}(\nu) = \lim_{n\to\infty} f(q_n).$$

Choose n_1, n_2 so large that
$$n > n_1 \Rightarrow |\mu - p_n| < \frac{\delta_1}{3}$$
$$n > n_2 \Rightarrow |\nu - q_n| < \frac{\delta_1}{3},$$

and let $n_3 = \max\{n_1, n_2\}$. Let $\delta = \delta_1/3$ and observe, for $n > n_3$,
$$|\mu - \nu| < \delta \Rightarrow |p_n - q_n| \le |p_n - \mu| + |\mu - \nu| + |\nu - q_n|$$
$$\Rightarrow |p_n - q_n| < \frac{\delta_1}{3} + \frac{\delta_1}{3} + \frac{\delta_1}{3} = \delta$$
$$\Rightarrow |f(p_n) - f(q_n)| < \frac{\epsilon}{3}.$$

But by the definition of $\overline{f}$, there are n_4, n_5 such that
$$n > n_4 \Rightarrow |f(p_n) - \overline{f}(\mu)| < \frac{\epsilon}{3}$$
$$n > n_5 \Rightarrow |f(q_n) - \overline{f}(\nu)| < \frac{\epsilon}{3}.$$

Thus, for $n > n_0 = \max\{n_3, n_4, n_5\}$,
$$|\overline{f}(\mu) - \overline{f}(\nu)| \le |\overline{f}(\mu) - f(p_n)| + |f(p_n) - f(q_n)| + |f(q_n) - \overline{f}(\nu)|$$
$$< \frac{\epsilon}{3} + \frac{\epsilon}{3} + \frac{\epsilon}{3} = \epsilon. \qquad \square$$

3.9.27. Corollary. *Let $\mathcal{F}$ be an equicontinuous family of functions f defined on the rationals of an interval $[a,b]$, and let $\overline{\mathcal{F}}$ be the family of extensions $\overline{f}$ of functions $f \in \mathcal{F}$ to all reals in $[a,b]$. The family $\overline{\mathcal{F}}$ is again equicontinuous.*

9. CAUCHY'S PROOF

For, every function $f \in \mathcal{F}$ is uniformly continuous and by the Theorem has a uniformly continuous extension $\overline{f}$. But if one examines the proof of the uniform continuity of $\overline{f}$, one notices that the modulus of continuity produced, δ, for a given ϵ depended only on the modulus δ_1 of the function f. If all the functions $f \in \mathcal{F}$ share the same modulus of continuity, the functions $\overline{f} \in \overline{\mathcal{F}}$ will likewise have a common modulus of continuity.

We wish, of course, to apply this to the functions
$$f_x(q) = x^q, \; q \text{ rational},$$
for $x > 0$. To do this we must show that f_x is uniformly continuous on the rational numbers in $[a, b]$.

3.9.28. Lemma. *Let a be a positive real number and define*
$$f_a(q) = a^q, \text{ for rational } q.$$
i. *if $a > 1$, f_a is strictly increasing*
ii. *if $a < 1$, f_a is strictly decreasing*
iii. *if $a = 1$, f_a is constant: $f_1(q) = 1^q = 1$.*

Proof. iii is trivial. And ii follows from i: If $a < 1$, then $1/a > 1$ and
$$p < q \Rightarrow f_{1/a}(p) < f_{1/a}(q) \Rightarrow \left(\frac{1}{a}\right)^p < \left(\frac{1}{a}\right)^q$$
$$\Rightarrow a^q < a^p \Rightarrow f_a(q) < f_a(p).$$
Thus, it suffices to prove assertion i.

Assume $a > 1$ and $p < q$, say $q = p + r$, where r is a positive rational number. Now
$$a^q = a^{p+r} = a^p \cdot a^r$$
and we will have $a^p < a^q$ if $a^r > 1$. But the function g defined by
$$g(x) = x^r,$$
is strictly increasing, $g(1) = 1^r = 1$, and $a > 1$, whence $g(a) = a^r > 1$. □

3.9.29. Theorem. *For $[a, b] \subseteq (0, \infty)$ and $x \in [a, b]$, define*
$$f_x(q) = x^q$$
for rational $q \in [-R, R]$. The family
$$\mathcal{F} = \{f_x \,|\, x \in [a, b]\}$$
is equicontinuous.

Proof. Let $\epsilon > 0$ be given.

Let $x \in [a, b]$ and let $p, q \in [-R, R]$ be rational numbers with $q < p$. Write $p = q + r$, r a positive rational number. Consider
$$|x^p - x^q| = |x^{q+r} - x^q| = |x^q| \cdot |x^r - 1|.$$

For fixed $q > 0$, $g(x) = x^q$ is strictly increasing as a function of x. Hence $|x^q| \le b^q$. And $h(q) = b^q$ is, by the Lemma, strictly increasing for $b > 1$, constant (and equal to 1) for $b = 1$, and strictly decreasing for $b < 1$. Thus
$$b^q \le M_0 = \max\{b^R, 1, b^{-R}\}.$$

For $q < 0$, $g(x) = x^q$ is strictly decreasing and $x^q \leq a^q$. In this case we have
$$a^q \leq M_1 = \max\{a^R, 1, a^{-R}\}.$$
Thus, for $M = \max\{M_0, M_1\}$, we have
$$|x^p - x^q| \leq M \cdot |x^r - 1|,$$
with the constant M depending only on a, b, R.

If n is a positive integer such that $r < 1/n$ (i.e., if $n > 1/r$), we have

for $x > 1$, $|x^r - 1| = x^r - 1 < x^{1/n} - 1$, by the Lemma

for $x - 1$, $|x^r - 1| = 1 - 1 = 0 = x^{1/n} - 1$

for $x < 1$, $|x^r - 1| = 1 - x^r \leq 1 - x^{1/n}$, since $x^r > x^{1/n}$ by the Lemma.

Thus, for $x \in [a, b]$,
$$|x^r - 1| \leq |x^{1/n} - 1|$$
and we can show $|x^r - 1|$ tends to 0 by showing $|x^{1/n} - 1|$ tends to 0.

We have
$$|x - 1| = \left|\left(x^{1/n}\right)^n - 1\right| = \left|x^{1/n} - 1\right| \cdot \left|\left(x^{1/n}\right)^{n-1} + \left(x^{1/n}\right)^{n-2} + \ldots + 1\right|$$
$$= \left|x^{1/n} - 1\right| \cdot \left(\left(x^{1/n}\right)^{n-1} + \left(x^{1/n}\right)^{n-2} + \ldots + 1\right).$$

Now each function $g(x) = x^\mu$, for positive rational exponents μ, is strictly increasing, whence the function
$$h(x) = \left(x^{1/n}\right)^{n-1} + \left(x^{1/n}\right)^{n-2} + \ldots + 1$$
is also strictly increasing and assumes a minimum on $[a, b]$ at a. By Lemma 3.9.28, the function
$$k(x) = a^x$$
is increasing for $a > 1$, decreasing for $a < 1$, and constant for $a = 1$. Thus, for $a < 1$,
$$a^{k/n} \geq a^1, \text{ for } k = 1, 2, \ldots, n-1$$
and $h(a) > a + a + \ldots + a = na$. And for $a > 1$,
$$h(a) \geq \left(1^{1/n}\right)^{n-1} + \ldots + 1 = n.$$

Thus, for $A = \min\{a, 1\}$,
$$|x - 1| = |x^{1/n} - 1| \cdot h(x) \geq |x^{1/n} - 1| \cdot h(a) \geq |x^{1/n} - 1| \cdot nA,$$
i.e.,
$$|x^{1/n} - 1| \leq \frac{|x - 1|}{nA} \leq \frac{|x| + 1}{nA} \leq \frac{b + 1}{nA}.$$
Choose n so large that $M \frac{b+1}{nA} < \epsilon$ and choose $\delta < \frac{1}{n}$. Observe
$$|p - q| < \delta \Rightarrow |r| < \delta$$
$$\Rightarrow r < \frac{1}{n}$$
$$\Rightarrow |x^{1/n} - 1| < \frac{b+1}{nA}$$

$$\Rightarrow |x^r - 1| < \frac{b+1}{nA}$$

$$\Rightarrow |x^p - x^q| < M\frac{b+1}{nA} = \epsilon. \qquad \square$$

We are all but finished establishing the continuity of exponentiation. Given intervals $[a,b] \subseteq (0,\infty)$ and $[-R,R] \subseteq (-\infty,\infty)$. Theorem 3.9.29 tells us that the family,

$$\mathcal{F} = \{f_x \,|\, x \in [a,b]\},$$

of functions $f_x = x^q$ defined on the rationals $q \in [-R,R]$ is equicontinuous. By Theorem 3.9.26 each f_x has a uniformly continuous extension $\overline{f}_x$ defined for all $\mu \in [-R,R]$. Moreover, the family

$$\overline{\mathcal{F}} = \{\overline{f}_x \,|\, x \in [a,b]\}$$

is equicontinuous by Corollary 3.9.27.

But, defining $f(x,\mu) = \overline{f}_x(\mu)$ for $\langle x,\mu\rangle \in [a,b] \times [-R,R]$, we know that the family $\mathcal{G}$ of functions

$$g_\mu(x) = f(x,\mu), \ \mu \in [-R,R],$$

is equicontinuous by Theorem 3.9.22. By Cauchy's "false lemma" (i.e., Theorem 3.9.8), f itself is uniformly continuous as a function of two variables in the rectangle $[a,b] \times [-R,R]$.

3.9.30. Definition. *Let μ be irrational. For $x > 0$, let a,b,R be given such that $x \in [a,b] \subseteq (0,\infty)$ and $\mu \in [-R,R]$. We define $x^\mu = f(x,\mu)$ for f as just defined.*

At this point, or perhaps just before the Definition, one would point out that the value of x^μ is the limit of values of x^q for rationals $q \to \mu$ and thus does not depend on the exact choice of a,b,R.

So we have finally defined x^μ for $x > 0$ for all real μ and verified that it is continuous in both variables on $(0,\infty) \times (-\infty,\infty)$, uniformly so on every closed rectangle $[a,b] \times [-R,R]$ contained in the domain. We are now ready to return to Cauchy's proof of the Binomial Theorem. Given the number of pages I have already devoted to Cauchy, I shall be brief. I can do this because the overall structure of the proof is familiar: it is the Euler-Bolzano proof, with some lemmas proven in greater generality and, perhaps, greater clarity than Bolzano provided.

By what we've been discussing, for fixed $x > -1$,

$$g(\mu) = (1+x)^\mu$$

is a continuous function of μ.

3.9.31. Lemma. *Let x be fixed, $|x| < 1$. The function*

$$\varphi(\mu) = \sum_{k=0}^{\infty} \binom{\mu}{k} x^k$$

is uniformly continuous on any interval $[-R,R]$.

Proof. Apply the Ratio Test:

$$\left|\frac{\binom{\mu}{k}x^k}{\binom{\mu}{k+1}x^{k+1}}\right| = \left|\frac{\binom{\mu}{k}}{\frac{\mu-k}{k+1}\binom{\mu}{k}} \cdot \frac{1}{x}\right| = \left|\frac{k+1}{\mu-k}\right| \cdot \left|\frac{1}{x}\right| \to \frac{1}{|x|}$$

as $k \to \infty$, But $1/|x| > 1$, whence the series for $\varphi(\mu)$ is absolutely convergent for all μ, and thus convergent. But for $\mu \in [-R, R]$, one readily sees

$$\left|\binom{\mu}{k}x^k\right| \leq \left|\binom{-R}{k}x^k\right|.$$

Let $\epsilon > 0$ be given. Choose n_0 so large that for $n > n_0$,

$$\sum_{k=n+1}^{\infty}\left|\binom{-R}{k}x^k\right| < \epsilon,$$

and observe

$$\left|\varphi(\mu) - \sum_{k=0}^{n}\binom{\mu}{k}x^k\right| = \left|\sum_{k=n+1}^{\infty}\binom{\mu}{k}x^k\right|$$

$$\leq \sum_{k=n+1}^{\infty}\left|\binom{\mu}{k}x^k\right|$$

$$\leq \sum_{k=n+1}^{\infty}\left|\binom{-R}{k}x^k\right| < \epsilon.$$

Thus convergence is uniform. By Cauchy's other "false lemma", φ is uniformly continuous on $[-R, R]$. □

Cauchy's proof of the Binomial Theorem is embedded in more theory, with some lemmas stated in greater generality than needed for this Theorem. For example, he does not just prove, as Bolzano did, that the Cauchy product of two convergent binomial series is convergent with limit equalling the product of the limits of the two series, but proves more generally that the Cauchy product of two absolutely convergent series also converges and equals the product of the two series. His proof of this is essentially the simplified proof we gave on page 219, above, but written with general absolutely convergent series replacing the binomial ones. I leave the rewriting as an exercise for the more industrious reader, and refer the less enthusiastic reader to Abel's general proof given on pages 272 – 273, below, which proof is not dissimilar. This is done in Chapter 6 of the *Cours d'analyse*[93], where he also cites a simple counterexample to the general convergence of the Cauchy product when only conditional convergence is assumed of the individual factors:

---

[93]Bradley and Sandifer, *op. cit.*, pp. 100 – 102 and 106.

3.9.32. Example. Let, for $n = 1, 2, 3, \ldots$, $a_n = b_n = (-1)^{n+1}/\sqrt{n}$. The series

$$\sum_{n=1}^{\infty} a_n = \sum_{n=1}^{\infty} b_n = 1 - \frac{1}{\sqrt{2}} + \frac{1}{\sqrt{3}} - \frac{1}{\sqrt{4}} + \ldots$$

converges by the Alternating Convergence Test. The general term of the Cauchy product is

$$c_n = \pm \left(\frac{1}{\sqrt{n \cdot 1}} + \frac{1}{\sqrt{(n-1)2}} + \frac{1}{\sqrt{(n-2)3}} + \ldots + \frac{1}{\sqrt{2(n-1)}} + \frac{1}{\sqrt{1 \cdot n}} \right).$$

In Note II of the Appendix, Cauchy proves a number of useful inequalities, culminating in the famous inequality of Gauss relating the arithmetical and geometrical means[94]: For m positive real numbers $A, B, C, \ldots, M$,

$$\sqrt[m]{ABC \cdots M} \leq \frac{A + B + C + \ldots + M}{m}.$$

For $m = 2$, it is proven simply:

$$0 \leq (\sqrt{A} - \sqrt{B})^2 = A - 2\sqrt{AB} + B,$$

whence

$$\sqrt{AB} \leq \frac{A + B}{2}.$$

Applying this to k and $n - k + 1$ with $0 < k < n$, we get

$$\frac{n+1}{2} = \frac{n - k + 1 + k}{2} \geq \sqrt{(n - k + 1)k},$$

whence

$$\frac{1}{\sqrt{(n - k + 1)k}} \geq \frac{2}{n + 1}.$$

Thus

$$|c_n| \geq \underbrace{\frac{2}{n+1} + \frac{2}{n+1} + \ldots + \frac{2}{n+1}}_{n} = \frac{2n}{n+1} > 1$$

for $n \geq 1$ and $\sum_{n=1}^{\infty} c_n$ diverges.

He applies his theorem on Cauchy products in Chapter 6 to $\varphi(\mu)$ and $\varphi(\nu)$.[95] That this product is $\varphi(\mu + \nu)$ depends on Vandermonde's Theorem,

$$\sum_{i=0}^{k} \binom{\mu}{i} \binom{\nu}{k-i} = \binom{\mu + \nu}{k},$$

which he has already proven in Chapter 4.[96] To conclude $\varphi(\mu) = (1 + x)^\mu$ [97], he now cites a general theorem proven in Chapter 5 [98]:

---

[94] Ibid., pp. 306 – 307.
[95] Ibid., p. 107.
[96] Ibid., pp. 67 – 69.
[97] Ibid., pp. 110 – 111.
[98] Ibid., pp. 73 – 75.

3.9.33. Lemma. *Let φ be continuous on some interval $[\alpha, \beta]$ and assume that for all real $\mu, \nu \in [\alpha, \beta]$, $\varphi(\mu + \nu) = \varphi(\mu) \cdot \varphi(\nu)$. Then, for all μ, $\varphi(\mu) = \varphi(1)^\mu$.*

3.9.34. Exercise. *Prove the Lemma.*

To conclude that

$$(1+x)^\mu = \sum_{k=0}^{\infty} \binom{\mu}{k} x^k = \varphi(\mu),$$

it suffices, since $\psi(\mu) = (1+x)^\mu$ is also continuous and satisfies $\psi(\mu + \nu) = \psi(\mu) \cdot \psi(\nu)$ [99], to show $\psi(1) = \varphi(1)$. But $\psi(1) = (1+x)^1 = 1 + x$ and

$$\varphi(1) = \sum_{k=0}^{\infty} \binom{1}{k} x^k = \binom{1}{0} x^0 + \binom{1}{1} x^1 + 0x^2 + 0x^3 + \ldots = 1 + x.$$

This finishes Cauchy's proof of the Binomial Theorem for arbitrary real exponents. It does not, however, finish our discussion of Cauchy and the Binomial Theorem.

10. The Exponential and Logarithmic Functions

As we saw in Chapter 2, the Binomial Theorem opens the door to many other power series. Both Bolzano and Cauchy continued their expositions to considerations of the exponential and logarithmic functions. Bolzano's proofs are from first principles and are a bit ponderous. Cauchy, however, reduces everything to the Binomial Theorem.

Corollary I.—If we replace μ by $\frac{1}{\alpha}$ and x by αx in equation (20)[100], where α denotes an infinitely small quantity, then for all values of αx contained between the limits -1 and $+1$, or what amounts to the same thing, for all values of x contained between the limits $-\frac{1}{\alpha}$ and $+\frac{1}{\alpha}$, we have

$$(1 + \alpha x)^{\frac{1}{\alpha}} = 1 + \frac{x}{1} + \frac{x^2}{1 \cdot 2}(1 - \alpha) + \frac{x^3}{1 \cdot 2 \cdot 3}(1 - \alpha)(1 - 2\alpha) + \ldots$$

$$\left(x = -\frac{1}{\alpha}, \; x = +\frac{1}{\alpha} \right).$$

This last equation ought to remain true, no matter how small the numerical value of α may be. If we denote as usual by the abbreviation lim placed in front of an expression that includes the variable α the limit towards which this expression converges as the numerical value of α decreases indefinitely, then in passing to the limit, we

---

[99]Obviously it holds for rational values of μ, ν. To conclude the same for irrational values, let $p_n \to \mu$, $q_n \to \nu$ and take limits:

$$\psi(\mu + \nu) = \lim_{n \to \infty} \psi(p_n + q_n) = \lim_{n \to \infty} \left(\psi(p_n) \cdot \psi(q_n) \right)$$

$$= \left(\lim_{n \to \infty} \psi(p_n) \right) \cdot \left(\lim_{n \to \infty} \psi(q_n) \right) = \psi(\mu) \cdot \psi(\nu).$$

[100]Cauchy's formula (20) expresses the binomial expansion for $(1+x)^\mu$.

10. THE EXPONENTIAL AND LOGARITHMIC FUNCTIONS

find

(21) $\lim(1+\alpha x)^{\frac{1}{\alpha}} = 1 + \dfrac{x}{1} + \dfrac{x^2}{1\cdot 2} + \dfrac{x^3}{1\cdot 2\cdot 3} + \ldots$ $(x = -\infty, x = +\infty)$

It remains to seek the limit of $(1+\alpha x)^{\frac{1}{\alpha}}\ldots$ [101]

The rest of Cauchy's discussion is so close to the modern that I would simply be repeating myself when interpreting Cauchy should I continue, so I break off here.

In modern terms, Cauchy defines the function $\exp(x)$ in terms of exponentiation:

3.10.1. Definition. *For real x,*

$$\exp(x) = \lim_{\alpha\to 0}(1+\alpha x)^{1/\alpha},$$

provided the limit exists.

By the Binomial Theorem, for any $\alpha \neq 0$, one has for $|x| < 1/|\alpha|$,

$$(1+\alpha x)^{1/\alpha} = \sum_{n=0}^{\infty}\binom{1/\alpha}{n}(\alpha x)^n$$

$$= 1 + \dfrac{1}{\alpha}\alpha x + \dfrac{\frac{1}{\alpha}\left(\frac{1}{\alpha}-1\right)}{2\cdot 1}\alpha^2 x^2 + \dfrac{\frac{1}{\alpha}\left(\frac{1}{\alpha}-1\right)\left(\frac{1}{\alpha}-2\right)}{3\cdot 2\cdot 1}\alpha^3 x^3 + \ldots$$

$$= 1 + x + \dfrac{1-\alpha}{2}x^2 + \dfrac{(1-\alpha)(1-2\alpha)}{6}x^3 + \ldots$$

But the binomial series is continuous as a function of μ. Thus, for any x,

$$\lim_{\alpha\to 0}(1+\alpha x)^{1/\alpha} = 1 + x + \dfrac{x^2}{2} + \dfrac{x^3}{6} + \ldots = \sum_{n=0}^{\infty}\dfrac{x^n}{n!}.$$

Substituting $x = 1$, one has

$$\lim_{\alpha\to 0}(1+\alpha)^{1/\alpha} = \sum_{n=0}^{\infty}\dfrac{1}{n!}.$$

We define this sum to be e and note that

$$(1+\alpha x)^{1/\alpha} = \left((1+\alpha x)^{1/(\alpha x)}\right)^x = \left((1+\beta)^{1/\beta}\right)^x \text{ for } \beta = \alpha x,$$

and, since $\beta \to 0$ as $\alpha \to 0$,

$$\lim_{\alpha\to 0}(1+\alpha x)^{1/\alpha} = \lim_{\beta\to 0}\left((1+\beta)^{1/\beta}\right)^x = \left(\lim_{\beta\to 0}(1+\beta)^{1/\beta}\right)^x,$$

by the continuity of exponentiation as a function of the base. But this last limit is e^x. Thus, there is a number e such that

$$\exp(x) = \lim_{\alpha\to 0}(1+\alpha x)^{1/\alpha} = e^x,$$

and

$$e^x = \sum_{n=0}^{\infty}\dfrac{x^n}{n!}.$$

---

[101] Bradley and Sandifer, *op. cit.*, pp. 111 – 112.

3. THE BINOMIAL THEOREM PROVEN

For good measure, Cauchy offers another treatment by starting with the series definition of $\exp(x)$:

3.10.2. Definition. *For real x,*

$$\exp(x) = \sum_{n=0}^{\infty} \frac{x^n}{n!},$$

provided the limit exists.

By the Ratio Test,

$$\left| \frac{x^n/n!}{x^{n+1}/(n+1)!} \right| = \frac{n+1}{|x|} \to \infty$$

for all $x \neq 0$, whence the series converges for all x. By his result on the equality of the Cauchy product and the product of the series, the $(n+1)$-th term of $\exp(x)\exp(y)$ is

$$\sum_{k=0}^{n} \frac{x^k}{k!} \cdot \frac{y^{n-k}}{(n-k)!}. \tag{121}$$

But the corresponding term of $\exp(x+y)$ is

$$\frac{(x+y)^n}{n!} = \frac{\sum_{k=0}^{n} \binom{n}{k} x^k y^{n-k}}{n!}$$

$$= \sum_{k=0}^{n} \frac{\frac{n!}{k!(n-k)!} x^k y^{(n-k)}}{n!} = \sum_{k=0}^{n} \frac{1}{k!(n-k)!} x^k y^{n-k},$$

which equals (121). Thus $\exp(x+y) = \exp(x)\exp(y)$ and by Lemma 3.9.33, $\exp(x) = \exp(1)^x = e^x$, where we again define $e = \exp(1)$.

Many advanced textbooks in the theory of the Calculus choose Definition 3.10.2 as the definition of exponentiation to base e, prove the existence of the natural logarithm $\ln(x)$ as an inverse to $\exp(x)$, and then define x^y to be $x^{y \ln(x)}$. Doing so saves the effort of proving the existence of continuous extensions to the reals of functions uniformly continuous on the rationals. It also eliminates the need to discuss equicontinuity. A full discussion of this approach would take us away from the Binomial Theorem and I refer the interested reader to any old-fashioned Advanced Calculus book or undergraduate Real Analysis text. Here I shall only discuss Cauchy's further application of the Binomial Theorem to the natural logarithm.

Cauchy obtains the power series for the natural logarithm through the simple device of differentiating $(1+x)^y$ with respect to y at $y = 0$, i.e., by evaluating

$$\lim_{\mu \to 0} \frac{(1+x)^\mu - (1+x)^0}{\mu}.$$

However, Cauchy does not treat differentiation in the *Cours d'analyse*, so this calculation is given without reference to the operation and it appears in the guise of a clever trick.

10. THE EXPONENTIAL AND LOGARITHMIC FUNCTIONS

For fixed x with $|x| < 1$ and any μ,
$$(1+x)^\mu - 1 = \mu x + \frac{\mu(\mu-1)}{2\cdot 1}x^2 + \frac{\mu(\mu-1)(\mu-2)}{3\cdot 2\cdot 1}x^3 + \cdots$$
whence, for $\mu \neq 0$,
$$\frac{(1+x)^\mu - 1}{\mu} = x + \frac{\mu-1}{2}x^2 + \frac{(\mu-1)(\mu-2)}{3\cdot 2\cdot 1}x^3 + \cdots \qquad (122)$$

Now the series (122) converges by the Ratio Test,
$$\left| \frac{(\mu-1)\cdots(\mu-n)}{(n+1)!} x^n \middle/ \frac{(\mu-1)\cdots(\mu-n-1)}{(n+2)!} x^{n+1} \right|$$
$$= \frac{n+2}{|\mu-n+1|} \cdot \frac{1}{|x|} = \frac{1+\frac{2}{n}}{\left|\frac{\mu+1}{n}-1\right|} \cdot \frac{1}{|x|} \to \frac{1}{|x|} > 1.$$

And convergence is uniform for μ bounded, whence the series is uniformly continuous as a function of μ. Thus
$$\lim_{\mu \to 0} \frac{(1+x)^\mu - 1}{\mu} = x + \frac{-1}{2}x^2 + \frac{(-1)(-2)}{3!}x^3 + \cdots$$
$$= x - \frac{x^2}{2} + \frac{x^3}{3} - \frac{x^4}{4} + \cdots \qquad (123)$$

The result of the differentiation ought to be
$$\frac{d}{dy}e^{y\ln(1+x)}\bigg|_{y=0} = e^{y\ln(1+x)} \cdot \ln(1+x)\bigg|_{y=0} = \ln(1+x).$$

Now, defining $\ln(x)$ to be the inverse to $\exp(x)$, so that
$$1 + x = \exp\big(\ln(1+x)\big),$$
one has
$$(1+x)^\mu = \exp\big(\mu\ln(1+x)\big) = 1 + \mu\ln(1+x) + \frac{\mu^2\big(\ln(1+x)\big)^2}{2!} + \cdots$$
and
$$\frac{(1+x)^\mu - 1}{\mu} = \ln(1+x) + \mu(\text{other terms}),$$
and again taking the limit as $\mu \to 0$ this yields
$$\lim_{\mu \to 0} \frac{(1+x)^\mu - 1}{\mu} = \ln(1+x).$$

With (123), this yields
$$\ln(1+x) = x - \frac{x^2}{2} + \frac{x^3}{3} - \frac{x^4}{4} + \cdots$$

11. Abel's Proof

Niels Henrik Abel (1802 – 1829), Norway's most celebrated mathematician, was trained in the spirit of the 17th and 18th centuries. Then he discovered Cauchy and it was a revelation. On 24 October 1826 he wrote to his old teacher Bernt Michael Holmboe (1795 – 1850)

> Cauchy is crazy and there is nothing to be done with him, even though at the moment he is the mathematician who knows how mathematics must be done. His works are excellent, but he writes in a very confusing way. At first I understood virtually nothing of what he wrote, but now it goes better.[102]

Bottazzini explains that

> Indeed (and this was the thing that most impressed the young Abel), for some time Cauchy had been working on a critical revision of the principles of analysis, whose manifesto was the publication of the *Cours d'analyse* in 1821, the lectures that Cauchy had given at the École Polytechnique. From it Abel had accepted the need for a new rigor in analysis. It was a problem that worried the Norwegian mathematician and appears frequently in his letters. "I want to dedicate all my efforts to bring a little more clarity into the prodigious obscurity that one incontestably finds today in analysis," he had earlier written to his professor Hansteen in Christiania [Oslo].[103]

Bottazzini then quotes a bit more from Abel's letter to Christoffer Hansteen (1784 – 1873) and then he quotes at length from a letter of 16 January 1826 to Holmboe, from which I borrow the following still longish excerpt:

> Divergent series are in their entirety an invention of the devil and it is a disgrace to base the slightest demonstration on them. You can take out whatever you want when you use them, and they are what has produced so many failures and paradoxes... I have become extremely sensitive to all this because, except for cases of the most extreme simplicity, for example geometric series, there is hardly anywhere in the whole of mathematics a single infinite series whose sum is determined in a rigorous manner. In other words, that which is the most important in mathematics is without foundation...

> I do not think you can show me many propositions where infinite series appear, where I cannot make fundamental objections against their demonstration. Do it, I reply. Even the binomial formula is not yet rigorously demonstrated.

---

[102]Bottazzini, *op. cit.*, p. 85. In 1902, on the centenary of his birth, a conference was organised in honour of Abel. Versions of the proceedings were published in several countries, most notably in Norway and France. Included in these proceedings was Abel's correspondence in Norwegian in the first of these volumes and in French translation along with the Norwegian originals in the second. Both editions can be found online.

[103]*Ibid.*, p. 86.

I have found that
$$(1+x)^m = 1 + mx + \frac{m(m-1)}{2}x^2 + \cdots$$
for all values of m when x is smaller than 1. If x is equal to $+1$, one obtains the same formula in the case where $m > -1$, and only then, but if $x = -1$, the formula is invalid, unless m is positive. For all the other values of x and m the series $1 + mx+$ etc. is divergent. The Taylor theorem, the basis of all of higher mathematics[104], is also poorly founded. I have found only one rigorous demonstration, that of Cauchy in his *Résumé des lecons sur le calcul infinitesimal*. He there demonstrates that
$$\phi(x+\alpha) = \phi x + \alpha \phi' x + \frac{\alpha^2}{2}\phi'' x + \cdots$$
whenever the series is convergent (but one would sooner use it in all cases).[105]

The first two paragraphs of this quotation are also translated by Abel's biographer, Oystein Ore, whose phrasing is a bit milder in tone:

> On the whole, divergent series are a deviltry, and it is a shame to base any demonstration upon them. By using them one can produce any result one wishes, and they are the cause of many calamities and paradoxes... My eyes have been opened in the most surprising manner. If you disregard the very simplest cases, there is in all of mathematics not a single infinite series whose sum has been stringently determined. In other words, the most important parts of mathematics stand without a foundation... I don't believe you can propose many theorems to me in which infinite series appear, where I cannot make justified objections against the proof. Try it, and I shall reply.—Even the binomial formula is not proved satisfactorily.[106]

Ore follows this quotation with the remark

> He was writing an article on the topic, he told Holmboe, but it embraced much more than the binomial series; it gave in reality a

---

[104]Lagrange had tried to avoid the use of limits by assuming all functions were represented by Taylor series and performing purely algebraic operations on them.

[105]Bottazzini, *op. cit.*, pp. 87 – 88. We shall discuss Taylor's Theorem in the next section, including Abel's misstatement of Cauchy's result.

[106]Oystein Ore, *Niels Henrik Abel; Mathematician Extraordinary*, University of Minnesota Press, Minneapolis, 1957, pp. 96 – 97. We now have three translations of the first sentence of this paragraph—my English translation of Hankel's (page 193, above), the English translation of Bottazzini's quote or Italian translation from (a French translation of?) the Norwegian original, and Ore's direct translation from the Norwegian. The referee points out that Ore, being a native Norwegian speaker, should be assumed to have offered the most accurate translation. My decision to retain all three is strictly pædagogical: it illustrates that translation is not an exact science and one must exercise caution in relying on such.

prototype for the treatment of infinite series... It became one of the classical memoirs in mathematics.[107]

Abel's memoir[108] was indeed a classic. And it may have concerned more than the binomial series, but this series is central to it, as proclaimed by the paper's title, "Untersuchungen über die Reihe:

$$1 + \frac{m}{1}x + \frac{m.(m-1)}{1.2}.x^2 + \frac{m.(m-1).(m-2)}{1.2.3}.x^3 + \ldots \text{u.s.w.''}$$

["Investigations on the series..."], Abel's announcement that the goal of his paper was to fill in the gaps and give a complete proof of the Binomial Theorem for all real and complex numbers x and m for which the series converged, and the fact that all the "embraced" extras were there to support this proof.

This is not to say that the extras in Abel's paper are of negligible importance. They are not and could stand on their own without reference to the Binomial Theorem. Indeed, they are of greater lasting importance than Abel's proof of the Binomial Theorem. The excerpts from the paper that appear in the English language anthologies bear this out in that they only cover the introductory portion of Abel's paper and the editors offer varying reasons for its inclusion in these anthologies.

In his famous source book, David Eugene Smith, for example, titles his selection "On the continuity of functions defined by power series", referring to Abel's Theorem IV, where the excerpt stops. Smith's introduction states in part:

> The theorem (which is fundamental in analytic function theory) may be stated in modern notation as follows. *If a real power series converges for some positive value of the argument, the domain of uniform convergence extends at least up to and including this point, and the continuity of the sum-function extends at least up to and including this point...*

> This theorem is of special interest, in that it was included in the scope of the investigation by Cauchy, referred to above. Cauchy correctly stated and in substance proved the theorem for the trivial case of the infinite geometric progression. Cauchy proceeded at once to state and claimed to prove a much more general theorem of which this would have been a special case. Cauchy's more general theorem is however false. Abel remarks indeed in this paper in a footnote (p. 316):

> In the above mentioned work of Mr. Cauchy (page. 131) one finds the following theorem:

---

[107]*Ibid.*

[108]Niels Henrik Abel, "Untersuchungen über die Reihe: $1 + \frac{m}{1}x + \frac{m.(m-1)}{1.2}.x^2 + \frac{m.(m-1).(m-2)}{1.2.3}.x^3 + \ldots$u.s.w.", *Journal für die reine und angewandte Mathematik* 1 (1826), pp. 311–339. The paper was originally written in French, but translated into German by the editor Crelle, a French version being first published in his collected works in an 1839 edition edited by Holmboe and again in 1881 in a second edition of the collected works edited by Ludvig Sylow and Sophus Lie. The German text has a number of typographical errors that are corrected in the later French publications.

11. ABEL'S PROOF

"If the different terms of the series

$$u_0 + u_1 + u_2 + u_3 + \ldots \text{etc.}$$

are functions of one and the same variable x, and indeed continuous functions with respect to this variable in the neighborhood of a particular value for which the series converges, then the sum s of the series is also a continuous function of x in the neighborhood of this particular value."

It appears to me that this theorem suffers exceptions.[109] Thus for example the series

$$\sin\phi - \frac{1}{2}\sin 2\phi + \frac{1}{3}\sin 3\phi - \ldots \text{etc.}$$

is discontinuous for each value $(2m+1)\pi$ of ϕ, where m is a whole number. It is well-known that there are many series with similar properties.

Abel was the first to note that Cauchy's announced theorem is not in general valid, and to prove the correct theorem for general power series.[110]

Smith adds a few additional remarks, then includes an excerpt featuring Theorems I – IV and their proofs, and finishing with the brief comment:

> The paper continues, giving an imperfect discussion of power series with variable coefficients, Theorem V, and in Theorem VI disposes of the product of two convergent series: Parts III and IV which form the main substance of the paper deal strictly with the binomial series.[111]

Perhaps Smith had read Hankel and accepted Bolzano as the author of the first rigorous demonstration of the Binomial Theorem, for he nowhere mentions that Abel proved the Theorem.

Garrett Birkhoff ostensibly includes his selection as a milestone in the history of series convergence, pairing it with an extract from Gauss's work on the hypergeometric series as examples of rigorous proofs of the convergence of power series. He discusses Cauchy's "false lemma", cites Abel as giving "the first really rigorous general study of the domain of convergence of power series" [112], and then discusses uniform convergence. He does not mention the Binomial Theorem in his commentary, but he does label the excerpt "Abel on the binomial series" and cites an additional line Smith had not included: "The aim of

---

[109] This affords a more dramatic example of disagreement among translations as described in footnote 106 on page 251, above. The original German publication clearly translates to "suffer exceptions". The French version, however, translates to "admits exceptions". Both versions inform us that there are exceptions, but the nuances of imparting this information are quite different.

[110] D.E. Smith (ed.), *A Source Book in Mathematics*, 1929; reprinted: Dover Publications, Inc., New York, 1959 (pp. 286 – 291). I have followed Smith's formatting convention here: small text for commentary and normal sized text for quoted material.

[111] *Ibid.*, p. 291.

[112] Garrett Birkhoff, *A Source Book in Classical Analysis*, Harvard University Press, Cambridge (Mass.), 1973 (pp. 61 – 62 for commentary and pp. 68 – 70 for an edited reprint of the translation (by Albert A. Bennett) from Smith's source book.

this memoir is to...solve completely the problem of summing the [binomial] series for all real and complex values of x and m for which it is convergent".[113]

Jacqueline Stedall's introduction to Abel's memoir approaches the subject from yet another angle:

> Neither Cauchy nor Bolzano managed to prove that Cauchy sequences are convergent: Bolzano tried but failed; Cauchy merely stated it as obvious. Nevertheless, Cauchy sequences turned out to be an extraordinarily powerful tool. In 1826, only five years after the publication of the *Cours d'analyse*, Abel used them with great skill to give the first rigorous proof of the binomial theorem...
>
> Abel's proof was very long: he allowed both x and m to be complex, but always separated them into real and imaginary parts, which led to some fearsome algebraic manipulation; further, his notation was somewhat unreliable because he more than once gave new meanings to letters already in use. Nevertheless, the underlying concepts of his arguments are as sophisticated as those in any modern proof. Here we give only the opening of Abel's paper to show his use of Cauchy sequences.[114]

Stedall's words "fearsome algebraic manipulation" may well be the most honest explanation of why the anthologists stop short of translating the full paper. One ought to include the paper, but no one would find it rewarding to read the "fearsome algebraic manipulation", so one's translation stops short of it, and the anthologists find different justifications for including those parts of the paper that are readable.

Abel's memoir divides into five parts. Part I is a short preface in which he explains that the treatment of the convergence of series requires greater care than was then customary, even the Binomial Theorem never having been adequately proven, and that the goal of his paper was to demonstrate rigorously the truth of this theorem.

The traditionally anthologised Part II includes some general lemmas on the convergence of series. It begins

> We first wish to establish a few necessary theorems on series. The excellent work of Cauchy *"Cours d'analyse de l'école polytechnique"*, which should be read by every analyst who loves rigour in mathematical investigations, will serve us thereby as a guide.
>
> D e f i n i t i o n. An arbitrary series
>
> $$v_0 + v_1 + v_2 + \ldots\ldots + v_m [+] \text{etc.}$$
>
> is called convergent if, for ever increasing values of m, the sum $v_0 + v_1 + v_2 + \ldots\ldots + v_m$ continually approaches a certain limit. This limit is called the s u m o f t h e s e r i e s: In the contrary

---

[113] *Ibid.*, p. 68.

[114] Stedall, *op. cit.*, p. 515. Stedall's excerpt (pp. 515 – 524) includes her introduction, a photoreproduction of the opening parts of Abel's paper, and a new translation of these parts.

case the series is called divergent, and therefore has no sum. From this definition it follows that, for a series to be convergent, it is necessary and sufficient that, for ever increasing values of m, the sum $v_m + v_{m+1} + \ldots\ldots + v_{m+n}$ continually approaches 0, whatever value n may have.

In any arbitrary [convergent] series, the general term v_m will thus continually approach 0.[115]

This is, of course, just the definition of convergence accompanied by a statement of Cauchy convergence as a criterion for the existence of the sum of a series. The Cauchy criterion is, in effect, Abel's choice of completeness axiom. He now cites in two parts an additional convergence criterion—a variation on the Ratio Test, which, as we saw, was implicit in d'Alembert and which Cauchy's "guide" explicitly stated and proved. Abel states the first without proof.

T h e o r e m I. If one denotes by $\rho_0, \rho_1, \rho_2, \ldots\ldots$ a sequence of positive quantities, and for ever increasing values of m the quotient $\frac{\rho_{m+1}}{\rho_m}$ approaches a limit α which is g r e a t e r t h a n 1: then the series

$$\varepsilon_0 \rho_0 + \varepsilon_1 \rho_1 + \varepsilon_2 \rho_2 + \ldots\ldots + \varepsilon_m \rho_m + \ldots\ldots,$$

in which ε_m is a quantity, which d o e s n o t a p p r o a c h 0 for ever increasing values of m, will necessarily be d i v e r g e n t.

T h e o r e m II. If in a series of positive quantities $\rho_0 + \rho_1 + \rho_2 + \ldots\ldots + \rho_m [+ \ldots\ldots]$, for ever increasing values of m, the quotient $\frac{\rho_{m+1}}{\rho_m}$ approaches a limit α, which i s s m a l l e r t h a n 1, then the series

$$\varepsilon_0 \rho_0 + \varepsilon_1 \rho_1 + \varepsilon_2 \rho_2 + \ldots\ldots + \varepsilon_m \rho_m [+ \ldots\ldots],$$

in which $\varepsilon_0, \varepsilon_1, \varepsilon_2$, etc. are quantities which d o n o t e x c e e d 1, will necessarily be convergent.

In fact one can, as a result of the assumptions, always take m large enough so that $\rho_{m+1} < \alpha \rho_m, \rho_{m+2} < \alpha \rho_{m+1}, \ldots\ldots, \rho_{m+n} < \alpha \rho_{m+n-1}$. From this it follows that $\rho_{m+k} < \alpha^k . \rho_m$, and with that

$$\rho_m + \rho_{m+1} + \ldots\ldots + \rho_{m+n} < \rho_m (1 + \alpha + \ldots\ldots + \alpha^n) < \frac{\rho_m}{1-\alpha},$$

and consequently all the more

$$\varepsilon_m \rho_m + \varepsilon_{m+1} \rho_{m+1} + \varepsilon_{m+2} \rho_{m+2} + \ldots\ldots + \varepsilon_{m+n} \rho_{m+n} < \frac{\rho_m}{1-\alpha}. \quad (124)$$

However $\rho_{m+k} < \alpha^k \rho_m$ and $\alpha < 1$, so it is clear that ρ_m, and hence also the sum

$$\varepsilon_m \rho_m + \varepsilon_{m+1} \rho_{m+1} + \varepsilon_{m+2} \rho_{m+2} + \ldots\ldots + \varepsilon_{m+n} \rho_{m+n},$$

---

[115] Abel, op. cit., pp. 312 – 313. The German version occasionally omits a few items like the bracketed + above or the occasional ellipsis. Here and below, on the basis of the French text, I have simply indicated the omissions by bracketed expressions.

will tend to 0.

Consequently the above series is convergent.[116]

The inequality I have labelled (124) is probably meant to be read

$$\left|\varepsilon_m\rho_m + \varepsilon_{m+1}\rho_{m+1} + \varepsilon_{m+2}\rho_{m+2} + \ldots + \varepsilon_{m+n}\rho_{m+n}\right| < \frac{\rho_m}{1-\alpha},$$

as he does not specify that $\varepsilon_0, \varepsilon_1, \varepsilon_2, \ldots$ are positive, unless this is implied by his use of the word "quantities" instead of "numbers"—but this is doubtful since he explicitly refers to "positive quantities" elsewhere. In any event, the i's of the proof are easily dotted and the t's crossed:

$$\left|\varepsilon_m\rho_m + \varepsilon_{m+1}\rho_{m+1} + \varepsilon_{m+2}\rho_{m+2} + \ldots + \varepsilon_{m+n}\rho_{m+n}\right|$$
$$\leq \left|\varepsilon_m\rho_m\right| + \left|\varepsilon_{m+1}\rho_{m+1}\right| + \left|\varepsilon_{m+2}\rho_{m+2}\right| + \ldots + \left|\varepsilon_{m+n}\rho_{m+n}\right|$$
$$\leq \left|\varepsilon_m\right|\rho_m + \left|\varepsilon_{m+1}\right|\rho_{m+1} + \left|\varepsilon_{m+2}\right|\rho_{m+2} + \ldots + \left|\varepsilon_{m+n}\right|\rho_{m+n}$$
$$\leq \rho_m + \rho_{m+1} + \ldots + \rho_{m+n}$$
$$< \frac{\rho_m}{1-\alpha},$$

the third inequality following from the second because all the ε_k's are assumed not to exceed 1 in absolute value.

Theorem III. If one denotes by $t_0, t_1, t_2 \ldots \ldots [,]t_m[, \ldots \ldots]$ a sequence of arbitrary quantities, and if the quantity $p_m = t_0 + t_1 + t_2 + \ldots \ldots + t_m$[117] is always smaller than a definite quantity δ, then one has

$$r = \varepsilon_0 t_0 + \varepsilon_1 t_1 + \varepsilon_2 t_2 + \ldots \ldots + \varepsilon_m t_m < \delta.\varepsilon_0,$$

where $\varepsilon_0, \varepsilon_1, \varepsilon_2[, \ldots \ldots]$ denote positive, decreasing quantities.

In fact one has

$$t_0 = p_0, \quad t_1 = p_1 - p_0, \quad t_2 = p_2 - p_1, \text{ etc.}$$

whence

$$r = \varepsilon_0 p_0 + \varepsilon_1(p_1 - p_0) + \varepsilon_2(p_2 - p_1) + \ldots \ldots + \varepsilon_m(p_m - p_{m-1}),$$

or too

$$r = p_0(\varepsilon_0 - \varepsilon_1) + p_1(\varepsilon_1 - \varepsilon_2) + \ldots \ldots + p_{m-1}(\varepsilon_{m-1} - \varepsilon_m) + p_m\varepsilon_m.$$

However, $\varepsilon_0 - \varepsilon_1, \varepsilon_1 - \varepsilon_2, \ldots \ldots$ are positive, so the quantity r is obviously smaller than $\delta.\varepsilon_0$.[118]

Here again it makes most sense to assume all inequalities refer to absolute values. If $t_0, t_1, \ldots$ were all positive, the result would be trivial. For then each $\varepsilon_i t_i \leq \varepsilon_0 t_i$ and

$$0 < r = \varepsilon_0 t_0 + \varepsilon_1 t_1 + \ldots + \varepsilon_m t_m \leq \varepsilon_0(t_0 + t_1 + \ldots + t_m) < \varepsilon_0 \delta.$$

---

[116] *Ibid.*, pp. 313 – 314.
[117] The German text inserts an ellipsis here that doesn't belong.
[118] Abel, *op. cit.*, p. 314.

However, no such restriction has been made on the t_i's and a different argument is needed:

$$|r| \leq |p_0(\varepsilon_0 - \varepsilon_1)| + |p_1(\varepsilon_1 - \varepsilon_2)| + \ldots + |p_{m-1}(\varepsilon_{m-1} - \varepsilon_m)| + |p_m \varepsilon_m|$$
$$\leq |p_0|(\varepsilon_0 - \varepsilon_1) + |p_1|(\varepsilon_1 - \varepsilon_2) + \ldots + |p_{m-1}|(\varepsilon_{m-1} - \varepsilon_m) + |p_m|\varepsilon_m$$
$$< \delta(\varepsilon_0 - \varepsilon_1) + \delta(\varepsilon_1 - \varepsilon_2) + \ldots + \delta(\varepsilon_{m-1} - \varepsilon_m) + \delta\varepsilon_m$$
$$< \delta\varepsilon_0.$$

That part of the argument equating the two expressions for r has come to be known as *Abel's Summation Formula* or *summation by parts*:

3.11.1. Lemma (Abel's Summation Formula). *Let $a_0, a_1, a_2, \ldots$ and $b_0, b_1, b_2, \ldots$ be sequences. Then*

$$\sum_{k=1}^{n} a_k b_k = \left(\sum_{k=1}^{n} a_k\right) b_{n+1} - \sum_{k=1}^{n} \left(\sum_{i=1}^{k} a_i\right)(b_{k+1} - b_k).$$

It is the discrete analogue to integration by parts, and is one of the elements of Abel's memoir of continuing value.

> D e f i n i t i o n . A function $f(x)$ is said to be a c o n t i n u o u s f u n c t i o n of x between the limits $x = a$,[119] $x = b$, if for an arbitrary value of x between these limits, the quantities $f(x - \beta)$ approach the limit $f(x)$ for ever decreasing values of β.
>
> T h e o r e m IV. If the series
>
> $$f(\alpha) = v_0 + v_1\alpha + v_2\alpha^2 + \ldots\ldots + v_m\alpha^m + \ldots\ldots$$
>
> converges for a given value δ of α, then it will also converge for each s m a l l e r value of α, and will be such that $f(\alpha - \beta)$ approaches the limit $f(\alpha)$ for ever decreasing values of β, assuming that α is less than or equal to δ.[120]

It is not clear whether the definition of continuity is to be for the open interval (a, b) or the closed $[a, b]$. The French text uses *"comprise entre"*, whence faithfulness to Cauchy's usage[121] would suggest he had the closed interval $[a, b]$ in mind. The expression "ever decreasing values of β" [*stets abnehmende Werthe von β*] obviously intends to be read $\beta \to 0$, and not merely that β decreases monotonically.

The statement and meaning of the Theorem are not so clear. Smith's interpretation, cited above, is that δ is positive, $f(\alpha)$ converges for $0 \leq \alpha \leq \delta$ and f is continuous on $[0, \delta]$. However, not assuming inequalities to mean inequalities of absolute values would suggest β is positive and only *left continuity* is intended. The Theorem is related to our earlier Lemma 3.6.16 (page 181, above)

---

[119]The German text gives the first "limit", i.e., endpoint of the interval, as o, presumably meaning 0. The French text in the collected works gives the more general a. Smith quietly corrected this in his source book; Stedall, who customarily incorporates the French corrections, uses 0.

[120]Abel, *op. cit.*, p. 314.

[121]*Cf.* page 228, above.

which established the (absolute) convergence of $f(x)$ for $|\alpha| < |\delta|$. Moreover, the proof of that Lemma, which is a simplified variant of Abel's proof of Theorem IV, albeit with absolute value explicitly noted, is readily modified to yield the uniform convergence of the series, whence the uniform continuity of the function on closed intervals $[-r, r]$ for $0 < r < |\delta|$. Continuity on $(-|\delta|, |\delta|)$ follows. What Theorem IV does beyond this is to establish continuity in the half-closed interval obtained by including the endpoint δ. This result, *sans* proof, goes back to Leibniz:

3.11.2. Theorem. *Suppose the series* $f(x) = \sum_{k=0}^{\infty} a_k x^k$ *converges for* $x = 1$. *Then*
$$\lim_{x \to 1^-} f(x) = f(1),$$
where $x \to 1^-$ *means x tends to 1 from the left, i.e., only values of $x < 1$ are considered in taking the limit:*
$$\forall \epsilon > 0 \, \exists \delta > 0 \, \forall x > 0 \Big(1 - \delta < x < 1 \Rightarrow |f(x) - f(1)| < \epsilon\Big).$$

Leibniz used this evident fact to justify defining $f(1)$ to be this limit, when it exists, in cases where $f(1)$ does not converge. For example, the so-called Grandi sequence $1 - 1 + 1 - 1 + \ldots$ has no limit, but for $|x| < 1$,
$$1 - x + x^2 - x^3 + \ldots = \frac{1}{1+x}$$
and $\lim_{x \to 1^-} \frac{1}{1+x} = \frac{1}{2}$, whence Leibniz accepted Grandi's value $1/2$ for this sum.[122] Leibniz rejected Christian Wolf's summations
$$1 - 2 + 4 - 8 + \ldots = \frac{1}{3}$$
$$1 - 3 + 9 - 27 + \ldots = \frac{1}{4},$$
which result from formally summing the geometric progressions because the values $1/3$ and $1/4$ were not limits of this form:
$$\lim_{x \to 1^-} \Big(1 - 2x + 4x^2 - 8x^3 + \ldots\Big) \neq \frac{1}{3}$$
$$\lim_{x \to 1^-} \Big(1 - 3x + 9x^2 - 27x^3 + \ldots\Big) \neq \frac{1}{4},$$
as, indeed, the limits did not exist.

Theorem 3.11.2 is not a fully satisfactory interpretation of Theorem IV, but is a useful one and will be the form first used by Abel following his proof of his Theorem VI, and by him again in Part IV of his paper in discussing the validity of the complex Binomial Theorem on the boundary of its circle of convergence.

---

[122]The series had been discovered by Johann Bernoulli (1667 – 1748) and given the value $1/2$ before its rediscovery by Guido Grandi (1671 - 1742) with whom Leibniz corresponded on the subject. Leibniz also had a probabilistic justification for this sum.

11. ABEL'S PROOF

The extent to which Abel wants to state Theorem 3.11.2 is perhaps better clarified by the proof itself.

Let
$$v_0 + v_1\alpha + \ldots\ldots + v_{m-1}\alpha^{m-1} = \varphi(\alpha)$$
$$v_m\alpha^m + v_{m+1}\alpha^{m+1} + \text{etc}\ldots\ldots = \psi(\alpha),$$

so
$$\psi(\alpha) = \left(\frac{\alpha}{\delta}\right)^m . v_m\delta^m + \left(\frac{\alpha}{\delta}\right)^{m+1} . v_{m+1}\delta^{m+1} [+\ldots\ldots] \text{etc.};$$

whence, by dint of Theorem (III.), $\psi(\alpha) < \left(\frac{\alpha}{\delta}\right)^m . p$, where p denotes the largest of the quantities, $v_m\delta^m$, $v_m\delta^m + v_{m+1}\delta^{m+1}$, $v_m\delta^m + v_{m+1}\delta^{m+1} + v_{m+2}\delta^{m+2}$ etc. Therefore, for each value of α which is less than or equal to δ, one can take m large enough so that[123]

$$\psi(\alpha) = w.$$

Now $f(\alpha) = \varphi(\alpha) + \psi(\alpha)$, thus $f(\alpha) - f(\alpha - \beta) = \varphi(\alpha) - \varphi(\alpha - \beta) + w$. Because $\varphi(\alpha)$ is an entire function of α [124], β can be taken small enough so that
$$\varphi(\alpha) - \varphi(\alpha - \beta) = w;$$

Thus in the same way
$$f(\alpha) - f(\alpha - \beta) = w,$$

from which the Theorem is proven.[125]

The proof contains a minor, readily correctible error. The terms $v_m\delta^m$, $v_m\delta^m + v_{m+1}\delta^{m+1}$, $v_m\delta^m + v_{m+1}\delta^{m+1} + v_{m+2}\delta^{m+2}$, ... are partial sums of the series $\psi(\delta)$. There are infinitely many of these and a largest one may not exist. However, by the Cauchy convergence of $f(\delta)$, these sums are bounded and a least upper bound p exists.

If we rewrite the proof in modern terms (and notation) and look to see what it proves, we come up with the following:

3.11.3. Theorem. *Let the series* $\sum_{k=0}^{\infty} a_k x^k$ *converge at some point* $x_0 \neq 0$.
i. *if* $x_0 > 0$, *the series converges uniformly on* $[0, x_0]$ *to a uniformly continuous function* f;
ii. *if* $x_0 < 0$, *the series converges uniformly on* $[x_0, 0]$ *to a uniformly continuous function* f.

---

[123] In an earlier footnote, Abel announces that he will use w to stand for a quantity that can be made smaller than any given quantity, no matter how small. His use of w is thus like Bolzano's use of Ω. In the French publications the letter used is ω.

[124] In modern usage, the phrase "entire function" refers to a function defined, continuous, differentiable, etc. in the entire complex plane. Here it just means that φ is a polynomial and is thus continuous.

[125] Abel, op. cit., pp. 314 – 315.

Proof. i. Let $\epsilon > 0$ be given. By the Cauchy convergence of $\sum_{k=0}^{\infty} a_k x_0^k$ we can find an m_0 so large that for all $m > m_0$ and all $n \geq 0$,

$$\left| \sum_{k=m}^{m+n} a_k x_0^k \right| = \left| \sum_{k=0}^{m+n} a_k x_0^k - \sum_{k=0}^{m-1} a_k x_0^k \right| < \frac{\epsilon}{3}.$$

Choose such an m.

For $0 < x < x_0$, the sequence

$$\varepsilon_i = \left(\frac{x}{x_0}\right)^{m+i}$$

is strictly decreasing with $1 > \varepsilon_0$. By Theorem III,

$$\left| \sum_{k=m}^{m+n} a_k x^k \right| = \left| \sum_{k=m}^{m+n} \left(\frac{x}{x_0}\right)^k a_k x_0^k \right| < \varepsilon_0 \cdot \frac{\epsilon}{3} < \frac{\epsilon}{3}.$$

The inequality obviously holds for $x = 0$ and $x = x_0$ as well, whence we have shown $\sum_{k=0}^{\infty} a_k x^k$ to be uniformly Cauchy convergent on $[0, x_0]$, whence the limit $f(x) = \sum_{k=0}^{\infty} a_k x^k$ is uniformly continuous on that interval.

Abel did not have the concept of uniform convergence nor the theorem on the continuity of the limit of a uniformly convergent series of continuous functions. Consequently he had to show directly that f was continuous. To this end he introduces the functions

$$\varphi(x) = a_0 + a_1 x + \ldots + a_{m-1} x^{m-1}$$
$$\psi(x) = a_m x^m + a_{m+1} x^{m+1} + \ldots$$

so that $f(x) = \varphi(x) + \psi(x)$. Now, for any $x \in [0, x_0]$, each finite sum $\sum_{k=m}^{m+n} a_k x^k$ is bounded in absolute value by $\epsilon/3$. It follows that

$$|\psi(x)| = \lim_{n \to \infty} \left| \sum_{k=m}^{m+n} a_k x^k \right| \leq \frac{\epsilon}{3}.$$

This holds for all $x, y \in [0, x_0]$.

But φ is a polynomial, hence uniformly continuous on $[0, x_0]$ and one can find $\delta > 0$ such that for all $x, y \in [0, x_0]$,

$$|x - y| < \delta \Rightarrow |\varphi(x) - \varphi(y)| < \frac{\epsilon}{3}.$$

Thus, for $|x - y| < \delta$,

$$|f(x) - f(y)| \leq |f(x) - \varphi(x)| + |\varphi(x) - \varphi(y)| + |\varphi(y) - f(y)|$$
$$\leq \frac{\epsilon}{3} + |\varphi(x) - \varphi(y)| + \frac{\epsilon}{3}$$
$$< \frac{\epsilon}{3} + \frac{\epsilon}{3} + \frac{\epsilon}{3} = \epsilon,$$

and f is indeed uniformly continuous on $[0, x_0]$.

ii. Similar. Or: reduce the negative case to the positive one by applying the result of part i to the series

$$\sum_{k=0}^{\infty}(-1)^k a_k x^k = \sum_{k=0}^{\infty} a_k(-x)^k. \qquad \square$$

Using the uniform continuity of φ is perhaps cheating as Abel had no concept of uniformity. One can be a little more faithful to his methods by fixing $y \in [0, x_0]$ and appealing to the pointwise continuity of φ at y to conclude the pointwise continuity of f at y.

The modernisation of Abel's proof is not hard for one trained in epsilontics. Abel and his contemporaries were not. Care must be taken in first choosing m and then δ (or, in Abel's notation, first choosing m and then β). In 1862 Joseph Liouville inserted into the *Journal de mathématiques pures et appliquées*[126] a note[127] written for him by Peter Gustav Lejeune Dirichlet offering a new proof of Abel's Theorem IV as given in the form of Theorem 3.11.2. This note is quite short and, as the result is important, I take the liberty of presenting an English version here:

PROOF OF A THEOREM OF ABEL.

[Note of Mr. LEJEUNE DIRICHLET communicated by Mr. LIOUVILLE]

This is to prove that if the series

$$a_0 + a_1 + a_2 + \cdots + a_n + \cdots$$

is convergent and its sum is A, the sum of the series

$$a_0 + a_1\rho + a_2\rho^2 + \cdots + a_n\rho^n + \cdots,$$

that is even more so convergent, will by taking the variable ρ positive and less than unity, tend to the limit A when ρ tends indefinitely to unity. Talking one day with my good and late lamented friend LEJEUNE DIRICHLET, I told him that I found quite difficult to explain (and understand) the demonstration that ABEL has given this important theorem: DIRICHLET began on the spot to write the note below before my eyes, just to help me; it was a great help and I am pleased to deliver it to the public. The mode of demonstration has found multiple applications and often helps me in my lectures at the Collège de France.

I transcribe verbatim the DIRICHLET note without adding anything, and of course without changing anything.

---

[126] Generally referred to as *Liouville's Journal*, Liouville being the editor and principal contributor.

[127] Lejeune-Dirichlet, "Demonstration d'un théoreme d'Abel", *Journal de mathématiques pures et appliquées* ser. 2, volume 7 (1862), pp. 253 – 255; reprinted in Dirichlet's collected works, volume 2, pp. 305 – 306. I follow the latter as much as possible in layout.

"It follows from the assumed convergence of the series
$$A = a_0 + a_1 + a_2 + \cdots + a_n + \text{etc.}$$
that the sum
$$s_n = a_0 + a_1 + \cdots + a_n$$
remains numerically less than some constant k and converges to the limit A, when n increases indefinitely. Now consider the series
$$S = a_0 + a_1\rho + a_2\rho^2 + \cdots + a_n\rho^n + \text{etc.}$$
the quantity ρ being assumed positive and less than unity, by replacing a_0, a_1, a_2, etc., by $s_0, s_1 - s_0, s_2 - s_1$, etc. It will take the form
$$S = s_0 + (s_1 - s_0)\rho + (s_2 - s_1)\rho^2 + \cdots + (s_n - s_{n-1})\rho^n + \text{etc.}$$
and this, after reordering, is transformed into
$$S = (1-\rho)(s_0 + s_1\rho + s_2\rho^2 + \cdots + s_n\rho^n + \cdots),$$
which offers no difficulty, since it amounts to adding to the sum of the first $n+1$ terms the term $-s_n\rho^{n+1}$, which vanishes for $n = \infty$."

"Now let's see to what limit S converges, when the positive variable $\varepsilon = 1 - \rho$ becomes infinitely small. We decompose S into two parts, a first including n terms and the other all the following terms, and let n grow as ε decreases, but slowly enough that the limit of $n\varepsilon$ is zero. The first part
$$(1-\rho)\left(s_0 + s_1\rho + \cdots + s_{n-1}\rho^{n-1}\right),$$
being numerically smaller than $n\varepsilon k$, converges to zero. As for the second
$$(1-\rho)\left(s_n\rho^n + s_{n+1}\rho^{n+1} + \cdots\right)$$
we can give it the form
$$P(1-\rho)\left(\rho^n + \rho^{n+1} + \cdots\right) = P\rho^n = P(1-\varepsilon)^n,$$
P designating a value between the largest and smallest of the quantities $s_n, s_{n+1}, \cdots$. Now these last converge to the limit A; it will be the same for P. And as on the other hand the factor $(1-\varepsilon)^n$, by virtue of the assumption made above, obviously converges to unity, it is proven that the limit of S, when the variable $\rho < 1$ indefinitely approaches unity, is the same sum A of the series considered in the first place."

I do not think anyone can now think to ask for further clarification. Dirichlet's proof of Abel's Theorem 3.11.2 avoids all reference to uniformity and is the proof of choice in some expositions[128].

---

[128] I cite two: Ferrar, *op. cit.*, pp. 79 – 80; and Walter Rudin, *Principles of Mathematical Analysis*, 2nd edition, McGraw-Hill Book Company, New York, 1964, p. 160. The reader who may ask for further clarification can find more detailed modern expositions of this proof therein.

11. ABEL'S PROOF

Theorems IV, 3.11.2, and 3.11.3 clearly have the nature of technical lemmas, proven to apply to some fixed purpose and not as ends in themselves. For, clearly, the proof can be modified to yield a much sharper result. Examining the proof of Lemma 3.6.16 shows it i. to be a simple variant of Abel's proof, and ii. to establish the uniform convergence of the series, whence the uniform continuity of the limit function f, in the interval $[-r, r]$ for any $0 < r < |x_0|$. Thus, the proper result, for $x_0 > 0$ reads:

3.11.4. Theorem. *Let* $\sum_{k=0}^{\infty} a_k x_0^k$ *converge. For any r with $0 < r < x_0$, the series* $\sum_{k=0}^{\infty} a_k x^k$ *converges uniformly to a function f uniformly continuous on the interval $[-r, x_0]$. Moreover, f is continuous on $(-x_0, x_0]$.*

The analogous result holds for $x_0 < 0$.

In words, a power series is continuous everywhere inside its radius of convergence, uniformly so on any closed subinterval of its interval of convergence; if the series converges at one of its endpoints, the continuity and uniform continuity extend to include that endpoint.

In general, nothing can be said about the other endpoint.

3.11.5. Example. *Let*

$$f(x) = \sum_{k=0}^{\infty} \frac{(-1)^k}{k+1} x^k.$$

Then $f(1) = 1 - 1/2 + 1/3 - 1/4 + \ldots$ is the alternating harmonic series and converges to $\ln 2$, but $f(-1) = 1 + 1/2 + 1/3 + 1/4 + \ldots$ is the divergent harmonic series. Thus f is convergent and continuous on $(-1, 1]$, uniformly convergent and uniformly continuous on $[-r, 1]$ for any $0 < r < 1$.

Abel was, of course, interested in the continuity of the binomial series, but of more critical concern in the eventual proof of the Binomial Theorem was the technical lemma in the form of Theorem 3.11.2.

Abel had two more lemmas to establish before turning to the proof of the Binomial Theorem itself, including his own "false lemma", his answer to Cauchy's "false" result on the continuity of a convergent series:

T h e o r e m V. *Let*

$$v_0 + v_1 \delta + v_2 \delta^2 + \ldots\ldots \text{etc.}$$

be a [convergent] series in which $v_0, v_1, v_2 [, \ldots\ldots]$ are continuous functions of one and the same variable quantity x between the limits $x = a$ and $x = b$. Then the series

$$f(x) = v_0 + v_1 \alpha + v_2 \alpha^2 + \ldots\ldots,$$

where $\alpha < \delta$ [129], *is convergent and is a continuous function of x between the same limits.*

---

[129] In the German version, δ has incorrectly been rendered β.

3. THE BINOMIAL THEOREM PROVEN

It has already been proven that the series $f(x)$ is convergent. That the function $f(x)$ is continuous can be proven as follows.

Let
$$v_0 + v_1\alpha + \ldots\ldots + v_{m-1}\alpha^{m-1} = \varphi(\alpha)$$
$$v_m\alpha^m + v_{m+1}\alpha^{m+1} + \ldots\ldots = \psi(\alpha),$$

so that
$$f(x) = \varphi(x) + \psi(x).$$

However,
$$\psi(x) = \left(\frac{\alpha}{\delta}\right)^m.v_m\delta^m + \left(\frac{\alpha}{\delta}\right)^{m+1}v_{m+1}\delta^{m+1} + \left(\frac{\alpha}{\delta}\right)^{m+2}v_{m+2}\delta^{m+2} + \text{etc.},$$

so one has, if one denotes by $\theta(x)$ the largest of the quantities $v_m\delta^m$, $v_m\delta^m + v_{m+1}\delta^{m+1}$, $v_m\delta^m + v_{m+1}\delta^{m+1} + v_{m+2}\delta^{m+2}$ etc., by dint of Theorem (III.):
$$\psi(x) < \left(\frac{\alpha}{\delta}\right)^m.\theta(x).$$

From here it follows that one can take m large enough so that $\psi(x) = w$, thus likewise
$$f(x) = \varphi(x) + w,$$
where w is smaller than any given quantity.

It is even the case that
$$f(x - \beta) = \varphi(x - \beta) + w,$$
whence
$$f(x) - f(x - \beta) = \varphi(x) - \varphi(x - \beta) + w.$$
By the expression for $\varphi(x)$ it is clear that one can choose β small enough that
$$\varphi(x) - \varphi(x - \beta) = w, \text{ and}$$
thus likewise
$$f(x) - f(x - \beta) = w.$$

Thus the function $f(x)$ is continuous.[130]

There is much to discuss here: the meaning of Theorem V, its relation to Theorem IV, and, of course, the error. The English language anthologists, Smith, Birkhoff, and Stedall, all stop their excerpts just before the statement of Theorem V. There are good reasons for this. Theorem IV can be given the readily intelligible gloss: "the power series is continuous wherever it converges". Theorem V has to be explained. Its proof is incorrect, the error must be explained, and a correct version of the Theorem supplied. If one is not going to discuss the full proof of the Binomial Theorem, this promises to be a large page count for comparatively little benefit. The reader might wish to skip ahead to page 272 and Theorem VI.

---

[130] *Ibid.*, pp. 315 – 316.

What Theorem V is intended to do is replace Cauchy's "false lemma" on the continuity of an infinite series by a true version that can be applied to the binomial series to establish Lemma 3.9.31. Abel is here considering

$$f(x,\alpha) = \sum_{k=0}^{\infty} v_k(x)\alpha^k,$$

like the function

$$g(\mu, x) = \sum_{k=0}^{\infty} \binom{\mu}{k} x^k$$

of Lemma 3.9.31. Theorem IV yielded the continuity of f in the second variable for each fixed x and Theorem V asserts this continuity in the variable x for each fixed α. The proof given is clearly intended to be analogous to that of Theorem IV. Unfortunately, it is incorrect. The problem lies with $\theta(x)$. One wants $|\psi(x)|$ to be small. For $0 < d < |\delta|$ and $|\alpha| \leq d$, $(\alpha/\delta)^m$ can be made small by choosing m large. But in making $(\alpha/\delta)^m \theta(x)$ small, m may depend on $\theta(x)$, i.e., on x. One needs a uniform upper bound on the values of $\theta(x)$ for $x \in [a,b]$, and such a bound need not exist.

Abel himself apparently had doubts about his proof, as he attempted to prove it anew in 1827. The result was again inadequate, the unfinished paper being published first only in 1881 in the second edition of his collected works. The Italian mathematician Felice Casorati (1835 – 1890) visited Berlin late in 1864 and had many conversations with Leopold Kronecker (1823 – 1891) who, according to Casorati

> was led to say that Abel in his paper on the binomial series (where he does not define continuity precisely enough), although correcting Cauchy's error, gives a demonstration that is not valid. This is because it [the demonstration] rests essentially on this, that when, by taking for x any value in a given interval (for example from -1 to $+1$), we can always assign an upper limit to the value of a function $\phi(x)$, the function $\phi(x)$ must have a maximum in this interval. Kronecker says that Abel did not consider that if this upper limit depends on x, we cannot assert the existence of the maximum. He sees the defect in Abel's demonstration, but says he cannot see the means of obtaining a rigorous proof.[131]

Presumably Kronecker had not given a counterexample to Theorem V or he would not have said he couldn't see how to obtain a rigorous proof. Counterexamples are, in fact, not hard to find. Bottazzini cites Pierre Dugac for an

---

[131]Bottazzini, *op. cit.*, p. 262, citing Erwin Neuenschwander, "Der Nachlass von Casorati in Pavia", *Archive for History of Exact Sciences* 19 (1978), pp. 1 – 89.

example.[132] I've not seen Dugac's paper, but can cite Henrik Kragh Sørensen[133] for the following example.

3.11.6. Example. *For $n \geq 2$, let v_n be the spike function defined on $[0,1]$ algebraically by*

$$v_n(x) = \begin{cases} 0, & 0 \leq x < \frac{1}{n+1} \\ 2^n \frac{x - \frac{1}{n+1}}{\frac{1}{n} - \frac{1}{n+1}}, & \frac{1}{n+1} \leq x < \frac{1}{n} \\ 2^n \frac{\frac{1}{n-1} - x}{\frac{1}{n-1} - \frac{1}{n}}, & \frac{1}{n} \leq x < \frac{1}{n-1} \\ 0, & \frac{1}{n-1} < x \leq 1 \end{cases}.$$

Then define $f_\alpha(x) = \sum_{n=2}^{\infty} v_n(x) \alpha^n$. It is easy to see that each f_α converges pointwise and that $f_{1/2}$ is not continuous at 0. (Cf. FIGURE 1, below.)

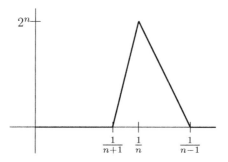

FIGURE 1. Graph of v_n

3.11.7. Exercise. *Prove the assertions of Example 3.11.6.*

In 1871 Paul du Bois-Reymond published a correct proof of a general result that can stand in for Theorem V. His paper is short and to the point:

> In C a u c h y's Cours d'Analyse de l'Ecole Polytechnique one finds put forward on p. 131 the theorem that the sum of a convergent series in which the individual terms are continuous functions of a variable x must likewise be a continuous function of this variable. The physical series teach us that this assertion, so generally expressed, is incorrect, and A b e l already opposed it in that he introduced as an example of a series discontinuous despite the continuity of its terms: $\sin x - \frac{\sin 2x}{2} + \cdots$ (C r e l l e's Journ. 1, p. 316

---

[132]P. Dugac, "Fondements de l'analyse", in: Jean Dieudonné (ed.), *Abrégé d'histoire des mathématiques 1700 – 1900*, vol. 1 (1978), pp. 355 – 392.

[133]Henrik Kragh Sørensen, "Throwing some light on the vast darkness that is analysis: Niels Henrik Abel's critical revision of the concept of absolute convergence", *Centaurus* 52 (2010), pp. 38 – 72; here: pp. 62 – 63, 69. Of particular interest for this section is Sørensen's dissertation, *The Mathematics of Niels Henrik Abel: Continuation and New Approaches in Mathematics During the 1820s*, University of Aarhus, Denmark, 2002; electronic edition 2010. The electronic edition is available online.

11. ABEL'S PROOF

footnote). Nevertheless relations must exist between the continuity of the terms of a series and the continuity of the sum, and such have already been uncovered especially by trigonometric series, while the following theorem has not yet to my knowledge been announced:

If in the infinite series $w_1\mu_1 + w_2\mu_2 + \cdots$ the μ are independent of x and the series $\mu_1 + \mu_2 + \cdots$ converges absolutely, if further the quantities $w_1 = w_1(x), w_2 = w_2(x), \ldots$ are for each index continuous functions of the argument x in the interval $x = a$ to $x = b$, if finally none of these functions including w_∞ becomes infinite for any value of x belonging to the interval $x = a$ to $x = b$ [by $w_p(x)$ is always understood $\lim_{\varepsilon=0} w_p(x \pm \varepsilon)$], then the sum of the series $w_1\mu_1 + w_2\mu_2 + \cdots$ is a continuous function of x in the interval $x = a$ to $x = b$.

Proof. We set $U_n(x) = w_1\mu_1 + w_2\mu_2 + \cdots w_n\mu_n$, and assume (without loss of generality) the μ are positive.

As starting point of the proof we take the equation:

$$U_n(x+\varepsilon) - U_n(x) = \Delta w_1 \mu_1 + \Delta w_2 \mu_2 + \cdots \Delta w_m \mu_m$$
$$+ w_{m+1}(x+\varepsilon)\mu_{m+1} + w_{m+2}(x+\varepsilon)\mu_{m+2} + \cdots w_n(x+\varepsilon)\mu_n$$
$$- \{w_{m+1}(x)\mu_{m+1} + w_{m+2}(x)\mu_{m+2} + \cdots w_n(x)\mu_n\},$$
$$\Delta w_p = w_p(x+\varepsilon) - w_p(x).$$

Let ΔW_m be a certain mean value[134] among the differences $\Delta w_1, \Delta w_2 .. \Delta w_m$, which however needs to equal none of them, likewise $W_{mn}^{(1)}$ a mean value of the quantities $w_{m+1}(x+\varepsilon), \ldots w_n(x+\varepsilon)$ and W_{mn} a mean value of the quantities $w_{m+1}(x), \ldots . w_n(x)$; so we can write:

$$U_n(x+\varepsilon) - U_n(x) = \Delta W_m(\mu_1 + \mu_2 + \cdots \mu_m)$$
$$+ \left(W_{mn}^{(1)} - W_{mn}\right)(\mu_{m+1} + \mu_{m+2} + \cdots \mu_n).$$

Now let n become infinite, so $W_{m\infty}^{(1)}, W_{m\infty}$ will, for each value of x, be finite completely determined quantities, and one has:

$$U_\infty(x+\varepsilon) - U_\infty(x) = \Delta W_m(\mu_1 + \mu_2 + \cdots \mu_m)$$
$$+ \left(W_{m\infty}^{(1)} - W_{m\infty}\right)(\mu_{m+1} + \mu_{m+2} + \cdots \text{in inf.}).$$

The function $U_\infty(x)$ is continuous everywhere, where $U_\infty(x \pm 0) - U_\infty(x)$ is null. We allow ε to vanish in the above equation. ΔW_m also vanishes and we find

$$U_\infty(x+\varepsilon) - U_\infty(x) =$$
$$(\mu_{m+1} + \mu_{m+2} + \cdots \text{in inf.}) \left\{\lim_{\varepsilon=0} W_{m\infty}^{(1)} - W_{m\infty}\right\};$$

---

[134]In the second appendix to the *Cours d'analyse* Cauchy defined a mean value to be any value in the range between the lowest and highest of a given set of values, and presented rules for dealing with such. A modern exposition of du Bois-Reymond's proof, such as that to be given shortly, needs make no mention of them.

since we cannot know a priori, whether the quantities $\lim_{\varepsilon=0} W_{m\infty}^{(1)}$ and $W_{m\infty}$ are equal to one another. That they are finite however, so the right side of the above equation will be so much smaller, the larger m, and can through increasing m be *made* smaller than any given quantity however small. With this the left side, which does not contain m, *is* smaller than any given quantity, however small; thus it is null. Q. E. D.

...

The series
$$\frac{\sin x}{1^\mu} + \frac{\sin 2x}{2^\mu} + \cdots$$
will by the above theorem necessarily be continuous if $\mu > 1$ for every value of x, because the series $\frac{1}{1^\mu} + \frac{1}{2^\mu} + \cdots$ converges for $\mu > 1$.[135]

The exposition is even less clear than Abel's! The clearest part of the paper is that which follows the proof. Here du Bois-Reymond cites examples of series to which his theorem applies as well as a series to which it does not apply. Of these I have only included the trigonometric series as it is the only one tangentially related to our discussion: If one alternates the signs of the terms and sets $\mu = 1$, the result is the discontinuous Fourier series Abel cited as an exception to Cauchy's Theorem; but if one takes $\mu > 1$, the series converges absolutely to a continuous function.

The rest of du Bois-Reymond's paper requires decipherment, starting with the statement of the theorem. The "interval $x = a$ to $x = b$" can be taken to mean the *closed* interval $[a, b]$. That was easy enough, but I confess that the bracketed expression "by $w_p(x)$ is always understood $\lim_{\varepsilon=0} w_p(x \pm \varepsilon)$" gave me trouble until I read Sørensen's "Throwing some light...", the following passage from which also helps to clarify the proof.

> One central piece of proof technique which was missing in ABEL's proof, namely a technique to elucidate the order of limit processes, is more explicit in DU BOIS-REYMOND's proof: First he introduced n and m and auxiliary quantities defined from them. Then, he considered the limit process $n \to \infty$ which took care of the least interesting part of the series. Then he let $\epsilon \to 0$ which brought into play the continuity of the terms. And, finally, he let $m \to \infty$ which proved the continuity of the sum. However, the final step was still convoluted in a way that may remind us of ABEL's proof. The assumptions that are necessary to ensure that $\lim_{\epsilon=0} W'_{m,\infty}$ is bounded were still hidden in the statement of the theorem where DU BOIS-REYMOND had stipulated that $\lim_{\epsilon=0} w_\infty(x \pm \epsilon)$ does not become infinite. The assumption—the equivalent of a uniform bound

---

[135]Paul du-Bois Reymond, "Notiz über einen Cauchy'schen Satz, die Stetigkeit von Summen unendlicher Reihen betreffend", *Mathematischen Annalen* 4 (1871), pp. 135 – 137. The bracketed expression in the statement of the theorem is not an editorial insertion, but a verbatim quote.

near x—was not completely clear and only with the turn towards uniform convergence would this be made precise.[136]

So, where du Bois-Reymond says that w_p does not become infinite for any value $x \in [a, b]$, he is saying, by virtue of "by $w_p(x)$ is always understood $\lim_{\varepsilon=0} w_p(x \pm \varepsilon)$", that $\lim_{\varepsilon \to 0} w_p(x \pm \varepsilon)$ does not become infinite for any value $x \in [a, b]$. For finite p, this is a trivial consequence of continuity. The real import is the statement for $p = \infty$, which likewise is not to be taken about the function w_∞ but about $\lim_{p \to \infty} w_p$. The limit w_∞ is, in fact, a red herring and need not exist; all that is necessary is the *uniform local boundedness* of $w_p(x \pm \varepsilon)$ as $p \to \infty$.

3.11.8. Definition. *A sequence of functions $w_1(x), w_2(x), w_3(x), \ldots$ defined on an interval $[a, b]$ is* locally uniformly bounded *on $[a, b]$ if for any $x \in [a, b]$ there are positive numbers n_0, B, δ such that, for all $y \in [a, b]$ and all n,*

$$n > n_0 \ \& \ |x - y| < \delta \Rightarrow |w_n(y)| < B.$$

With this definition we can offer a precise formal statement of du Bois-Reymond's result.

3.11.9. Theorem. *Let $w_1(x), w_2(x), w_3(x), \ldots$ be a sequence of functions defined on a closed interval $[a, b]$ and let $\mu_1, \mu_2, \mu_3, \ldots$ be a sequence of numbers. If*
i. *the series $\mu_1 + \mu_2 + \mu_3 + \ldots$ converges absolutely,*
ii. *the functions $w_1(x), w_2(x), w_3(x), \ldots$ are all continuous on $[a, b]$, and*
iii. *the sequence $w_1(x), w_2(x), w_3(x), \ldots$ is locally uniformly bounded on $[a, b]$, then the series*

$$f(x) = \sum_{k=1}^{\infty} w_k(x) \mu_k$$

converges and forms a continuous function on $[a, b]$.

Before embarking on the proof, let me quickly note that the assumption that the limit

$$w_\infty(x) = \lim_{p \to \infty} w_p(x)$$

exists and is locally bounded is insufficient. The functions $w_n(x) = v_n(x)$ of Example 3.11.6 converge to the constant function $w_\infty(x) = 0$ and $f_{1/2}$ is still discontinuous at 0.

Also, if the functions $w_1(x), w_2(x), w_3(x), \ldots$ are all continuous then local uniform boundedness implies: for all $x \in [a, b]$ there are $B, \delta > 0$ such that, for all $y \in [a, b]$ and all n,

$$|x - y| < \delta \Rightarrow |w_n(y)| < B.$$

For, let $n_0, \delta_1, B_1 > 0$ satisfy

$$n > n_0 \ \& \ |x - y| < \delta_1 \Rightarrow |w_n(y)| < B_1,$$

---

[136]Sørensen, "Throwing some light...", *op. cit.*, p. 62.

and, by the continuity of $|w_1(x)| + |w_2(x)| + \ldots + |w_{n_0}(x)|$ at x, choose $\delta_2 > 0$ such that

$$|x - y| < \delta_2 \Rightarrow |w_1(y)| + |w_2(y)| + \ldots + |w_{n_0}(y)| < B_2 + 1,$$

where $B_2 = |w_1(x)| + |w_2(x)| + \ldots + |w_{n_0}(x)|$. Choosing $B = \max\{B_1, B_2\}$ and $\delta = \min\{\delta_1, \delta_2\}$ does the trick.

Third, by a more advanced result of the theory of the Calculus, known as the Heine-Borel Theorem, one can further conclude the existence of a bound B such that for all $y \in [a, b]$ and all n,

$$|w_n(y)| < B.$$

Thus, to prove Theorem 3.11.9 it suffices to prove the following weaker-looking result:

3.11.10. Theorem. *Let $w_1(x), w_2(x), w_3(x), \ldots$ be a sequence of functions defined on a closed interval $[a, b]$ and let $\mu_1, \mu_2, \mu_3, \ldots$ be a sequence of numbers. If*
i. *the series $\mu_1 + \mu_2 + \mu_3 + \ldots$ converges absolutely,*
ii. *the functions $w_1(x), w_2(x), w_3(x), \ldots$ are all continuous on $[a, b]$, and*
iii. *the functions $w_1(x), w_2(x), w_3(x), \ldots$ are uniformly bounded: there is a $B > 0$ such that, for all n and all $x \in [a, b]$,*

$$|w_n(x)| < B,$$

then the series

$$f(x) = \sum_{k=1}^{\infty} w_k(x)\mu_k$$

converges and forms a continuous function on $[a, b]$.

[This may not be faithful to du Bois-Reymond, but it is faithful to the period. Bolzano in the 1840s, Dirichlet in the 1850s, and Eduard Heine in 1872 all used the Heine-Borel Theorem at least implictly.]

Proof of Theorem 3.11.10. We combine Abel's and du Bois-Reymond's proofs.

Let $\epsilon > 0$. Let m be a positive integer and define

$$\varphi_m(x) = w_1(x)\mu_1 + \ldots + w_m(x)\mu_m$$
$$\psi_m(x) = w_{m+1}(x)\mu_{m+1} + w_{m+2}(x)\mu_{m+2} + \ldots,$$

so that $f(x) = \varphi_m(x) + \psi_m(x)$. Note that, for any $x \in [a, b]$,

$$|\psi_m(x)| = \left|\sum_{k=m+1}^{\infty} w_k(x)\mu_k\right| \leq \sum_{k=m+1}^{\infty} |w_k(x)| \cdot |\mu_k|$$
$$\leq \sum_{k=m+1}^{\infty} B|\mu_k| = B \sum_{k=m+1}^{\infty} |\mu_k|.$$

By the absolute convergence of the μ-series, we can choose m_0 so large that for $m > m_0$,
$$\sum_{k=m+1}^{\infty} |\mu_k| < \frac{\epsilon}{3B},$$
whence
$$|\psi_m(x)| < B\frac{\epsilon}{3B} = \frac{\epsilon}{3}.$$
Thus, for any $x, y \in [a, b]$, and $m > m_0$,
$$|\psi_m(x) - \psi_m(y)| \leq |\psi_m(x)| + |\psi_m(y)| < \frac{\epsilon}{3} + \frac{\epsilon}{3} = \frac{2\epsilon}{3}.$$
Now $\varphi_m(x)$ is continuous, whence for each $x \in [a, b]$ there is a $\delta > 0$ such that for all $y \in [a, b]$,
$$|x - y| < \delta \Rightarrow |\varphi_m(x) - \varphi_m(y)| < \frac{\epsilon}{3}.$$
Putting this together we have, for $|x - y| < \delta$,
$$\begin{aligned}|f(x) - f(y)| &= |\varphi_m(x) + \psi_m(x) - \varphi_m(y) - \psi_m(y)| \\ &\leq |\varphi_m(x) - \varphi_m(y)| + |\psi_m(x) - \psi_m(y)| \\ &< \frac{\epsilon}{3} + \frac{2\epsilon}{3} = \epsilon,\end{aligned}$$
i.e.,
$$|x - y| < \delta \Rightarrow |f(x) - f(y)| < \epsilon,$$
with δ depending only on x, ϵ. □

A decade after du Bois-Reymond's paper appeared, Sylow and Lie edited a new edition of Abel's collected works in which appeared for the first time Abel's 1827 paper with two new attempts to prove Theorem V. Sylow observed that Abel's original argument worked if one assumed a uniform bound for all terms $|v_m(x)\delta^m|$ in some neighbourhood of x.

Whatever replacement one chooses for Theorem V, one can now draw the conclusion relevant to the proof of the Binomial Theorem, namely, that, for any x with $|x| < 1$, the series
$$f(\mu) = \sum_{k=0}^{\infty} \binom{\mu}{k} x^k$$
is a continuous function of μ on any interval $[-R, R]$ for $R > 0$. I leave to the reader the task of verifying this for Sylow's formulation and give the details for du Bois-Reymond's as it is slightly more devious. It uses the trick of writing, for $0 < |x| < 1$, $x = yz$ with $0 < y < 1$ and $|z| < 1$, e.g., by choosing $y = x^{2/3}, z = x^{1/3}$. Making such a choice,
$$f(\mu) = \sum_{k=0}^{\infty} \binom{\mu}{k} y^k \cdot z^k,$$

with $w_k(\mu) = \binom{\mu}{k} y^k$ and $\sum_{k=0}^{\infty} z^k$ absolutely convergent. Moreover, for any $\mu \in [-R, R]$,
$$\left|\binom{\mu}{k} y^k\right| \leq \left|\binom{-R}{k}\right| y^k = \frac{R(R+1)\cdots(R+k-1)}{k!} y^k.$$
But
$$\left|\binom{-R}{k+1} y^{k+1}\right| / \left|\binom{-R}{k} y^k\right| = \frac{R+k}{k+1} y < 1,$$
if
$$k > \frac{Ry - 1}{1 - y}.$$
Thus, if we choose $k_0 > \dfrac{Ry - 1}{1 - y}$, the sequence $\left|\binom{\mu}{k}\right| y^k$ is decreasing for $k > k_0$ and
$$B = \max\left\{\left|\binom{-R}{0} y^0\right|, \ldots, \left|\binom{-R}{k_0+1} y^{k_0+1}\right|\right\}$$
bounds the absolute value of $\left|\binom{\mu}{k} y^k\right|$ for all k and all $\mu \in [-R, R]$.

Let us now return, finally, to Abel's paper.

Theorem VI is the final lemma before the proof of the Binomial Theorem, his version of the equality of the product of two series with their Cauchy product. The reader will recall that Bolzano proved the result directly for the binomial series. Cauchy proved that if the two series converged absolutely, then so did their Cauchy product and it equalled the product of the two series. Abel now gives pretty much the same proof and then follows it up with the observation that by Theorem 3.11.2, if each of two series and their Cauchy product converge, then the product of the two equals the Cauchy product.

T h e o r e m VI. Let ρ_0, ρ_1, ρ_2 etc., $\rho_0^1, \rho_1^1, \rho_2^1$ etc. denote the numerical values[137] of the respective terms of two convergent series
$$v_0 + v_1 + v_2 + \ldots\ldots = p \text{ and}$$
$$v_0^1 + v_1^1 + v_2^1 + \ldots\ldots = p^1.$$
If the series
$$\rho_0 + \rho_1 + \rho_2 + \ldots\ldots \text{ and}$$
$$\rho_0^1 + \rho_1^1 + \rho_2^1 + \ldots\ldots$$
are likewise still convergent, then too the series
$$r_0 + r_1 + r_2 + \ldots\ldots + r_m [+ \ldots\ldots]$$
the general term of which is
$$r_m = v_0 v_m^1 + v_1 v_{m-1}^1 + v_2 v_{m-2}^1 + \ldots\ldots [+] v_m v_0^1,$$
converges and has as its sum
$$\left(v_0 + v_1 + v_2 + \ldots\ldots\right) \times \left(v_0^1 + v_1^1 + v_2^1 + \ldots\ldots\right).$$

---

[137] I.e., the absolute values.

Proof. Setting
$$p_m = v_0 + v_1 + \ldots\ldots + v_m,$$
$$p_m^1 = v_0^1 + v_1^1 + \ldots\ldots + v_m^1,$$
one easily sees that
$$r_0 + r_1 + r_2 + \ldots\ldots + r_{2m} =$$
$$p_m p_m^1 + \left(p_0 v_{2m}^1 + p_1 v_{2m-1}^1 + \ldots\ldots + p_{m-1} v_{m+1}^1 (=t) \right. \text{ (a)}$$
$$\left. + p_0^1 v_{2m} + p_1^1 v_{2m-1} + \ldots\ldots + p_{m-1}^1 v_{m+1} (=t^1) \right).$$

If one now sets
$$\rho_0 + \rho_1 + \rho_2 + \ldots\ldots = u,$$
$$\rho_0^1 + \rho_1^1 + \rho_2^1 + \ldots\ldots = u^1,$$
it is clear that, without reference to signs,
$$t < u(\rho_{2m}^1 + \rho_{2m-1}^1 + \ldots\ldots + \rho_{m+1}^1),$$
$$t^1 < u^1(\rho_{2m} + \rho_{2m-1} + \ldots\ldots + \rho_{m+1}).$$

Since, however, the series
$$\rho_0 + \rho_1 + \rho_2 + \ldots\ldots, \quad \rho_0^1 + \rho_1^1 + \rho_2^1 + \ldots\ldots$$
are convergent, so the quantities t and t^1 will, for ever increasing values of m, approach the limit null. If one thus sets m infinitely large in equation (a), then
$$r_0 + r_1 + r + r_3 + \text{etc.} =$$
$$(v_0 + v_1 + v_2 + \text{etc.})(v_0^1 + v_1^1 + v_2^1 + \text{etc.}).^{138}$$

This completes the proof of Theorem VI. I have followed the French text in correcting a couple of typographic errors, but stuck to the German original as regards notation. The German text continues, without forming a new paragraph, to prove the general result when absolute convergence is not assumed. The French text more sensibly begins a new paragraph. This conclusion should be separated from the proof of Theorem VI both to indicate that the proof is over and because Abel deemed it important enough to stress it in the first paragraph of his paper:

If, e.g., two series are multiplied together, one writes
$$\left(u_0 + u_1 + u_2 + u_3 + \text{etc.} \right)\left(v_0 + v_1 + v_2 + v_3 + \text{etc.} \right)$$
$$= u_0 v_0 + \left(u_0 v_1 + u_1 v_0 \right) + \left(u_0 v_2 + u_1 v_1 + u_2 v_0 \right) + \text{etc.}$$
$$\ldots\ldots + \left(u_0 v_n + u_1 v_{n-1} + v_2 v_{n-2} + \ldots\ldots + u_n v_0 \right) + \text{etc.}$$

---

[138] Abel, *op. cit.*, p. 317.

3. THE BINOMIAL THEOREM PROVEN

This equation is completely correct if the series

$$u_0 + u_1 + \ldots \ldots \text{ and } v_0 + v_1 + \ldots \ldots$$

are finite. If however they are infinite, they must first of all necessarily c o n v e r g e, because a divergent series has no sum, and then too the series in the second member must likewise c o n v e r g e. Only with these restrictions is the above expression correct. If I am not mistaken this restriction has till now not been taken into account. This will happen in the present essay.[139]

His proof of this assertion, so prominently announced in the paper's preface, casually follows the proof of Theorem VI with no fanfare:

Assume t_0, t_1, t_2 etc., t_1^1, t_2^1, t_3^1 etc. are two series of positive and negative quantities whose individual terms approach the null without end. It follows from Theorem (II.) that the series

$$t_0 + t_1\alpha + t_2\alpha^2 + \text{etc.}, \quad t_0^1 + t_1^1\alpha + t_2^1\alpha^2[+]\text{etc.},$$

where α denotes a quantity smaller than 1, must be convergent. This holds even if one gives each term its numerical[140] value, thus in consequence of the preceding theorem:

$$\left(t_0 + t_1\alpha + t_2\alpha^2 + \ldots \ldots\right)\left(t_0^1 + t_1^1\alpha + t_2^1\alpha^2 + \ldots \ldots\right) =$$
$$t_0 t_0^1 + (t_1 t_0^1 + t_0 t_1^1)\alpha + (t_2 t_0^1 + t_1 t_1^1 + t_0 t_2^1)\alpha^2 + \text{etc.} \quad \text{(b)}$$
$$\ldots \ldots + (t_m t_0^1 + t_{m-1} t_1^1 + t_{m-2} t_2^1 + \ldots \ldots + t_0 t_m^1)\alpha^m + \text{etc.}$$

If one now assumes that the three series

$$t_0 + t_1 + t_2 + \text{etc.},$$
$$t_0^1 + t_1^1 + t_2^1 + \text{etc., and}$$
$$t_0 t_0^1 + (t_1 t_0^1 + t_0 t_1^1) + (t_2 t_0^1 + t_1 t_1^1 + t_0 t_2^1) + \text{etc.}$$

are convergent, one finds, by dint of Theorem (IV.), if one lets α approach the unit in equation (b):

$$(t_0 + t_1 + t_2 + \text{etc.})(t_0^1 + t_1^1 + t_2^1 + \text{etc.}) =$$
$$t_0 t_0^1 + (t_1 t_0^1 + t_0 t_1^1) + (t_2 t_0^1 + t_1 t_1^1 + t_0 t_2^1) + \text{etc.}[141]$$

A brief aside: There is one more result about the Cauchy product cited in most textbooks. Cauchy and Abel proved that if two series converged absolutely then their Cauchy product also converged absolutely and equalled the product of the two series. And Cauchy gave an example of two conditionally convergent series whose Cauchy product did not converge.[142] In 1875 Franz

---

[139] *Ibid.*, p. 311.
[140] I.e., absolute.
[141] *Ibid.*, pp. 317 – 318.
[142] Cited in Example 3.9.32 on page 244, above.

11. ABEL'S PROOF

Mertens showed that if both series converged and only one of them was absolutely convergent, then their Cauchy product still converged and equalled the product of the series.[143]

Proof of Mertens's Theorem. Mertens's proof is fairly simple. Let $u_0, u_1, u_2, \ldots$ and $v_0, v_1, v_2, \ldots$ be sequences of real numbers such that

$$\sum_{k=0}^{\infty} u_k \text{ converges absolutely and } \sum_{k=0}^{\infty} v_k \text{ converges.}$$

Further, let $w_0, w_1, w_2, \ldots$ be the terms of their Cauchy product,

$$w_k = u_0 v_k + u_1 v_{k-1} + \ldots + u_k v_0.$$

One must show that the differences

$$\Delta_{2n} = w_0 + w_1 + \ldots + w_{2n} - (u_0 + u_1 + \ldots + u_n)(v_0 + v_1 + \ldots + v_n)$$
$$\Delta_{2n+1} = w_0 + w_1 + \ldots + w_{2n+1} - (u_0 + u_1 + \ldots + u_{n+1})(v_0 + v_1 + \ldots + v_n)$$

tend to 0. To this end, let $\epsilon > 0$ be given and observe that a little algebra yields

$$\Delta_{2n} = u_0(v_{n+1} + v_{n+2} + \ldots + v_{2n}) + u_1(v_{n+1} + v_{n+2} + \ldots + v_{2n-1}) +$$
$$\ldots + u_{n-1} v_{n+1}$$
$$+ u_{n+1}(v_0 + v_1 + \ldots + v_{n-1}) + u_{n+2}(v_0 + v_1 + \ldots + v_{n-2}) + \ldots + u_{2n} v_0,$$

$$\Delta_{2n+1} = u_0(v_{n+1} + v_{n+2} + \ldots + v_{2n+1}) + u_1(v_{n+1} + v_{n+2} + \ldots + v_{2n}) +$$
$$\ldots + u_n v_{n+1}$$
$$+ u_{n+2}(v_0 + v_1 + \ldots + v_{n-1}) + u_{n+3}(v_0 + v_1 + \ldots + v_{n-2}) + \ldots + u_{2n+1} v_0.$$

Now, by the convergence assumptions, there are bounds A, B such that

$$|u_0| + |u_1| + \ldots + |u_m| < A$$
$$|v_0 + v_1 + \ldots + v_m| < B$$

for any m. By Cauchy convergence, there is a number n_0 so large that for all $n > n_0$ and all m,

$$|u_{n+1}| + |u_{n+2}| + \ldots + |u_{n+m}| < \frac{\epsilon}{A+B}$$
$$|v_{n+1} + v_{n+2} + \ldots + v_{n+m}| < \frac{\epsilon}{A+B}.$$

Thus

$$|\Delta_{2n}| \leq |u_0| \cdot |v_{n+1} + v_{n+2} + \ldots + v_{2n}| + |u_1| \cdot |v_{n+1} + v_{n+2} + \ldots + v_{2n-1}| +$$
$$\ldots + |u_{n-1}| \cdot |v_{n+1}|$$
$$+ |u_{n+1}| \cdot |v_0 + v_1 + \ldots + v_{n-1}| + |u_{n+2}| \cdot |v_0 + v_1 + \ldots + v_{n-2}| +$$
$$\ldots + |u_{2n}| \cdot |v_0|$$

---

[143] F. Mertens, "Ueber die Multiplikationsregel für zwei unendliche Reihen", *Journal für die reine und angewandte Mathematik* 79 (1875), pp. 182 – 184.

$$< |u_0| \cdot \frac{\epsilon}{A+B} + |u_1| \cdot \frac{\epsilon}{A+B} + \ldots + |u_{n-1}| \cdot \frac{\epsilon}{A+B} +$$
$$|u_{n+1}| \cdot B + |u_{n+2}| \cdot B + \ldots + |u_{2n}| \cdot B$$
$$< (|u_0| + |u_1| + \ldots + |u_{n-1}|) \frac{\epsilon}{A+B} + (|u_{n+1}| + |u_{n+2}| + \ldots + |u_{2n}|) B$$
$$< A \cdot \frac{\epsilon}{A+B} + \frac{\epsilon}{A+B} \cdot B = \epsilon.$$

Likewise $|\Delta_{2n+1}| < \epsilon$. □

Returning to Abel's paper, we see that with Theorem VI he has all the ingredients necessary to complete the by then standard Eulerian proof of the Binomial Theorem. With Theorem IV he also has the tool to allow him to discuss the possible validity of the binomial identity at the endpoints ± 1 of the interval of convergence. Abel, however, did not stop there, but turned his attention immediately in part III to the complex case, i.e., the question of the validity of the equation

$$(1+x)^m = \sum_{k=0}^{\infty} \binom{m}{k} x^k,$$

where x and m are allowed to be complex numbers, i.e., numbers of the form $a + b\sqrt{-1}$. Part III handled the case of $|x| = \sqrt{a^2 + b^2} < 1$, i.e., complex x in the interior of the circle of convergence of the power series, and Part IV dealt with the thornier issue of the behaviour of the series on the boundary. (A short fifth part considered a few other series.)

Abel was not the first to extend the Binomial Theorem to complex arguments. In Chapter 9 of the *Cours d'analyse*, Cauchy had proven the Theorem for complex x with $|x| < 1$ and real exponents in a readable sort of way. Abel's approach of splitting everything into real and imaginary parts resulted in the "fearsome algebraic manipulation" that modern translators prefer not to deal with. There is, in fact, little to be gained by rising to the challenge of cutting a swath through Abel's algebraic jungle. The interesting case—$|x| < 1$—is handled more easily by Cauchy's methods. And, consider this: We have thus far considered only 7 of the 29 pages of Abel's paper. The page count of this book—we are already on page 276 of what was intended to be a short (< 250 pages) account—is rapidly increasing and an editorial decision has to be made. This is what I have decided: I am going to stick to real numbers in this book; the proper treatment of the Binomial Theorem in the complex case is best handled by means of Taylor series in the complex plane as taught in the basic undergraduate course in Complex Analysis.

I will finish this section only with the consideration of the behaviour of the binomial series at $x = \pm 1$. I note that in discussing the application of Abel's Theorem V, I slipped into writing μ, à la Bolzano, for the exponent instead of using Abel's m. As the current convention is that m usually denotes an integer, I am more comfortable using μ in this context and shall continue to do so here.

Thus, consider the question: when does

$$(1+x)^\mu = \sum_{k=0}^{\infty} \binom{\mu}{k} x^k,$$

for $x = \pm 1$? The analyses of both cases start out routinely, but each requires a twist we have not yet considered. Let us first consider the case $x = 1$.

3.11.11. Lemma. *Let μ be a real number not a nonnegative integer. The series*

$$\sum_{n=0}^{\infty} \binom{\mu}{n} \qquad (125)$$

converges for $\mu > -1$ and diverges for $\mu \leq -1$.

Proof. Write $a_n = \binom{\mu}{n}$. The Ratio Test is, of course, inconclusive. But one can still look at the ratio:

$$\left| \binom{\mu}{n+1} \bigg/ \binom{\mu}{n} \right| = \frac{|\mu - n|}{n+1}.$$

For $\mu \leq -1$, $|\mu - n| = n + |\mu|$, whence

$$\left| \frac{a_{n+1}}{a_n} \right| = \frac{n + |\mu|}{n+1} \geq \frac{n+1}{n+1} = 1,$$

and $|a_{n+1}| \geq |a_n|$, whence $a_n \not\to 0$. Thus (125) does not converge.

For $\mu > 1$ and $n > \mu$, the ratio is

$$\left| \frac{a_{n+1}}{a_n} \right| = \frac{n - \mu}{n+1} < \frac{n+1}{n+1} = 1,$$

and the sequence $|a_n|, |a_{n+1}|, \ldots$ is strictly decreasing for $n > \mu$. Also, as

$$a_{n+1} = \binom{\mu}{n+1} = \frac{\mu - n}{n+1} \binom{\mu}{n} = \frac{\mu - n}{n+1} a_n,$$

the factor $\mu - n$ being negative for $n > \mu$, the sequence $a_n, a_{n+1}, \ldots$ alternates in sign for $n > \mu$. It remains only to apply the Alternating Convergence Test, but to do so requires us to show $a_n \to 0$.

This is the non-routine part of the proof. Consider

$$\left| \frac{a_n}{a_{n+1}} \right| = \frac{n+1}{n-\mu}, \quad \text{for } n > \mu$$

$$= \frac{n - \mu + \mu + 1}{n - \mu} = 1 + \frac{\mu + 1}{n - \mu}$$

$$> 1 + \frac{\mu + 1}{n + 1}.$$

If we assume

$$\left| \frac{a_n}{a_{n+k}} \right| > 1 + (\mu + 1) \left(\frac{1}{n+1} + \frac{1}{n+2} + \ldots + \frac{1}{n+k} \right), \qquad (126)$$

then
$$\left|\frac{a_n}{a_{n+k+1}}\right| = \left|\frac{a_n}{a_{n+k}}\right| \cdot \left|\frac{a_{n+k}}{a_{n+k+1}}\right|$$
$$> \left(1 + (\mu+1)\left(\frac{1}{n+1} + \frac{1}{n+2} + \ldots + \frac{1}{n+k}\right)\right) \cdot \left(1 + \frac{\mu+1}{n+k+1}\right)$$
$$> 1 + (\mu+1)\left(\frac{1}{n+1} + \frac{1}{n+2} + \ldots + \frac{1}{n+k}\right) + \frac{\mu+1}{n+k+1},$$

whence (126) holds when k is replaced by $k+1$, and induction yields (126) for all k. But, for fixed $n > \mu$,
$$\lim_{k\to\infty}\left|\frac{a_n}{a_{n+k}}\right| \geq (\mu+1)\sum_{k=1}^{\infty}\frac{1}{n+k} = +\infty,$$

the series being the divergent harmonic series whence $a_{n+k} \to 0$ as $k \to \infty$. $\square$

The case $x = -1$ likewise involves a trick.

3.11.12. Lemma. *Let μ be a real number not a nonnegative integer. The series*
$$\sum_{n=0}^{\infty}(-1)^n\binom{\mu}{n} \tag{127}$$
converges for $\mu \geq 0$ and diverges for $\mu < 0$.

Proof. Let $a_n = (-1)^n\binom{\mu}{n}$ denote the general term of the series and again consider
$$\left|\frac{a_{n+1}}{a_n}\right| = \frac{|n-\mu|}{n+1}.$$

Also note that
$$a_{n+1} = \binom{\mu}{n+1}(-1)^{n+1} = \frac{\mu-n}{n+1}\binom{\mu}{n}(-1)^{n+1} = -\frac{\mu-n}{n+1}a_n,$$

whence for $n > \mu$, a_n and a_{n+1} have the same sign.

For $\mu < 0$ and $n > 0$,
$$a_n = \frac{\mu(\mu-1)\cdots(\mu-n+1)}{n!}(-1)^n$$

is the product of $2n$ negative numbers, whence positive. Also
$$|\mu - n| = n - \mu = n + |\mu| > n,$$

whence
$$a_{n+1} > \frac{n}{n+1}a_n.$$

Let $c = a_1/2$ so that $a_1 > c/1$. A simple induction shows that $a_n > c/n$ for all $n \geq 1$:
$$a_{n+1} > \frac{n}{n+1}a_n > \frac{n}{n+1}\cdot\frac{c}{n} = \frac{c}{n+1}.$$

Thus
$$\sum_{n=0}^{\infty}a_n \geq \sum_{n=1}^{\infty}a_n \geq \sum_{n=1}^{\infty}\frac{c}{n} = c\sum_{n=1}^{\infty}\frac{1}{n},$$

which last series is the divergent harmonic series.

For $\mu > 0$, the proof is moderately devious. Choose $n_0 > \mu$ and note again that for $n > n_0$, a_{n+1} and a_n have the same sign. Let us assume this sign is positive, i.e., a_n is positive for $n > n_0$. Then
$$(n+1)a_{n+1} = (n-\mu)a_n < na_n.$$
Thus the sequence
$$n_0 a_{n_0}, (n_0+1)a_{n_0+1}, \ldots$$
is a decreasing sequence of positive reals and, by the Monotone Convergence Theorem, has a limit γ. Let
$$c_n = na_n - (n+1)a_{n+1}$$
and note that
$$\sum_{k=0}^{p} c_{n_0+k} = \left(n_0 a_{n_0} - (n_0+1)a_{n_0+1}\right) + \left((n_0+1)a_{n_0+1} - (n_0+2)a_{n_0+2}\right) +$$
$$\ldots + \left((n_0+p)a_{n_0+p} - (n_0+p+1)a_{n_0+p+1}\right)$$
$$= n_0 a_{n_0} - (n_0+p+1)a_{n_0+p+1}$$
$$\to n_0 a_{n_0} - \gamma \text{ as } p \to \infty,$$
i.e., $\sum c_n$ converges. But, for $n > n_0$,
$$c_n = na_n - (n+1)a_{n+1} = na_n - (n-\mu)a_n = \mu a_n,$$
whence $\sum a_n = \frac{1}{\mu} \sum c_n$ converges.

If the terms a_n are all negative for $n > n_0$, simply replace the series $\sum a_n$ by $\sum(-a_n)$. □

12. Taylor's Theorem

A sort of coda to our history of the Binomial Theorem is the subsequent proof of Taylor's Theorem and the modern reversal of the dependence: Taylor's Theorem is no longer founded upon the Binomial Theorem, but quite the reverse holds—provided one's formulation of Taylor's Theorem is sufficiently strong. We should take a look at this, but, insofar as it is a sequel to our main story and not part of the main narrative, I shall be brief.

The rigorous proof and, indeed, adequate formulation of Taylor's Theorem begins with Lagrange, who provided a key step, but not for this purpose. Lagrange disliked the uses of motion (fluxional calculus), infinitesimals, and limits, and set out to develop the Calculus purely algebraically by assuming every function to be expandable into a power series, reporting on his development in 1797 in *Théorie des fonctions analytiques, contenant les principes du calcul différentiel dégagés de toute considération d'infiniment petits ou d'évanouissans, de limites ou de fluxions, et réduits a l'analyse algébrique des quantités finies* [*Theory of analytic functions containing the principles of the differential calculus freed of all consideration of the infinitely small or evanescent quantities, of limits or of fluxions, and reduced to the algebraic analysis of finite quantities*], usually referred to simply as the *Théorie des fonctions analytiques*.

From an expansion,
$$f(x) = a_0 + a_1(x-a) + a_2(x-a)^2 + \ldots,$$
of a function f around a point a, he *defined* the derivative of f at a to be the coefficient a_1 of the linear term, introducing the notation $f'(a)$ for this value. The dependence of $f'(a)$ on a being functional, he abstracted a function f' and called it the *derived function* of f. As a function f' would also be expandable into a power series and have its own derived function f'', which would have its own derived function f''', etc. Ignoring questions of convergence, he determined algebraically that
$$a_0 = f(a), \quad a_1 = f'(a), \quad a_2 = \frac{f''(a)}{2}, \quad a_3 = \frac{f'''(a)}{3!}, \quad \ldots,$$
i.e., that
$$f(x) = f(a) + f'(a)(x-a) + \frac{f''(a)}{2!}(x-a)^2 + \frac{f'''(a)}{3!}(x-a)^3 + \ldots$$
This he took to be a proof of Taylor's Theorem.

Fortunately, Lagrange did not stop here. Like d'Alembert with the binomial series, he also considered the practical problem of how closely the series, truncated to $n+1$ terms, approximated the function and established, if not completely rigorously, a result which, in modern terms, is stated as follows.

3.12.1. Theorem (Taylor's Theorem with the Lagrange Form of the Remainder). *Let f be n times continuously differentiable on $[a,b]$ and $n+1$ times differentiable on (a,b). There is some $c \in (a,b)$ such that*
$$f(b) = f(a) + f'(a)(b-a) + \frac{f''(a)}{2}(b-a)^2 + \ldots$$
$$+ \frac{f^{(n)}(a)}{n!}(b-a)^n + \frac{f^{(n+1)}(c)}{(n+1)!}(b-a)^{n+1}.$$

The condition of *continuous differentiability* is simply the existence of a continuous derivative. The n-fold continuous differentiability assumed here is simply the assumption of the existence and continuity on $[a,b]$ of $f, f', f'', \ldots, f^{(n)}$. The further assumption of $(n+1)$-fold differentiability on (a,b) assumes also the existence, but not the continuity, of $f^{(n+1)}$ on the open interval.

The Theorem as cited is readily seen to be an $(n+1)$-th order version of the Mean Value Theorem. Indeed, for $n=0$ it is the Mean Value Theorem itself and for $n=1$ the second order version, Theorem 2.10.10 of page 143, above. Some textbooks in fact refer to Theorem 3.12.1 as the *Extended Mean Value Theorem* and prove it by reduction to Rolle's Theorem using the cleverly chosen auxiliary function,
$$h(t) = \left(f(b) - \sum_{k=0}^{n} \frac{f^{(k)}(t)}{k!}(b-t)^k \right) - \left(\frac{b-t}{b-a}\right)^n \left(f(b) - \sum_{k=0}^{n} \frac{f^{(k)}(a)}{k!}(b-a)^k \right).$$

3.12.2. Exercise. *Prove Theorem 3.12.1 by applying Rolle's Theorem to the function $h(t)$.*

This is not quite how Lagrange's proof went—or, I should say, how his proofs went. In 1799 and 1801 Lagrange lectured again on the Calculus, publishing his lectures with a new treatment of Theorem 3.12.1 in 1806 in *Leçons sur le calcul des fonctions* [*Lessons on the calculus of functions*].

Lagrange's formulation of Theorem 3.12.1 was not as sharp as that above. His proofs, however, were better motivated, if incompletely given and relying on inadequately established lemmas.

Because he assumes Taylor's Theorem already established for all functions, Lagrange assumes that the derivatives of all orders exist and are continuous, not just the first $n+1$ and n, respectively. He is in fact inferring the result for all values of n. With our modern recognition that not all functions have derivatives of all orders, we find it necessary to see what is established for those functions possessing derivatives only up to order $n+1$. And examination of his proof reveals a result slightly less sharp than Theorem 3.12.1 in that he must assume $f^{(n+1)}$ to be continuous on $[a,b]$ as well. The clever application of Rolle's Theorem yielding the Mean Value Theorem without assuming the continuity of the derivative came over half a century after Lagrange.

There are also cosmetic differences in the formulations. The now familiar Definition-Theorem-Proof style of writing was not the expositional standard. He does not isolate and label his theorems in boldface or italic. In his *Théorie des fonctions analytiques* he begins by stating the problem, working through the solution for the cases $n = 0, 1, 2$ (i.e., deriving the Mean Value Theorem and the Mean Value Theorems of the Second and Third Orders), and only afterwards summarising his conclusion by stating what he has proven—explicitly up to the third order and writing "&c." to express the result for higher orders.

Translating from the French is time-consuming and not all the words in the copy of the 1797 edition I downloaded are legible, so I shall simply paraphrase his treatment. Lagrange proves the Theorem in the special case $a = 0, b = x$, noting that every element in the interval $[0, x]$ can be written in the form $x - xz$ for some $z \in [0, 1]$. By Taylor's Theorem,

$$f(x) = f(x - xz) + xzf'(x - xz) + \frac{x^2 z^2}{2} f''(x - xz) + \ldots$$

In particular

$$f(x) = f(x - xz) + xP, \qquad (128)$$

where P can be viewed as a function of z (x being held fixed). Differentiating (128) with respect to z, we have

$$0 = -xf'(x - xz) + xP',$$

whence

$$P' = f'(x - xz).$$

Now P is an antiderivative of $f'(x-xz)$, whence
$$P = \int_0^z f'(x-xz)dz + C$$
for some constant C. From (128), however, we see that $P = 0$ when $z = 0$, whence $C = 0$. Thus
$$f(x) = f(x-xz) + x\int_0^z f'(x-xz)dz. \tag{129}$$
Setting $z = 1$, (129) reads
$$f(x) = f(0) + x\int_0^1 f'(x-xz)dz. \tag{130}$$
He now pretty much observes that, if
$$N = \min\{f'(z)|z \in [0,1]\} \quad \text{and} \quad M = \max\{f'(z)|z \in [0,1]\},$$
which numbers exist by virtue of the Extreme Value Theorem (using the continuity of f'), then one has
$$\int_0^1 Ndz \le \int_0^1 f'(x-xz)dz \le \int_0^1 Mdz, \tag{131}$$
i.e.,
$$N \le \int_0^1 f'(x-xz)dz \le M.$$
By the Intermediate Value Theorem (again using the continuity of f'), there is some $u \in (0,1)$ strictly between the points where f' takes on the values N and M such that
$$f'(u) = \int_0^1 f'(x-xz)dz.$$
Thus (130) now becomes
$$f(x) = f(0) + xf'(u)$$
for some $u \in (0,x)$. A change of variables yields
$$f(z+x) = f(z) + xf'(z+u),$$
i.e., the Mean Value Theorem.[144]

The Mean Value Theorems of the Second and Third Orders are established analogously. Taken as a proof of Taylor's Theorem, the proof is circular. However, it does establish the remainder term for functions which equal their Taylor expansions. Two years after publishing this proof, in 1799, Lagrange gave another proof not presupposing Taylor's Theorem and thus pointing the way to a proof of this Theorem. As I will shortly be presenting Cauchy's sharper result, I will not present the details here.[145]

Lagrange's proofs used several results he did not supply adequate proofs for, the Extreme Value Theorem and Intermediate Value Theorem being the most

---
[144] Actually, a slightly weaker version with the stronger hypothesis that f' is continuous.
[145] I refer the interested reader to Chabert, *op. cit.*, pp. 409 – 411, or to Smoryński, *Adventures in Formalism, op. cit.*, pp. 131 – 135.

obvious. His proof of (131) depended on a not quite correct proof of the lemma asserting that if the derivative of a function is positive on an interval then the function is strictly increasing. Today we prove this by appeal to the Mean Value Theorem.[146] But Lagrange only proved the Mean Value Theorem as a corollary to this lemma. In 1806 André Marie Ampère (1775 – 1836) attempted to use Theorem 3.12.1 to give a rigorous proof of Taylor's Theorem, defining the derivative of a function by a crucial property of the derivative Lagrange had singled out. His proof also was not rigorously correct, but Ampère was one of Cauchy's teachers and Cauchy would successfully complete the work of Lagrange and Ampère in providing rigour in the Differential Calculus.

Before discussing Cauchy, however, let us see what the Lagrange remainder can tell us about the convergence of the binomial series to $(1+x)^\mu$.

Let μ be given and consider the Lagrange remainder. For some c between 0 and x,

$$\left|(1+x)^\mu - \sum_{k=0}^{n} \binom{\mu}{k} x^k\right| = \left|\frac{x^{n+1}}{(n+1)!}\left(\frac{d^{n+1}}{dx^{n+1}}(1+x)^\mu\bigg|_{x=c}\right)\right|$$

$$= \left|\frac{x^{n+1}}{(n+1)!}\mu(\mu-1)\cdots(\mu-n)(1+c)^{\mu-n-1}\right|$$

$$= \left|\binom{\mu}{n+1}x^{n+1}\right|(1+c)^{\mu-n-1}.$$

Now, if $|x| < 1$, we know that the series $\sum \binom{\mu}{k} x^k$ converges, whence the k-th term goes to 0 as $k \to \infty$. Thus

$$\lim_{n \to \infty} \binom{\mu}{n+1} x^{n+1} = 0.$$

And, if $c \geq 0$, we will have $1 + c \geq 1$ and $\mu - n - 1$ will be negative for large n, whence

$$(1+c)^{\mu-n-1} \leq 1$$

and we see that for $0 \leq c < 1$, the remainder term can be made as small as we please by choosing n large enough. Thus, for $0 \leq x < 1$,

$$\lim_{n \to \infty} \sum_{k=0}^{n} \binom{\mu}{k} x^k = (1+x)^\mu$$

and the Binomial Theorem is proven in this case.

When $-1 < x < 0$, however, one has $-1 < c < 0$ and $1 + c < 1$, so for large n, $(1+c)^{\mu-n-1}$ grows exponentially and it is not clear that $\binom{\mu}{n+1}x^{n+1}$ decreases rapidly enough to counter this exponential growth.

3.12.3. Exercise. *Apply Taylor's Theorem with the Lagrange form of the remainder to the series for e^x, $\sin x$, $\cos x$ and $\ln(1+x)$. Does the Lagrange form suffice to establish convergence of these series to their given functions?*

---

[146] *Cf.* Corollary 2.10.7 of page 140 in Chapter 2, above.

We can say that, modulo a few key foundational lemmas, Lagrange can be credited with a partial proof of Taylor's Theorem and that a rigorous proof of a special case of the Binomial Theorem could have been had before Bolzano finally proved the Binomial Theorem in 1816, or Cauchy independently did so in 1821, or Abel is generally credited with having done in 1826. Even before Abel's Eulerian proof, however, Cauchy had improved on Lagrange's theorem and given a Taylor-based proof of the Binomial Theorem. This was done in his 1823 *Calcul infinitésimal* and again in his 1829 *Leçons sur le calcul différentiel*. He did this by improving the estimate of the error term.

3.12.4. Theorem (Taylor's Theorem with the Cauchy Form of the Remainder). *Let f be n times continuously differentiable on $[a,b]$ and $n+1$ times differentiable on (a,b). There is some $\theta \in (0,1)$ such that*

$$f(b) = f(a) + f'(a)(b-a) + \frac{f''(a)}{2}(b-a)^2 + \ldots$$
$$+ \frac{f^{(n)}(a)}{n!}(b-a)^n + \frac{(1-\theta)^n(b-a)^{n+1}}{n!}f^{(n+1)}\bigl(a+\theta(b-a)\bigr).$$

Proof. Fix b and consider the function $\varphi(t)$ denoting the remainder

$$f(b) - f(t) - \frac{b-t}{1}f'(t) - \ldots - \frac{(b-t)^n}{n!}f^{(n)}(t), \tag{132}$$

obtained by subtracting the first n terms of the Taylor expansion of f around t at b, for $t \in [a,b]$. f was assumed n times continuously differentiable and $n+1$ times differentiable, whence φ is continuous on $[a,b]$ and differentiable on (a,b). The Mean Value Theorem applies:

$$\varphi(b) - \varphi(t) = (b-t)\varphi'(c) \text{ for some } c \in (t,b) \subseteq (a,b).$$

Writing $c = t + \theta(b-t)$ for some $\theta \in (0,1)$, a little algebra yields

$$\varphi(t) = \varphi(b) - (b-t)\varphi'\bigl(t + \theta(b-t)\bigr). \tag{133}$$

Now (132) quickly yields

$$\varphi(b) = f(b) - f(b) - 0 - \ldots - 0 = 0. \tag{134}$$

On the other hand, differentiating (132) yields a telescoping sum as in the proof of Theorem 3.12.1 resulting in

$$\varphi'(t) = -\frac{(b-t)^n}{n!}f^{(n+1)}(t). \tag{135}$$

Combining (133) – (135) we have

$$\varphi(t) = -(b-t)\left(-\frac{\bigl(b-t-\theta(b-t)\bigr)^n}{n!}f^{(n+1)}\bigl(t+\theta(b-t)\bigr)\right)$$
$$= \frac{(b-t)(b-\theta b - t + \theta t)^n}{n!}f^{(n+1)}\bigl(t+\theta(b-t)\bigr)$$
$$= \frac{(1-\theta)^n(b-t)^{n+1}}{n!}f^{(n+1)}\bigl(t+\theta(b-t)\bigr). \qquad \square$$

12. TAYLOR'S THEOREM

After establishing the remainder terms, Cauchy applied the results of the *Cours d'analyse* to various functions to prove them to equal their Taylor series where they converged. In particular, he gave the first correct Taylor-based proof of the Binomial Theorem. Such a proof is now merely an ugly but routine estimate:

$$\left| (1+x)^\mu - \sum_{k=0}^n \binom{\mu}{k} x^k \right| = \left| \frac{(1-\theta)^n x^{n+1}}{n!} \left(\frac{d^{n+1}}{dt^{n+1}} (1+t)^\mu \bigg|_{t=\theta x} \right) \right|$$

$$= \frac{(1-\theta)^n |x|^{n+1}}{n!} |\mu(\mu-1)\cdots(\mu-n)|(1+\theta x)^{\mu-n-1}$$

$$= \left| (n+1) \binom{\mu}{n+1} x^n \right| |x| \left(\frac{1-\theta}{1+\theta x} \right)^n (1+\theta x)^{\mu-1}.$$

But

$$(n+1) \binom{\mu}{n+1} x^n \to 0 \text{ as } n \to \infty \text{ for } |x| < 1$$

because it is n-th degree term of the convergent series for the derivative of $(1+x)^\mu$. Also

$$\left(\frac{1-\theta}{1+\theta x} \right)^n \to 0 \text{ as } n \to \infty \text{ for } 0 < \theta < 1$$

because

$$0 < \frac{1-\theta}{1+\theta x} < 1 \text{ for } |x| < 1,$$

as is quickly established:

$$\frac{1-\theta}{1+\theta x} \geq 1 \Rightarrow 1 - \theta \geq 1 + \theta x \Rightarrow 0 \geq \theta(1+x)$$

which is impossible as θ and $1+x$ are positive. The extra factor of $|x|$ is obviously bounded by 1, and $(1+\theta x)^{\mu-1}$ is bounded by $(1+|x|)^{\mu-1}$ for $\mu > 1$, by $(1-|x|)^{\mu-1}$ for $\mu < 1$, and by 1 for $\mu = 1$. Thus

$$\left| (1+x)^\mu - \sum_{k=0}^n \binom{\mu}{k} x^k \right|$$

is the product of two terms that tend to 0 as $n \to \infty$ and two terms that are bounded, and hence itself tends to 0 as $n \to \infty$.

With this last proof we have completed our story. The central importance of the Binomial Theorem in the theory of the Calculus disappeared in 1823 with Cauchy's proof of and reliance on the Mean Value Theoem in the *Calcul infinitésimal*. Further developments of the theory of Taylor series took place in the complex plane, beginning already in 1821 in the *Cours d'analyse* and continuing from there in works by Cauchy, Riemann, Weierstrass, and countless others. Cauchy proved the convergence of its binomial series to $(1+x)^\mu$ for complex x with $|x| < 1$ and real μ already in the *Cours d'analyse*, and Abel extended this result to complex μ using his "fearsome algebraic manipulation". But Cauchy's more general treatment led to the very pleasing complex form of Taylor's Theorem:

3.12.5. Theorem (Taylor's Theorem). *Let the function f of a complex variable be differentiable throughout the open disc of radius $R > 0$ centred at a point z_0,*
$$D(z_0, R) = \{z \in \mathbb{C} \,|\, |z - z_0| < R\}.$$
Then: for all $z \in D(z_0, R)$,
$$f(z) = \sum_{k=0}^{\infty} \frac{f^{(k)}(z_0)}{k!}(z - z_0)^k.$$

To apply this to the real Binomial Theorem, for example, one has but to show that
$$f(z) = (1 + z)^\mu$$
is differentiable for all complex z with $|z| < 1$. This, of course, requires a definition of exponentiation in the complex domain,
$$(1 + z)^\mu = e^{\mu \ln(1+z)},$$
and the establishment of the usual laws of exponentiation, etc. There is a fair amount of work involved, but it is more routine than what we've been doing and very little of the work is a deviation from the course of developing the general theory of Complex Analysis. The upshot of all this is that Cauchy demoted Newton's Binomial Theorem from its central rôle in Real Analysis to that of an aside in a more rigorously developed theory of Complex Analysis.

APPENDIX A

Using the Calculator

1. Horner's Method on the Calculator

In one of my earlier books[1] I worked out a number of examples of the application of Horner's Method and performed substitutions in another chapter of that book. The algebra wasn't difficult, but I nonetheless found it easier to program the calculator to do the work for me. In this appendix I propose to automate the entire task. Doing so is a digression from our main path and the reader can omit this material without harm, but it is a pleasant diversion and may be of some service to the reader who hasn't yet made full use of the power of the modern graphing calculator. As to the choice of calculator, I shall use both the *TI-83 Plus* and the *TI-89 Titanium*. Because the *TI-89 Titanium* has so much more built-in functionality, I shall treat the two separately, dealing first in sections 1 to 3 with the *TI-83 Plus* and then in section 4 with the *TI-89 Titanium*. Owners of the latter ought not to skip the discussion for the former, as the ideas behind the methods are still relevant to the use of this calculator.

The first thing to do is to decide how one is going to represent a polynomial, e.g.,
$$P(x) = 5x^4 + 3x^3 - 6x^2 + 2,$$
in the calculator. On the *TI-83 Plus*, I find the most convenient representation is as the list,
$$\{5, 3, -6, 0, 2\}$$
of its coefficients. With this decision, we will represent polynomials by various lists with names like ∟POLY or ∟P*something*.

Our first task is to program the basic variable substitution taking us from a polynomial $P(x)$ to $Q(y) = P(y + a)$. We can do this either via iterating synthetic division or by means of the binomial coefficients. We shall do both, first the approach via synthetic division.

We assume someone has stored the polynomial P in a list ∟POLY1 and the number a, for which we wish to divide $P(x)$ by $x - a$, in the variable A. The following program will perform synthetic division, storing the quotient in the variable ∟POLY2 and the remainder $P(a)$ in the variable Z.

PROGRAM:SYNTHDIV
:dim(∟POLY1)→N
:N→dim(∟POLY2)

---

[1]Craig Smoryński, *History of Mathematics; A Supplement*, Springer Science + Business Media, LLC, New York, 2008.

```
:∟POLY1(1) →∟POLY2(1)
:For(I,2,N)
:∟POLY1(I)+A*∟POLY2(I−1)→∟POLY2(I)
:End
:∟POLY2(N)→Z
:N−1→dim(∟POLY2)
:DelVar N
:DelVar I.
```

I have not deleted ∟POLY1 because we will not want to lose it during calls to the program in the program HORNER, below.

To find $Q(y) = P(y + a)$, one again assumes $P(x)$ and a stored in the variables ∟POLY1 and A, and one will produce $Q(y)$ as a list ∟QPOLY of its coefficients:

```
PROGRAM:POLYSUB
:dim(∟POLY1)→M
:M→dim(∟QPOLY)
:∟POLY1(1)→∟QPOLY(1)
:For(J,2,M)
:prgmSYNTHDIV
:Z→∟QPOLY(M−J+2)
:If J<M
:∟POLY2→∟POLY1
:End
:DelVar M
:DelVar J
:DelVar ∟POLY2
:∟QPOLY.
```

I have added the last line so that the calculator will place a scrollable version of ∟QPOLY on the screen instead of the message Done.

Programming the substitution via the use of binomial coefficients is marginally more straightforward. Recall that, if we write

$$P(x) = a_1 x^n + a_2 x^{n-1} + \ldots + a_n x + a_{n+1}$$

and

$$Q(y) = b_1 y^n + b_2 y^{n-1} + \ldots + b_n y + b_{n+1},$$

where $y = x - a$ for some a and $Q(y) = P(x)$, then equation (22) told us

$$b_k = \sum_{i=1}^{k} a_i \binom{n-i+1}{k-i} a^{k-i}.$$

We have but to find the lists,

$$\{a_1, a_2, \ldots, a_k\}, \left\{ \binom{n}{k-1}, \binom{n-1}{k-2}, \ldots, \binom{n-k+1}{0} \right\}, \{a^{k-1}, a^{k-2}, \ldots, a, 1\},$$

multiply them, and sum the result to obtain b_k, as is done in the program POLYSUB2, below.

1. HORNER'S METHOD ON THE CALCULATOR

```
   PROGRAM:POLYSUB2
 1 :dim(∟POLY1)→N
 2 :N→dim(∟APOWS)
 3 :1→∟APOWS(1)
 4 :For(I,2,N)
 5 :A*∟APOWS(I−1)→∟APOWS(I)
 6 :End
 7 :N→dim(∟QPOLY)
 8 :For(K,1,N)
 9 :seq((N−I) nCr (K−I),I,1,K)→∟BIN
10 :seq(∟POLY1(I),I,1,K)→∟ACOEF
11 :seq(∟APOWS(I),I,K,1,⁻1)→∟POW
12 :∟ACOEF*∟BIN*∟POW→∟PRODS
13 :sum(∟PRODS)→∟QPOLY(K)
14 :End
15 :DelVar N
16 :DelVar I
17 :DelVar K
18 :DelVar ∟POLY1
19 :DelVar ∟APOWS
20 :DelVar ∟BIN
21 :DelVar ∟ACOEF
22 :DelVar ∟POW
23 :DelVar ∟PRODS
24 :∟QPOLY .
```

The added line numbers, which do not get entered into the calculator, are there to help me explain the steps of the program. Line 1 puts $n+1$ into N. The indices of the list go from 1 to $n+1$ and the degrees of the terms from 0 to n. This is the chief complication of the program. Lines 2 to 6 generate the list $\{1, a, a^2, \ldots, a^n\}$ of the powers of a. Line 7 creates a variable ∟QPOLY in which to store the coefficients of Q. If one has run the program before, ∟QPOLY already exists and has entries of its own. They will all be replaced and will not get in the way. Lines 8 to 14 do the heavy work of the program generating $b_1, b_2, \ldots, b_{n+1}$ in succession. For $1 \le k \le n+1$, line 9 creates the list

$$\left\{ \binom{n}{k-1}, \binom{n-1}{k-2}, \ldots, \binom{n-k+1}{0} \right\}.$$

[Bear in mind that N contains $n+1$, whence N−I will be $n+1-i = n-i+1$ and i goes from 1 to k, yielding $n, n-1, \ldots, n-k+1$ for the upper numbers of the binomial coefficients.] Line 10 picks out the subsequence,

$$\{a_1, a_2, \ldots, a_k\},$$

of the coefficients of $P(x)$, and line 11 picks out the powers of a up to $k-1$ in reverse order:
$$\{a^{k-1}, a^{k-2}, \ldots, a, 1\}.$$
Line 12 produces the list of products of these three lists,
$$\left\{a_1\binom{n}{k-1}a^{k-1}, a_2\binom{n-1}{k-2}a^{k-2}, \ldots, a_k\binom{n-k+1}{0}\right\},$$
and line 13 sums them up yielding b_k. Lines 15 to 23 perform the cleanup operations, and the last line puts a scrollable version of ∟QPOLY on the home screen.

Reviewing these programs we see that POLYSUB2 is comparable in length to POLYSUB and SYNTHDIV combined, and both it and the earlier pair simply involve the iteration of basic list operations. Writing the pair with a program call, however, requires a bit of extra care. In writing a single program, one would automatically avoid the conflicting use of variables. For example, one would not use the same counter in nested For loops. POLYSUB calls SYNTHDIV within a For loop and SYNTHDIV used the counter I in its For loop. Had I not been extra careful and used J instead of my preferred I in POLYSUB, the program would not have run as intended. The problem here is that on the *TI-83 Plus* only global variables are used. Writing the corresponding program on the *TI-89 Titanium* one could declare the counter variable i to be local in synthdiv and avoid a conflict in using it again in polysub. Indeed, on the *TI-89 Titanium*, one would program synthdiv and polysub as functions and thus be required to declare the variable to be local.

Incidentally, one can shorten POLYSUB2 by eliminating lines 2 – 6 and 19 entirely, and replacing line 11 by

seq(A^I,I,K−1,0,−1)→∟POW.

This would be a bad idea because of all the extra multiplications performed. Each a^i requires $i-1$ multiplications, whence the new command calls for
$$\sum_{i=1}^{k-1}(i-1) = \frac{(k-1)(k-2)}{2}$$
multiplications. And, this being done for $k=1$ to $n+1$, this requires
$$\sum_{k=1}^{n+1}\frac{(k-1)(k-2)}{2} = \frac{(n+1)n(n-1)}{6}$$
multiplications to produce all the powers of a; while retaining lines 2 – 6, 19, and the original 11 requires only $n-1$ multiplications for this task.

Fascinating as these considerations are, they digress from the purpose at hand, which is to program the calculator to solve polynomial equations.

Before programming the general procedure, a few difficulties have to be addressed. Two problems are obvious. First, there is the issue of the initial estimate for a root of the polynomial. There are methods, unrelated to the procedure at hand, for finding a rather large interval in which all the roots, if any exist, must lie. Discussing such methods would take us too far afield, so I will do what most expositors do and assume the initial estimate as given. The second issue concerns roots at or too near a local maximum or minimum value of the polynomial. The Chinese-Horner algorithm proceeds by finding the (say) left-endpoints of successively smaller intervals in which the successive substituted polynomials change signs. If the curve is too flat near the root, there may not be any sign changes at all.

A.1.1. Exercise. *Try to find the root of*
$$125000x^3 - 322500x^2 - 97650x + 492993,$$
given the initial estimate $a = 1$.

Like Qín's polynomial, the polynomial of the Exercise has a simple root that can be solved for exactly in a few steps using interval subdivision and synthetic division. But there will be no sign changes and one will have to vary the procedure. Moreover, even if one does this, unless one hits on the root exactly, one will not be certain whether one is zeroing in on a root or on the location of a minimum very close to 0. Thus, our program will be based on the assumption that the polynomial in question experiences a sign change near the root we are looking for.

There are inessential assumptions that can be made to simplify the programming task. For example, we shall assume à la the Chinese that the root sought is positive. Should one want a negative root for
$$P(x) = a_0 x^n + a_1 x^{n-1} + \ldots + a_{n-1} x + a_n,$$
one can simply run the program for
$$P(x) = (-1)^n a_0 x^n + (-1)^{n-1} a_1 x^{n-1} + \ldots + (-1) a_{n-1} x + a_0$$
and change the sign of the final result. Similarly, since $-P(x)$ is an increasing function whenever $P(x)$ is a decreasing one, we need only write our program for polynomials that are monotone increasing near the root.

I shall leave it to the reader to provide the necessary changes to be made to the program to accommodate the first of these two variations, i.e., building into the program the capability of finding negative roots. I would do the same for finding roots in the decreasing case, but, as our premier example, namely Qín's polynomial, is decreasing near his root and our program must handle his example directly, I incorporate this modification myself.

So that the user will not have to read the program before use to see what must be stored in which variables, our program HORNER will be interactive, instructing the user to enter the list of coefficients as a list from the highest to lowest degree, and then to enter two values r and c for an initial estimate $r \times 10^c$ to the root of the polynomial. It is assumed that the actual root lies between $r \times 10^c$ and $(r+1) \times 10^c$ and that $P(r \times 10^c)$ and $P((r+1) \times 10^c)$ have opposite signs.

The program begins with the interactive part requesting the necessary data:
    PROGRAM:HORNER
    :ClrHome
    :Disp "PLEASE ENTER"

:Disp "COEFFICIENT LIST"
:Disp "FOR POLYNOMIAL"
:Disp "(HIGHEST TO"
:Disp "LOWEST DEGREE):"
:Input "⌊P=",⌊P
:ClrHome
:Disp "PLEASE ENTER"
:Disp "ESTIMATE R∗10^C"
:Disp "OF ROOT WITH"
:Disp "R,C INTEGERS,"
:Disp "R NONNEGATIVE"
:Input "R=",R
:Input "C=",C
:ClrHome.

The program will generate two lists, a list of the digits of the root and a list of powers of 10 associated with these digits. The first few elements of the first list will be the digits of R. Thus, if R contains the positive integer 325, the first list will be $\{3, 2, 5, \ldots\}$. For a single digit number, we can start the list off by simply taking the singleton. For a multidigit number, we note that the number of digits in $r > 0$ is $[\log(r)] + 1$, where log is the logarithm to base 10 and $[\cdot]$ denotes the greatest integer function. This determines the length of the list to be formed, the entries being obtained by toying with powers of 10 and the greatest integer function:

:If R<10
:Then
:0→F
:{R}→L₁
:Else
:int(log(R))→F
:seq(int(R/10^I),I,0,F)→L₁
:seq((L₁(I)−10L₁(I+1)),I,1,F)→L₂
:augment(L₂,{L₁(F+1)})→L₂
:seq(L₂(I),I,F+1,1,⁻1)→L₁
:End.

L_1 is the initial segment of the list of digits of the root sought. L_2 was merely an auxiliary used in constructing L_1. The rest of the program is largely an iteration finding the remaining digits to complete L_1. Now, this requires a decision: How many digits do we want? The *TI-83 Plus* handles up to 14 digits and displays 10, so any number from 10 to 14 is reasonable. For a nice round figure I have chosen 10. Thus we begin by extending the existing L_1 to length 10.[2] We also generate the sequence of powers of 10, $\{10^{f+c}, 10^{f+c-1}, \ldots, 10^{f+c-9}\}$ (f being the number stored in F), and store the result in L_2:

:10→dim(L₁)
:seq(10^(F+C−I),I,0,9)→L₂.

---

[2] It is assumed R has fewer than 10 digits.

1. HORNER'S METHOD ON THE CALCULATOR

Preparatory for the calls to POLYSUB and SYNTHDIV, we copy ∟P and R∗10^C into ∟POLY1 and A, respectively. For our purposes, we do not need to keep ∟P or R any longer. Indeed, we could have input the list ∟P directly into ∟POLY1 in the interactive portion of the program. However, one might imagine wanting to rewrite HORNER to find several roots of a given polynomial more exactly, so I have written the program in this slightly odd manner.

:∟P→ ∟POLY1
:10^C→D
:R∗D→A.

Next come a few lines of code to accommodate decreasing polynomials. Recall that, after running SYNTHDIV, Z stores the value of $P(a)$:

:prgmSYNTHDIV
:If Z>0
:⁻∟POLY1→∟POLY1.

The main body of the program consists of a pair of nested loops. The outer loop is a For loop generating successive digits in L_1. The inner loop evaluates the current polynomial Q at successive values of y until a nonnegative value is generated. Then, either one has found a zero or the zero lies between y and the previous value of y. This is handled by a Repeat loop which repeats instructions until the stated condition is true. The reason for this choice is that the obvious approach of using Goto commands inside conditionals inside For loops never seems to work for me.[3]

:For(K,F+2,10)
:D/10→D
:prgmPOLYSUB
:0→L
:⁻1→Z
:∟QPOLY→∟POLY1

---

[3] When I first learned some programming in the late 1960s, FORTRAN was the programming language of choice and Goto commands were the standard mechanism for exiting loops or dealing with other branching commands. The higher the level of the language, however, the less appropriate they seem to be. The referee has called me out on this:

> There are many for whom this "obvious approach" has not worked! It was roundly denounced in a famous article by Edsger W. Dijkstra in 1968: "Goto Statement Considered Harmful", *Communications of the ACM* 11, no. 3, March 1968, pp. 147 – 148, which spurred a debate about structured programming in the computer science community. I would suggest to rephrase the reasons for the author's choice with a reference to structured programming in general and Dijkstra's article in particular.

Dijkstra, one of the more entertainingly outspoken pioneers of computer programming, offers cogent reasons why Goto commands should be given minimal use and do not belong at all in higher level languages. Whether they should or should not be used in one's calculator programs I cannot honestly say. I use them myself in this book and the programs work, but, as Dijkstra says, "The **go to** statement as it stands is just too primitive; it is too much an invitation to make a mess of one's program." (p. 147). The upshot is that one should use Goto commands sparingly if at all and only in the simplest contexts; otherwise it is highly likely one's programs will not run as intended.

```
:Repeat 0≤Z
:L+1→L
:D*L→A
:prgmSYNTHDIV
:End
:If Z=0
:Then
:L→L₁(K)
:L*D→A
:Else
:L−1→L₁(K)
:(L−1)*D→A
:End
:End.
```

We can now generate the root with a single line,

:sum(L_1*L_2)→Y,

and finish the program with a bunch of cleaning up commands as well as the convenient one placing Y on the screen:

```
:DelVar A
:DelVar C
:DelVar D
:DelVar F
:DelVar K
:DelVar L
:DelVar R
:DelVar Z
:DelVar ∟P
:DelVar ∟POLY1
:DelVar ∟POLY2
:DelVar ∟QPOLY
:ClrList L₁,L₂
:Y.
```

Now that we have the program, the first thing to do is to run it a few times.

A.1.2. Exercise. *Store the coefficients of Qín's polynomial in the list* L_3 *and run* HORNER, *entering* L_3 *when the program asks for* ∟P *and entering* 8 *and* 2 *for* R *and* C, *respectively. My necessity of using* Repeat *instead of* For *has as an accidental consequence the success of the program for the choices* 80 *and* 1, *and* 800 *and* 0, *for* R *and* C *as well. Run these. All three pairs give the value* 840, *but the last pair takes the longest time. Can you explain this?*

We also applied Horner's method to $P(x) = x^2 - 224$ in TABLE 3, above. The reader might wish to run HORNER for this polynomial using 14 for R and 0 for C.

Another recommendation is to run the program for the polynomial $P(x) = x^3 - 2x - 5$, which Newton used to introduce his own procedure for calculating roots. Starting with the estimate 2, the program produces 2.094551481 in short order.

The words "in short order" are, of course, relative. The several seconds this last run takes to produce 10 significant digits in the solution are negligible in comparison with the time it would take to work out, say, 2.094 by hand. But the calculator's built-in zero finder will deliver greater precision in much less time. Indeed, if one graphs the curve on the *TI-83 Plus* using the ZDecimal setting, then hits the CALC button, chooses the zero option, and successively enters 2, 3, 2.5 as LeftBound, RightBound, and Guess, respectively, an answer 2.0945515 will appear on the screen in less than a second—certainly less time than it took to press all the buttons or to quit the screen and press ANS to read the more accurate 2.094551482 or to check the hidden digits to learn that the stored answer is 2.0945514815423.

This does not mean that programming HORNER was an unnecessary waste of time, or that the calculator's built in functionality can outperform the old technique. It merely means that we have to do better. The program can be sped up and, after additional programming, can also be made more accurate.

There are two obvious methods that can be used to speed the program up. One is to rewrite it as an assembly language program for the calculator. This is a major task I will not take on in the present book. The other is to home in on the next digit more rapidly by starting the search for it using the negative ratio of the constant to the linear coefficient. For simplicity's sake I have based HORNER on the more straightforward but less efficient method of testing one possible digit at a time, always starting at 1 and working one's way up to 10 looking for the first change of sign, rather than estimating where the sign change occurred and starting from there. Thus, except for its reliance on synthetic division to evaluate the polynomial and perform the substitutions, it did not so much simulate the Chinese-Horner algorithm as a slightly refined version of the primitive bisection method by which an interval is divided into 10 equal parts instead of just two.

The new program HORNER2 will be a bit longer than HORNER, and will be more faithful to the instructions to Qín's diagram mentioned in footnote 54 on page 48, above. However, entering it in the calculator need not be too tedious. After beginning a new program and naming it HORNER2, the program looks like

PROGRAM:HORNER2
:

One can now press the RCL button, then the PRGM button, then choose the program HORNER from the EXEC submenu, and press ENTER. The entire program for HORNER gets copied into HORNER2, scrolling down in the display window as it does so. All one has to do is to replace the pair of nested loops— For and Repeat— from the beginning For command to the final End command. And, as I've added two new variables G and P, one will want to insert DelVar G and DelVar P commands somewhere amongst the existing DelVar commands.

The outer loop proceeds from one digit of the root to the next, starting with $-a_n/a_{n-1}$ as an estimate of the root of the substituted polynomial Q. To find these coefficients, one needs n and $n-1$. The variable N being already in use, I chose P and in it I will store the dimension of the list ∟POLY1 ($= n+1$, the length of the list $\{a_0, a_1, \ldots, a_{n-1}, a_n\}$ of coefficients) before beginning the loop. The variable D represents the power of 10, $d = 10^{c-k}$, corresponding to the next digit, and we only want that digit. This means we will want to take $b = (-a_n/a_{n-1})/d$ as our estimate of the digit and $[b]$ as an integral such estimate. However, although b will be positive and thus > 0 (as we assume $a_{n-1} \neq 0$ and Q is strictly increasing, and as a_n is assumed to be negative), it may still lie outside the interval unless we are already *very* close to the root. Hence we choose the minimum of $[b]$ and 10 as our starting digit. With this, the structure of the outer loop is fairly clear. Replace the entire block by the commands:

:dim(∟POLY1)→P
:For(K,F+2,10)
:D/10→D
:prgmPOLYSUB
:⁻∟QPOLY(P)/∟QPOLY(P−1)→L
:min({10,int(L/D)})→L
:L∗D→A
:∟QPOLY→∟POLY1
:prgmSYNTHDIV
:If Z=0
:Then
:L→L₁(K)
:Else
:prgmHORNAUX
:End
:End.

The call to the auxiliary program HORNAUX is clearly an expositional device that says "put the inner loop here". A separate program isn't required, but it is convenient for explanatory purposes. The current value in A may be to the left of the root or to the right of the root. In the first case, we keep adding 1 to L until the corresponding value of Z is $= 0$ and L is the digit, or Z is > 0 and L−1 is the digit. In the second case, we subtract 1 from L until Z is negative or 0, in which case L is the desired next digit. The two cases are not quite symmetric, but we can unify their treatment slightly by adding a new variable G which will store 1 if Z is negative and −1 if Z is positive. As we have tested to make sure Z is not 0 before calling HORNAUX, we could do this using the command

:⁻Z/abs(Z)→G.

Another possibility is to use a conditional construction,

:If Z<0
:Then
:1→G
:Else

:$^{-}1\to$G
:End.

Computer languages, however, are designed by computer scientists, not traditional mathematicians or logicians, so, in a desperate attempt to win their approval, I use tests instead. Z<0 gives the value 1 if the number stored in Z is negative and 0 otherwise; while Z>0 gives 1 if the number is positive and 0 otherwise. Thus I chose

:(Z<0)−(Z>0)→G.

With this said, the definition of HORNAUX proceeds thus:

PROGRAM:HORNAUX
:(Z<0)−(Z>0)→G
:L−G→L
:Repeat 0≤(G∗Z)
:L+G→L
:L∗D→A
:prgmSYNTHDIV
:End
:If Z≤0
:Then
:L→L$_1$(K)
:L∗D→A
:Else
:L−G→L$_1$(K)
:(L−G)∗D→A
:End

The DelVar G command can be placed at the end of HORNAUX or amidst the other clean-up commands of HORNER2.

HORNER2 improves on the performance of HORNER, albeit not dramatically so. The number of digits tested is cut down, as one can verify by inserting the command

:Disp {K,L}

into HORNAUX directly following the line :L+G→L, and doing the same in the appropriate spot in the original program HORNER and running the two programs on identical inputs.

A.1.3. Exercise. *To better see how the program is executing, one might consider instead placing the following commands immediately after the call to* SYNTHDIV:

:Disp {K,L,Z}
:For (Q,1,50)
:0→H
:End.

Run HORNER2 *for* $P(x) = x^2 - 224, r = 14, c = 0$. *[The* Disp *command displays the number* K *of the current digit the program is trying to find.* L *gives the value being considered for that digit, and* Z *the value of the polynomial at* the root with L in the K-th digit. The For loop is there merely to slow down the scrolling of the display to give one time to read the current values. The scroll may be slowed down even more by replacing 50 by a larger number, say 75 or 100. One could replace the For loop by a single Pause command, which stops program execution until one presses the ENTER button. But by the end one will have done this 20 times, which is a bit tiresome.]

A.1.4. Exercise. *Replacing the* Disp *command by* Output *commands allows one to display items in exact positions on the screen with no scrolling. Find places in* HORNER2 *to insert the following commands to give a better onscreen representation of the process:*

:Output(1,1,K), :Output(1,10,L),
:Output(2,1,R), :Output(2,K,L_1(K)).

Where would be the best place to insert a slow-down loop as in the preceding exercise? How would you change these commands in order to display the decimal point in its proper position on the second line of the display?

Oops! I started explaining how the program was too slow and in these exercises I am adding commands that slow the program down even more. However, being occupied with the feedback, one doesn't notice how slow the computational part of the program is. If one does not want to see how the program is doing, but to treat it as a black box and just get an answer quickly, one notices how far from instantaneous the process is and one must make it more efficient.

The real cause of the inefficiency of the program, however, is that one is not making full use of the estimate $-a_n/a_{n-1}$. While it may not initially be all that close to the desired digit, when one gets nearer and nearer to the root, the successive estimates provide greater and greater accuracy and the program is ignoring this, taking only 1 of many available digits, thus resulting in much duplication of effort.

For polynomials that are not too flat near the root, the number of correct decimals obtainable at any stage in the iteration can grow quite rapidly. Applying the technique to Newton's polynomial, the number of secured digits doubles at each stage. Thus, the number of correct digits obtainable via the use of $-a_n/a_{n-1}$ is

$$1, 2, 4, 8, 16, 32, 64, 128, 256, 512,$$

while our algorithm gives

$$1, 2, 3, 4, 5, 6, 7, 8, 9, 10.$$

With the limited built-in precision of the *TI-83 Plus*, we cannot take full advantage of the extra accuracy obtained by not limiting ourselves to using only one digit of $-a_n/a_{n-1}$. But, if we did, for Newton's polynomial we could stop at K = 4 and exceed the desired 10-digit accuracy we sought.[4] To draw the full benefit of the increasing accuracy offered by $-a_n/a_{n-1}$ as the number of iterations increases, we need *multiple precision*—thus we must either transfer

---

[4]In theory, we would have 15 digits, but we would lose one as the calculator only stores 14 of them. Moreover, some rounding may affect the last digit or so.

over to the *TI-89 Titanium* or program a means of obtaining such precision on the *TI-83 Plus*.

The switch to the *TI-89 Titanium* will be made two sections hence. In the immediately following section I propose to consider the problem of extending the precision of the *TI-83 Plus* for operations on nonnegative integers[5].

2. Exact Calculations on the *TI-83*

The key to increasing precision is simply to represent numbers by their lists of digits in some base. A list can have up to 999 entries, whereas a number can have only 14 digits in base 10. Obviously, though the accuracy will still be limited, it will be a lot less limited than the default amount. And the use of larger bases will allow greater accuracy yet: A number consisting of, say, 10 digits in base 100 contains twice as many digits in base 10 than does a number consisting of 10 digits in base 10. For our purposes, a base of $10^6 = 1000000$ is convenient. The sum or product of two such *micresimal* digits will have at most 12 decimal digits and the calculator's built-in operations will give accurate answers in base 10, which can then be readily represented in base 10^6. Moreover, as the number of decimals in each micresimal digit is divisible by 2 and 3, it is a convenient base should we want to find square or cube roots. (For 4th or 5th roots, one might prefer using 10^4 or 10^5, respectively, as the base.)

Our first program will simply convert a number into a list representing it in base 10^6. The following works for integers of up to 14 decimal digits.

```
PROGRAM:CONVERT
:log(Z)→N
:int(N/6)+1→N
:N→dim(LZ)
:For(K,1,N)
:int(Z/1000000)→A
:Z−1000000∗A→LZ(K)
:A→Z
:End
:seq(LZ(I),I,N,1,⁻1)→LZ
:DelVar A
:DelVar I
:DelVar K
:DelVar N
:DelVar Z
:LZ
```

When we first learned the arithmetic operations, we began by memorising the addition and multiplication tables and then learned how to handle multidigit numbers through repeated reference to these tables, the lining up of digits, and occasional carrying operations. The performance of the inverse functions of

---
[5]While it is true that students often find dealing with minus signs confusing, computationally the signs offer no challenge. Hence I limit my discussion here to the nonnegative case.

subtraction and division also required borrowing operations. With base 10^6, the tables are unwieldy, too large even to store on the calculator. Instead we find the sum or product of two micresimal "digits" by using the calculator's built-in arithmetic operations on numbers written in base 10. The treatment of "multidigit" numbers will then mimick the familiar techniques.

To add or subtract two numbers represented as lists ∟X and ∟Y by using the built-in operations ∟X+∟Y and ∟X−∟Y, the numbers must be presented with the same number of micresimal digits, i.e., the lists must have the same lengths. This is achieved by *padding* the shorter list by placing extra 0's in the front. And, as subtraction could result in a number with leading 0's, it will be handy to be able to *unpad* such a list. Thus, in addition to programs performing the basic arithmetic operations on lists ∟X and ∟Y, we will need four auxiliary programs XPAD, XUNPAD, XCARRY, and XBORROW operating on single lists ∟Z.[6]

The XPAD program is actually unnecessary, consisting of a single command:

PROGRAM:XPAD
:augment(seq(0,I,1,M),∟Z)→∟Z.

This assumes one is given a list ∟Z and a number M of 0 entries one wants to prefix to the list. The seq function produces the list of 0's and augment performs the concatenation. The only reason for making a call to XPAD from within another program rather than to list its command is to increase the readability of the program for anyone who reads it.

XUNPAD requires one to count the number of leading 0's in the list ∟Z, then produce only the rest of the list.

PROGRAM:XCOUNTER
:dim(∟Z)→N
:0→K
:While ∟Z(K+1)=0
:K+1→K
:If K=N
:Return
:End

PROGRAM:XUNPAD
prgmXCOUNTER
:If N=K
:Then
:{0}→∟Z
:Else
:seq(∟Z(I),I,K+1,N)→∟Z
:End
:DelVar N
:DelVar K.

Other than XPAD, XCOUNTER, and XUNPAD, each of the programs to follow assumes the inputs ∟X and ∟Y or ∟Z to be unpadded[7], leaves ∟X and ∟Y unchanged, and produces unpadded output. The only penalties for the use of padded inputs are padded outputs, which can be unpadded as necessary, and the occasional unnecessary computation in the calculator as lists are multiplied by 0, extra 0's are added, etc.

---

[6]The non-mnemonic prefixing of the program names by "X" is simply a device to push them to the bottom of the list of programs so that the important ones will appear on the screen when the PRGM button is pressed.

[7]I.e., no leading 0's unless the list is {0} itself.

2. EXACT CALCULATIONS ON THE TI-83

The XCARRY and XBORROW programs proceed from the last entry of a list $_LZ$ to the second, in each case adding or borrowing as necessary.

PROGRAM:XCARRY
:dim($_L$Z)→N
:For(I,N,1,⁻1)
:If $_L$Z(I)≥1000000
:Then
:int($_L$Z(I)/1000000)→R
:$_L$Z(I)−1000000∗R→$_L$Z(I)
:If I=1
:Then
:augment({R},$_L$Z)→$_L$Z
:Else
:$_L$Z(I−1)+R→$_L$Z(I−1)
:End
:End
:End
:DelVar N
:DelVar I
:DelVar R.

PROGRAM:XBORROW
dim($_L$Z)→N
:If N>1
:Then
:For(I,N,2,⁻1)
:If $_L$Z(I)<0
:Then
:$_L$Z(I)+1000000→$_L$Z(I)
:$_L$Z(I−1)−1→$_L$Z(I−1)
:End
:End
:prgmXUNPAD
:End
:DelVar N
:DelVar I

The two programs are not quite symmetric. The addition or subtraction of two numbers will always result in the carrying or borrowing of a 1. With multiplication, however, the number to be carried could easily be greater than 1.

The addition and subtraction programs are fairly straightforward. They begin with a preparation step padding any short argument. I place this preparatory step into a program of its own, thus overall taking up slightly less RAM:

PROGRAM:XPREP
:$_L$X→$_L$X1
:$_L$Y→$_L$Y1
:dim($_L$X)→A
:dim($_L$Y)→B
:If A<B
:Then
:B−A→M
:$_L$X→$_L$Z
:prgmXPAD
:$_L$Z→$_L$X1
:End
:If B<A
:Then
:A−B→M
:$_L$Y→$_L$Z

PROGRAM:SUM
:prgmXPREP
:$_L$X1+$_L$Y1→$_L$Z
:prgmXUNPAD
:prgmXCARRY
:DelVar A
:DelVar B
:DelVar M
:DelVar $_L$X1
:DelVar $_L$Y1

PROGRAM:DIFF
:prgmXPREP
:$_L$X1−$_L$Y1→$_L$Z
:prgmXUNPAD
:prgmXBORROW
:DelVar A
:DelVar B
:DelVar M
:DelVar $_L$X1
:DelVar $_L$Y1

```
:prgmXPAD
:ʟZ→ʟY1
:End.
```

There are two options for dealing with multiplication. We can simulate the usual cascade of multidigit multiplication by storing the results of the cascade in a matrix. For example the cascade part of the multiplication carried out in TABLE 2, below,

TABLE 2.

$$\begin{array}{rr} 123 & 456789 \\ \times\, 123 & 456789 \\ \hline 56185047 & 208656190521 \\ 15129\quad 56185047 & \\ \hline 15129\quad 112370094 & 208656190521 \end{array}$$

would be represented by the matrix with entries

$$\begin{bmatrix} 0 & 123 \cdot 456789 & 456789 \cdot 456789 \\ 123 \cdot 123 & 456789 \cdot 123 & 0 \end{bmatrix},$$

which the calculator stores internally as

$$\begin{bmatrix} 0 & 56185047 & 208656190521 \\ 15129 & 56185047 & 0 \end{bmatrix},$$

though the upper right entry will appear on the screen as 2.086561905E11. One could then take the cumulative sum of the matrix,

$$\begin{bmatrix} 0 & 56185047 & 208656190521 \\ 15129 & 112370094 & 208656190521 \end{bmatrix},$$

read off the bottom row as the list

$$\{15129, 112370094, 208656190521\},$$

and run the XCARRY program to obtain

$$\{15241, 598750, 190521\}$$

as the product.

Alternatively, and perhaps conceptually a bit simpler, we can multiply the two lists as if they were polynomials in x,

$$(123x + 456789)(123x + 456789),$$

and collect terms of like degree by adding the appropriate products: The coefficient of

x^2 is $123 \cdot 123$
x is $123 \cdot 456789 + 456789 \cdot 123$
1 is $456789 \cdot 456789$.

The program MULT follows the second approach.

```
PROGRAM:MULT
:dim(LX)→K
:dim(LY)→M
:K+M−1→N
:seq(0,I,1,N)→LZ
:For(I,1,K)
:For(J,1,M)
:LX(I)∗LY(J)+LZ(I+J−1)→LZ(I+J−1)
:End
:End
:prgmXCARRY
:DelVar K
:DelVar M
:DelVar N
:DelVar I
:DelVar J.
```

A.2.1. Exercise. *Use the program MULT to calculate* $123456789 \cdot 123456789$, *and compare the result with that obtained above.*

A.2.2. Exercise. *In* (12) *on page 7, above, one finds* $\sqrt{17}$ *is approximately* 4.1231056256176605498214. *Use the list*

$$\{41231, 56256, 176605, 498214\}$$

and the program MULT to show

$$4.1231056256176605498214^2 < 17,$$

and the list,

$$\{41231, 56256, 176605, 498215\}$$

and MULT to show

$$4.1231056256176605498215^2 > 17.$$

A.2.3. Exercise. (*For those with greater familiarity with programming and matrices*) *Write a program MULT2 using matrix operations to carry out multiplication via the cascading display as described above.*

The long division program is more of a pain to write than those for the simpler arithmetic operations. In addition to using the other operations, it will require one frequently to make comparisons to see if the number represented by the list LX is less than the number represented by LY. This program proceeds by unpadding the two lists and setting the truth value T to 0. If LX is shorter than LY, T is reset to 1 and the program quits. If LX is longer than LY, the program quits, leaving 0 in T. If the lists have equal length, the program looks for the first entry at which the two lists differ and puts 1 into T if LX has the smaller of the two values at that point, leaving 0 in T otherwise, and the program quits. If the two lists never differ, the lists are equal and the value of T is never changed.

We will actually need two comparisons, one that checks for strict inequality of the numbers represented by two lists as just described, and one that checks for weak inequality. I incorporate the two into a single program by adding a simple test that must be passed to test for the weak inequality:

```
PROGRAM:XLESS
:∟X→∟Z
:prgmXUNPAD
:∟Z→∟X
:∟Y→∟Z
:prgmXUNPAD
:∟Z→∟Y
:0→T
:dim(∟X)→X
:dim(∟Y)→Y
:If X<Y
:Then
    :1→T
    :Goto 1
:End
:If Y<X
    :Goto 1
:If Str1="WEAK"
:Then
    :If min(∟X=∟Y)
    :Then
        :1→T
        :Goto 1
    :End
:End
:For(I,1,X)
    :If ∟X(I)<∟Y(I)
    :Then
        :1→T
        :Goto 1
    :End
    :If ∟Y(I)<∟X(I)
        :Goto 1
:End
:Lbl 1
:DelVar I
:DelVar X
:DelVar Y
:DelVar Str1.
```

2. EXACT CALCULATIONS ON THE TI-83

For the sake of consistency I should again choose ∟X and ∟Y as the input variables for the program DIVIDE. However, this program will call DIFF for different choices of ∟X, so we might as well choose the mnemonic names ∟NUM and ∟DENOM for the numerator and denominator, respectively. The output variables will similarly have mnemonic names: ∟QUOT, and ∟REM for the quotient and remainder, respectively.

```
     PROGRAM:DIVIDE
 1   :∟NUM→∟X
 2   :∟DENOM→∟Y
 3   :"STRICT"→Str1
 4   :prgmXLESS
 5   :If T
 6   :Then
 7       :{0}→∟QUOT
 8       :∟NUM→∟REM
 9   :Else
10       :dim(∟NUM)→C
11       :dim(∟DENOM)→W
12       :∟NUM→∟NUM1
13       :dim(∟NUM1)→V
14       :seq(0,I,1,C)→∟QUOT
15       :If W=1
16       :Then
17           :prgmXDIVAUX1
18       :Else
19           :prgmXDIVAUX2
20       :End
21   :End
22   :SetUpEditor ∟NUM,∟DENOM,∟QUOT,∟REM
23   :prgmXCLEANUP.
```

If the denominator is larger than the numerator, the quotient is 0 and the remainder is just the numerator. Lines 1 – 8 handle this.

Lines 9 – 20 handle the case in which the denominator is small enough to divide the numerator. The numerator changes in the process of long division, so we introduce a new variable ∟NUM1 to stand in for ∟NUM, keeping the latter for "archival" purposes. During the process we need the lengths of ∟NUM, ∟DENOM, and the changing length of ∟NUM1, whence the new variables C, W, and V. There is nothing mnemonic about these names; if anything they represent alphabetically the stage in the development process in which I needed a new real variable. The quotient in a long division is usually written directly above the numerator and is shorter. If we temporarily allow leading 0's, it has the same length as ∟NUM, hence line 14: We start with all entries of QUOT being 0 and will make changes as necessary as we go along.

The real work will be done in auxiliary programs XDIVAUX1 and XDIVAUX2. The former handles the especially simple case in which the denominator consists of a single micresimal digit. Lines 15 – 20 determine which program is needed and makes the necessary call.

Line 22 is a convenience. It puts the crucial lists into the List Editor, allowing one to survey the quotient and remainder quickly. Line 23 calls a cleaning up program which will delete the no longer needed variables.

The auxiliary programs XDIVAUX1, XDIVAUX2, and auxiliaries to the latter will be presented piecemeal. In entering and debugging this mini-suite of programs, I found it convenient to use a simple do-nothing program,

> PROGRAM:XCLEANUP
> :DelVar W

as a temporary replacement until I was finished. This allowed me to check the values of other variables when something went wrong. The reader, should he diligently do the ensuing exercises as he enters more and more auxiliaries, will either have to omit line 23 from this program or follow my lead in using a temporary version like the above.

The basic long division algorithm for base 10^6 is familiar. It works the same way as long division in base 10. And, as with base 10, the procedure is simplest when the divisor has only one digit—in this case a single micresimal digit. For example, to divide $\{252, 1000, 6\}$ by $\{24\}$, one would divide 252 by 24 to get a quotient of 10 and a remainder of 12. The 10 replaces the first entry of ∟QUOT which is now $\{10, 0, 0\}$. One would next bring down the 1000 to divide $\{12, 1000\}$ by $\{24\}$, i.e., $12 \times 10^6 + 1000$ by 24, getting 500041 and a remainder of 16. ∟QUOT now becomes $\{10, 500041, 0\}$ and one divides $16 \times 10^6 + 6$ by 24 to get 666666 and a remainder of 22. The end result is that ∟QUOT is $\{10, 500041, 666666\}$ and ∟REM is $\{22\}$. Each division involved division of a number with at most 12 decimal digits by one of at most 6 such, whence the calculations of the integral parts of the fractions on the calculator,

> int(252/24)
> int((12∗10^6+1000)/24)
> int((16∗10^6+6)/24),

will all be exact.

[One can check the answer easily. Successively enter

> 24∗{10,500041,666666}→∟Z
> ∟Z(3)+22→∟Z(3)
> prgmXCARRY
> ∟Z

and

> {252 1000 6}

will appear on the right of the screen.]

The program XDIVAUX1 will handle the case of division by a single digit number. This is fairly straightforward. Our example left unexplained only the matter of keeping track of which digit of the quotient one is generating during a given step. This will be handled by the variable Q.

2. EXACT CALCULATIONS ON THE TI-83

```
PROGRAM:XDIVAUX1
:1→Q
:∟NUM1(1)→N
:∟DENOM(1)→D
:While Q≤C
   :If N<D
   :Then
      :If Q=C
      :Then
         :{N}→∟REM
         :Q+1→Q
      :Else
         :Q+1→Q
         :N*10^6+∟NUM1(Q)→N
      :End
   :Else
      :int(N/D)→A
      :A→∟QUOT(Q)
      :N−A*D→B
      :If Q=C
      :Then
         :{B}→∟REM
         :Q+1→Q
      :Else
         :Q+1→Q
         :B*10^6+∟NUM1(Q)→N
      :End
   :End
:End
:∟QUOT→∟Z
:prgmXUNPAD
:∟Z→∟QUOT
```

[*Note.* In XDIVAUX1 we do not yet swap ∟NUM1 for another numerator and could have simply used ∟NUM itself in the three lines where elements of the list are used. This will change with XDIVAUX2.]

A.2.4. Exercise. *Store* $\{252, 1000, 6\}$ *in* ∟NUM *and* $\{24\}$ *in* ∟DENOM *and run the program* DIVIDE *to check the result of our earlier division by hand. Repeat with* $\{142, 257, 10008\}$ *stored in* ∟NUM.

We have, of course, not completed the program DIVIDE insofar as XDIVAUX2 is as yet undefined. However, so long as ∟DENOM consists of a single micresimal digit, XDIVAUX2 will never be called and the program DIVIDE will run fine.

The program XDIVAUX2 runs along the same lines as XDIVAUX1, but is a bit more complicated. To see why, we must consider a couple of additional examples of long division in base 10^6. Suppose, by way of example, we wish to divide $\{2, 5, 8\}$ by $\{1, 2\}$, i.e., 2000005000008 by 1000002. We would note that 1 goes into 2 twice and that $2 * \{1,2\}$ is $\{2,4\}$, which is less than $\{2,5\}$. This allows us to subtract $\{2,4\}$ from $\{2,5\}$ to get a remainder of $\{1\}$. We then bring down the $\{8\}$ and divide $\{1,8\}$ by $\{1,2\}$. The whole procedure is summarised in FIGURE 1, below.

$$
\begin{array}{r}
2,\ \ 1 \\
1,2\ \overline{)\ 2,\ \ 5,\ \ 8} \\
2\ \ 4 \\ \hline
1\ \ 8 \\
1\ \ 2 \\ \hline
6
\end{array}
$$

FIGURE 1.

The process doesn't always work smoothly and adjustments have to be made. For example, in trying to divide $\{2,5,8\}$ by $\{1,3\}$, 1 still goes into 2 twice, but $2 * \{1,3\}$ is $\{2,6\}$ which is larger than $\{2,5\}$. So we reduce 2 to 1 and note that $1 * \{1,3\}$ is less than $\{2,5\}$. When we subtract and bring down the $\{8\}$, we have now to divide $\{1,2,8\}$ by $\{1,3\}$. We can use our calculators here and note

$$\text{int}((1*10^{\wedge}12+2*10^{\wedge}6+8)/(1*10^{\wedge}6+3))$$

evaluates to 999999 and carry on from there. The final result is given in FIGURE 2, below.

$$
\begin{array}{r}
1, \\
\cancel{2,}\ \ \ 999999 \\
1,3\ \overline{)\ 2,\ \ 5,\ \ 8} \\
\cancel{2}\ \ \cancel{6} \\ \hline
1\ \ 3 \\
1\ \ 2\ \ 8 \\
1\ \ 1\ \ 999997 \\ \hline
11
\end{array}
$$

FIGURE 2.

Another, more serious, problem arises when we try dividing $\{500000, 0, 0\}$ by $\{2, 50000\}$. Using only the 2 to estimate the first micresimal digit yields 250000. But $250000*\{2, 50000\}$ is $\{512500, 0\}$, which is too large. We could try reducing the digit by 1 to get 249999, but $249999 * \{2, 50000\}$ is $\{512497, 950000\}$. This is again too large... Evidently this could take a long time. Using both digits of $\{2, 50000\}$ and the calculator to estimate the quotient, we enter

$$\text{int}(500000*10^{\wedge}6/(2*10^{\wedge}6+50000))$$

and get 243902, which is correct. Our estimate using $\{2\}$ as divisor was off the mark by 6098, i.e., it would have taken 6098 steps to get the right digit if we tried successively 250000, 249999, 249998, ...

A.2.5. Exercise. *Complete the division of* $\{500000, 0, 0\}$ *by* $\{2, 50000\}$ *to find the quotient* $\{243902, x\}$ *and remainder* $\{y\}$. *Can you divide* $\{500000, 0, 0, 0\}$ *by* $\{2, 0, 50000\}$?

Will one have to use all three micresimal digits of $\{2, 0, 50000\}$ in this Exercise? Supposedly, as 200000050000 has only 8 significant digits (20000005 × 10^{12}) and 50000000000000000 has only 1 (5 × 10^{17}), the calculator will evaluate

$$\text{int}(500000*10^{\wedge}12/(2*10^{\wedge}12+50000))$$

exactly and there will be no problem. And we would expect the same were the divisor $\{2, 0, 50001\}$. But what if the divisor were $\{200000, 0, 50001\}$? And look at the numerator 1000002000008 of our earlier example. The calculator can divide its 13 digits by the 7 of 1000003 and get the integral part correct. But what if the numerator were $\{100, 2, 8\}$ with 15 digits and we were dividing by $\{100, 0, 3\}$?

Here, some numerical work is necessary to verify that we can estimate the desired digit of the quotient using the calculator with its built-in limitations. Basically, we can construct our algorithm so that we use divisions involving at most two-micresimal-digit numbers where the calculator will yield one micresimal digit accurately, and this estimate will be exceedingly close to the micresimal digit desired. We can do this largely on the basis of the following lemmas.

A.2.6. Lemma. *Let* $\alpha = A \cdot 10^6 + B$ *and* $\beta = a \cdot 10^6 + b$ *be two double-micresimal-digit numbers with* $a < A$, *i.e., assume* $0 < a < A < 10^6$ *and* $0 \leq B, b < 10^6$. *Then*

$$0 < \frac{\alpha+1}{\beta} - \frac{\alpha}{\beta+1} \leq 1,$$

with equality holding only for $A = B = 999999$, $a = 1$, *and* $b = 0$.

Proof. First, observe

$$\frac{\alpha}{\beta+1} < \frac{\alpha}{\beta} < \frac{\alpha+1}{\beta}, \quad \text{whence } 0 < \frac{\alpha+1}{\beta} - \frac{\alpha}{\beta+1}.$$

For the rest, observe

$$\frac{\alpha+1}{\beta} - \frac{\alpha}{\beta+1} = \frac{\alpha\beta + \alpha + \beta + 1 - \alpha\beta}{\beta(\beta+1)} = \frac{\alpha + \beta + 1}{\beta(\beta+1)}$$

$$\alpha + \beta + 1 = A \cdot 10^6 + B + a \cdot 10^6 + b + 1 \tag{136}$$

$$\beta(\beta + 1) = (a \cdot 10^6 + b)(a \cdot 10^6 + b + 1)$$
$$= a^2 \cdot 10^{12} + 2ab \cdot 10^6 + b^2 + a \cdot 10^6 + b. \tag{137}$$

We now consider the various cases.

Case 1. $b = 0$ & $a = 1$. In this case $\beta = 10^6$, whence $\beta(\beta+1) = 10^{12} + 10^6$. The largest possible value of $\alpha + \beta + 1$ occurs when $A = B = 999999$ and

$$\alpha + \beta + 1 = 999999 \cdot 10^6 + 999999 + 10^6 + 1$$
$$= 999999 \cdot 10^6 + 10^6 + 10^6$$
$$= (999999 + 1) \cdot 10^6 + 10^6$$

$$= 10^{12} + 10^6.$$

Thus $\alpha + \beta + 1 = \beta(\beta + 1)$ for $A = B = 999999$ and $\alpha + \beta + 1 < \beta(\beta + 1)$ for smaller values of A or B. That is,

$$\frac{\alpha + 1}{\beta} - \frac{\alpha}{\beta + 1} = \frac{\alpha + \beta + 1}{\beta(\beta + 1)} \leq 1,$$

with equality only at $A = B = 999999$.

Case 2. $b = 0$ & $a > 1$. Thus $\beta = a \cdot 10^6$ and $\beta(\beta + 1) = a^2 \cdot 10^{12} + a \cdot 10^6$. Observe

$$\begin{aligned}
\alpha + \beta + 1 &= A \cdot 10^6 + B + a \cdot 10^6 + 1, \quad \text{by (136)} \\
&= A \cdot 10^6 + B + 1 + a \cdot 10^6 \leq A \cdot 10^6 + 10^6 + a \cdot 10^6 \\
&\leq (A + 1) \cdot 10^6 + a \cdot 10^6 \\
&\leq 10^6 \cdot 10^6 + a \cdot 10^6 = 10^{12} + a \cdot 10^6 \\
&< a^2 \cdot 10^{12} + a \cdot 10^6, \text{ since } a > 1 \\
&< \beta(\beta + 1).
\end{aligned}$$

Case 3. $b > 0$ & $a = 1$. Then

$$\begin{aligned}
\alpha + \beta + 1 &= A \cdot 10^6 + B + a \cdot 10^6 + b + 1 \\
&< 10^{12} + B + 10^6 + b + 1 \\
&< 10^{12} + 2 \cdot 10^6 + b < \beta(\beta + 1).
\end{aligned}$$

Case 4. $b > 0$ & $a > 1$. Then

$$\begin{aligned}
\alpha + \beta + 1 &= A \cdot 10^6 + B + a \cdot 10^6 + b + 1 = A \cdot 10^6 + a \cdot 10^6 + B + b + 1 \\
&\leq A \cdot 10^6 + a \cdot 10^6 + (10^6 - 1) + (10^6 - 1) + 1 \\
&< A \cdot 10^6 + a \cdot 10^6 + 2 \cdot 10^6 \\
&< 10^{12} + a \cdot 10^6 + 2ab \cdot 10^6 < \beta(\beta + 1). \qquad \square
\end{aligned}$$

A.2.7. Lemma. *Let $\alpha = A \cdot 10^{12} + B \cdot 10^6$ and $\beta = a \cdot 10^6 + b$ be two double-micresimal-digit numbers, $0 < A < a < 10^6$ and $0 \leq B, b < 10^6$. Then*

$$0 < \frac{\alpha + 10^6}{\beta} - \frac{\alpha}{\beta + 1} < 1.$$

Proof. Again, observe

$$\begin{aligned}
\frac{\alpha + 10^6}{\beta} - \frac{\alpha}{\beta + 1} &= \frac{\alpha\beta + \beta \cdot 10^6 + \alpha + 10^6 - \alpha\beta}{\beta(\beta + 1)} \\
&= \frac{\beta \cdot 10^6 + \alpha + 10^6}{\beta(\beta + 1)} \\
&= \frac{(a \cdot 10^6 + b)10^6 + A \cdot 10^{12} + B \cdot 10^6 + 10^6}{\beta(\beta + 1)} \\
&= \frac{(a \cdot 10^{12} + b \cdot 10^6) + A \cdot 10^{12} + (B + 1) \cdot 10^6}{\beta(\beta + 1)}
\end{aligned}$$

2. EXACT CALCULATIONS ON THE *TI-83*

$$\leq \frac{(a \cdot 10^{12} + b \cdot 10^6) + (A+1) \cdot 10^{12}}{\beta(\beta+1)}$$

$$\leq \frac{a \cdot 10^{12} + b \cdot 10^6 + a \cdot 10^{12}}{\beta(\beta+1)}$$

$$< \frac{2(a \cdot 10^{12} + b \cdot 10^6)}{\beta(\beta+1)} = \frac{2\beta \cdot 10^6}{\beta(\beta+1)} < \frac{2 \cdot 10^6}{\beta+1}$$

$$< \frac{\beta}{\beta+1} < 1,$$

since $a \geq 2$ and thus $\beta \geq 2 \cdot 10^6$. □

These two lemmas will handle the individual steps in the division process in which the numerator ∟NUM1 and the denominator ∟DENOM differ in their first micresimal digit. The case in which these digits agree requires its own lemmas, which we will present shortly. First, however, let us consider the structure of XDIVAUX2 and the additional auxiliary programs that it will call to handle the cases covered by these lemmas. XDIVAUX2 is a more complex version of XDIVAUX1, with some of the analogous features hidden in the new auxiliaries.

```
PROGRAM:XDIVAUX2
:∟DENOM(1)→E
:E∗10^6+∟DENOM(2)→D
:Repeat V<W
    :∟NUM1(1)→F
    :If E<F
        :prgmXDIVA2A
    :If E>F
        :prgmXDIVA2B
    :If E=F
        :prgmXDIVA2C
:End
:∟QUOT→∟Z
:prgmXUNPAD
:∟Z→∟QUOT
:∟QUOT→∟X
:∟DENOM→∟Y
:prgmMULT
:∟NUM→∟X
:∟Z→∟Y
:prgmDIFF
:∟Z→∟REM.
```

The new auxiliary programs XDIVA2A, XDIVA2B, and XDIVA2C will determine micresimals in the cases $a < A$, $a > A$, and $a = A$, respectively, where A, a are as in Lemmas A.2.6 and A.2.7. How does this work? Well, if the current numerator is $A \cdot 10^{6n} + B \cdot 10^{6(n-1)} + \ldots$ and the denominator is

$a \cdot 10^{6m} + b \cdot 10^{6(m-1)} + \ldots$, then for, say, $a < A$, $\alpha = A \cdot 10^6 + B$, $\beta = a \cdot 10^6 + b$,

$$\frac{\alpha}{\beta+1} \cdot 10^{6(n-m)} \leq \frac{A \cdot 10^{6n} + B \cdot 10^{6(n-1)} + \cdots}{a \cdot 10^{6m} + b \cdot 10^{6(m-1)} + \cdots} \leq \frac{\alpha+1}{\beta} \cdot 10^{6(n-m)},$$

and the first micresimal digit of the ratio will lie between the greatest integers of

$$\frac{\alpha}{\beta+1} \quad \text{and} \quad \frac{\alpha+1}{\beta}.$$

This digit will thus lie between

$$\left[\frac{\alpha}{\beta+1}\right] \quad \text{and} \quad \left[\frac{\alpha+1}{\beta}\right].$$

As

$$0 < \frac{\alpha+1}{\beta} - \frac{\alpha}{\beta+1} \leq 1,$$

the greatest integers in the two numbers can differ by at most 1, i.e., the first micresimal digit of the quotient is either

$$\left[\frac{\alpha+1}{\beta}\right] \quad \text{or} \quad \left[\frac{\alpha+1}{\beta}\right] - 1.$$

Thus XDIVA2A will calculate $[(\alpha+1)/\beta]$ and check if this number times the denominator is not larger than the numerator. If it isn't, $[(\alpha+1)/\beta]$ is put into ∟QUOT(Q), Q is given an appropriate new value, and the difference between ∟NUM1 and the product becomes the new ∟NUM1; if it is too large, $[(\alpha+1)/\beta] - 1$ is put into ∟QUOT(Q) and the other steps are taken.

XDIVA2B is analogous, with only minor differences. I present the two programs side-by-side for the sake of comparison.

	PROGRAM:XDIVA2A	PROGRAM:XDIVA2B
1		:If W=V
2		:Then
3		:1→V
4		:Return
5		:End
6	:F*10^6+∟NUM1(2)+1 →N	:F*10^12+(∟NUM1(2)+1)*10^6 →N
7	:C−V+W→Q	:C−V+W+1→Q
8	:int(N/D)→A	:int(N/D)→A
9	:A*∟DENOM→∟Z	:A*∟DENOM→∟Z
10	:prgmXCARRY	:prgmXCARRY
11	:∟NUM1→∟Y	:∟NUM1→∟Y
12	:If W<V	:If W<V−1
13	:Then	:Then
14	:augment(∟Z,seq(0,J,1,V−W)) →∟X	:augment(∟Z,seq(0,J,1,V−W−1)) →∟X
15	:Else	:Else
16	:∟Z→∟X	:∟Z→∟X
17	:End	:End

2. EXACT CALCULATIONS ON THE TI-83

18	:"WEAK"→Str1	:"WEAK"→Str1
19	:prgmXLESS	:prgmXLESS
20	:If T	:If T
21	:Then	:Then
22	:A→ʟQUOT(Q)	:A→ʟQUOT(Q)
23	:ʟX→ʟY	:ʟX→ʟY
24	:ʟNUM1→ʟX	:ʟNUM1→ʟX
25	:prgmDIFF	:prgmDIFF
26	:ʟZ→ʟNUM1	:ʟZ→ʟNUM1
27	:Else	:Else
28	:A−1→ʟQUOT(Q)	:A−1→ʟQUOT(Q)
29	:(A−1)∗ʟDENOM→ʟZ	:(A−1)∗ʟDENOM→ʟZ
30	:prgmXCARRY	:prgmXCARRY
31	:If W<V	:If W<V−1
32	:Then	:Then
33	:augment(ʟZ,seq(0,J,1, V−W))→ʟY	:augment(ʟZ,seq(0,J,1, V−W−1))→ʟY
34	:Else	:Else
35	:ʟZ→ʟY	:ʟZ→ʟY
36	:End	:End
37	:ʟNUM1→ʟX	:ʟNUM1→ʟX
38	:prgmDIFF	:prgmDIFF
39	ʟZ→ʟNUM1	:ʟZ→ʟNUM1
40	:End	:End
41	:dim(ʟNUM1)→V	:dim(ʟNUM1)→V .

As one can see, the two programs are virtually identical. The differences are mainly due to the fact that in the second program one is using one more digit of the numerator in performing the local division. Thus the numerator must have more digits than the denominator—hence lines 1 – 5 escaping the program should this fail to occur and returning to XDIVAUX2, line 3 further forcing the escape from the Repeat loop in that program. Likewise, except for line 41 which assigns V the value of the new numerator ʟNUM1 as determined in line 21 or line 39, every occurrence of V in XDIVA2A is replaced by V−1 in XDIVA2B. The only other difference is line 6 determining $\alpha + 1$ in accordance with Lemma A.2.6 and $\alpha + 10^6$ in accordance with Lemma A.2.7 in XDIVA2A and XDIVA2B, respectively.

Line 7 determines where in ʟQUOT the next possibly nonzero digit is to be placed. In XDIVAUX1 one proceeded from Q to Q+1. In subtracting the multiple of the denominator from the numerator, however, several leading 0's might occur, as in the decimal division of FIGURE 3, below. The number of such 0's is C−V and one moves to C−V+W to determine Q in XDIVA2A and to C−V+W+1 in the second program.

The rest of each program is fairly straightforward, the only point perhaps requiring explanation is the presence of the augment commands in lines 14 and 33. They simply append extra 0's to the end of the multiple of the denominator

$$\begin{array}{r}002005\\102\overline{)204566}\\204000\\\hline 000566\\510\\\hline 56\end{array}$$

FIGURE 3.

to line it up properly against ∟NUM1 for performing the subtraction—as with the three 0's at the end of 204000 in FIGURE 3.

These programs are fairly straightforward, but are wasteful of space. Why fill up RAM with two nearly identical programs that could be handled by a single program with a toggle like we used in XLESS? A good exercise here might be to rewrite XDIVAUX2 and the auxiliaries thus far given with an eye to the more efficient use of the limited RAM storage available. Aside from combining XDIVA2A and XDIVA2B into a single program, can you see any other ways of conserving space?

Once XDIVA2A has been entered, DIVIDE can handle some additional divisions.

A.2.8. Exercise. *Run* DIVIDE *on the following pairs of inputs:*
i. ∟NUM: $\{500000, 0, 0\}$
   ∟DENOM: $\{2, 50000\}$
ii. ∟NUM: $\{2, 0, 4, 5, 6, 6\}$
   ∟DENOM: $\{1, 0, 2\}$.
Write a program XCHECKDV *which will multiply* ∟QUOT *and* ∟DENOM*, and then add* ∟REM *and compare the result with* ∟NUM *to check that the long divisions have been carried out correctly.*

And once XDIVA2B has been entered, DIVIDE can handle even more divisions.

A.2.9. Exercise. *Run* DIVIDE *on the following pairs of inputs:*
i. ∟NUM: $\{500000, 0, 0, 0\}$
    ∟DENOM: $\{2, 50000\}$
ii. ∟NUM: $\{2, 0, 6, 4, 0, 0\}$
    ∟DENOM: $\{3, 5, 2\}$
iii. ∟NUM: $\{2, 0, 6, 4, 0, 0\}$
    ∟DENOM: $\{3, 5\}$.

If one tries repeating this last exercise using
   ∟NUM: $\{2, 0, 6, 4, 0, 0, 0\}$
   ∟DENOM: $\{3, 5, 2\}$
at this stage in the development of DIVIDE, one will get the error message

2. EXACT CALCULATIONS ON THE *TI-83*

```
ERR:UNDEFINED
1:Quit
2:Goto .
```

Choosing the second option brings one to the line

```
:prgmXDIVA2C
```

of XDIVAUX2. We have yet to program the case in which the numerator and denominator have the same leading digit. Lemmas A.2.6 and A.2.7 do not apply and we must supply new lemmas to cover this case—lemmas in plural because there are several subcases. These are determined by the first digit at which the numerator and denominator disagree—which is larger and where the difference occurs.

A.2.10. Lemma. *Let* $k > 0$, $\alpha = A_1 \cdot 10^{6k} + A_2 \cdot 10^{6(k-1)} + \ldots + A_k \cdot 10^6 + A_{k+1}$ *and* $\beta = a_1 \cdot 10^{6k} + a_2 \cdot 10^{6(k-1)} + \ldots + a_k \cdot 10^6 + a_{k+1}$. *Suppose for* $i = 1, 2, \ldots, k+1$ *we always have* $0 \leq a_i, A_i < 10^6$ *with* a_i, A_i *integral and* $a_1, A_1 \neq 0$, *and for* $i = 1, 2, \ldots, k$ *we have* $a_i = A_i$. *If* $A_{k+1} > a_{k+1}$, *then*

$$\left[\frac{\alpha}{\beta+1}\right] = \left[\frac{\alpha}{\beta}\right] = \left[\frac{\alpha+1}{\beta}\right] = 1.$$

Proof. We have

$$\beta + 1 = a_1 \cdot 10^{6k} + \ldots + a_k \cdot 10^6 + a_{k+1} + 1$$
$$\leq a_1 \cdot 10^{6k} + \ldots + a_k \cdot 10^6 + A_{k+1}$$
$$\leq A_1 \cdot 10^{6k} + \ldots + A_k \cdot 10^6 + A_{k+1} = \alpha,$$

whence

$$1 \leq \frac{\alpha}{\beta+1}.$$

On the other hand,

$$\alpha + 1 < 2A_1 \cdot 10^{6k} = 2a_1 \cdot 10^{6k} \leq 2\beta,$$

whence

$$\frac{\alpha+1}{\beta} < 2.$$

Thus

$$\left[\frac{\alpha+1}{\beta}\right] \leq 1. \qquad \square$$

A.2.11. Lemma. *Let* $k > 1$, $\alpha = A_1 \cdot 10^{6(k+1)} + A_2 \cdot 10^{6k} + \ldots + A_k \cdot 10^{6 \cdot 2} + A_{k+1} \cdot 10^6$ *and* $\beta = a_1 \cdot 10^{6k} + a_2 \cdot 10^{6(k-1)} + \ldots + a_k \cdot 10^6 + a_{k+1}$. *Suppose for* $i = 1, 2, \ldots, k+1$ *we always have* $0 \leq a_i, A_i < 10^6$ *with* a_i, A_i *integral and* $a_1, A_1 \neq 0$, *and for* $i = 1, 2, \ldots, k$ *we have* $a_i = A_i$. *If* $A_{k+1} < a_{k+1}$, *then, for nonnegative integral* $x < 10^6$,

$$\left[\frac{\alpha}{\beta+1}\right] = \left[\frac{\alpha}{\beta}\right] = \left[\frac{\alpha+x}{\beta}\right] = 999999.$$

Proof. We have
$$\alpha + 10^6 = A_1 \cdot 10^{6(k+1)} + A_2 \cdot 10^{6k} + \ldots + A_k \cdot 10^{6 \cdot 2} + A_{k+1} \cdot 10^6 + 10^6$$
$$= a_1 \cdot 10^{6(k+1)} + a_2 \cdot 10^{6k} + \ldots + a_k \cdot 10^{6 \cdot 2} + (A_{k+1} + 1) \cdot 10^6$$
$$\leq a_1 \cdot 10^{6(k+1)} + a_2 \cdot 10^{6k} + \ldots + a_k \cdot 10^{6 \cdot 2} + a_{k+1} \cdot 10^6$$
$$\leq 10^6 \cdot \beta.$$

Thus
$$\frac{\alpha}{\beta + 1} < \frac{\alpha}{\beta} \leq \frac{\alpha + x}{\beta} < \frac{\alpha + 10^6}{\beta} \leq 10^6,$$

i.e.,
$$\left[\frac{\alpha}{\beta + 1}\right] \leq \left[\frac{\alpha}{\beta}\right] \leq \left[\frac{\alpha + x}{\beta}\right] \leq 999999.$$

If
$$\frac{\alpha}{\beta + 1} < 999999 = 10^6 - 1,$$
then
$$\alpha < (10^6 - 1)(\beta + 1) = 10^6 \beta - \beta + 10^6 - 1.$$
But $\alpha = 10^6 \cdot \beta + (A_{k+1} - a_{k+1}) \cdot 10^6$, whence
$$10^6 \cdot \beta + (A_{k+1} - a_{k+1}) \cdot 10^6 < 10^6 \cdot \beta - \beta + 10^6 - 1$$
$$\beta + A_{k+1} \cdot 10^6 + 1 < a_{k+1} \cdot 10^6 + 10^6$$
$$< (a_{k+1} + 1) \cdot 10^6$$
$$< 10^{12}.$$
However, $k > 1$, so
$$\beta = a_1 \cdot 10^{6k} + \ldots + a_{k+1} \geq a_1 \cdot 10^{12} \geq 10^{12}$$
and we have
$$10^{12} \leq \beta < \beta + A_{k+1} \cdot 10^6 + 1 < 10^{12},$$
a contradiction. □

A.2.12. Lemma. *Let $\alpha = a \cdot 10^{12} + B \cdot 10^6$ and $\beta = a \cdot 10^6 + b$, with $0 \leq a, B, b < 10^6$, a, B, b integral and $a \neq 0$, and suppose $b > B$. Then*
$$0 < \frac{\alpha + 10^6}{\beta} - \frac{\alpha}{\beta + 1} < 2.$$

The proof is similar to that of Lemma A.2.7 and I leave it to the reader as an exercise. On the basis of these lemmas, we construct the following program:

```
        PROGRAM:XDIVA2C
1   :1→K
2   :While ∟NUM1(K)=∟DENOM(K) and K<W
3       :K+1→K
4   :End
5   :If ∟NUM1(K)≥∟DENOM(K)
6   :Then
7       :C−V+W→Q
```

2. EXACT CALCULATIONS ON THE TI-83

```
8        :1→LQUOT(Q)
9        :If W<V
10       :Then
11           :augment(LDENOM,seq(0,J,1,V−W))→LY
12       :Else
13           :LDENOM→LY
14       :End
15       :LNUM1→LX
16       :prgmDIFF
17       :LZ→LNUM1
18   :Else
19       :If V=W
20       :Then
21           :1→V
22           :Return
23       :End
24       :If K>2
25       :Then
26           :C−V+W+1→Q
27           :999999→LQUOT(Q)
28           :999999∗LDENOM→LZ
29           :prgmXCARRY
30           :If W<V−1
31           :Then
32               :augment(LZ,seq(0,J,1,V−W−1))→LY
33           :Else
34               :LZ→LY
35           :End
36           :LNUM1→LX
37           :prgmDIFF
38           :LZ→LNUM1
39       :Else
40           :F∗10^12+(LNUM1(2)+1)∗10^6→N
41           :C−V+W+1→Q
42           :int(N/D)→A
43           :0→T
44           :Repeat T
45               :A∗LDENOM→LZ
46               :prgmXCARRY
47               :LNUM1→LY
48               :If W<V−1
49               :Then
50                   :augment(LZ,seq(0,J,1,V−W−1))→LX
51               :Else
52                   :LZ→LX
53               :End
```

```
54                  :"WEAK"→Str1
55                  :prgmXLESS
56                  :If T=0
57                      :A−1→A
58              :End
59              :A→∟QUOT(Q)
60              :∟X→∟Y
61              :∟NUM1→∟X
62              :prgmDIFF
63              :∟Z →∟NUM1
64          :End
65      :End
66  :dim(∟NUM1)→V
```

Lines 1 to 4 locate the position where the lists ∟NUM1 and ∟DENOM first differ and store this value, $k+1$, in the variable K. Lines 5 to 17 then use the information of Lemma A.2.10 to put 1 in the proper place in ∟QUOT and generate a new ∟NUM1. Lines 18 to 64 treat the case where ∟DENOM has the larger entry where the two lists first disagree. If ∟DENOM has the same length as ∟NUM1, a 0 is left in ∟QUOT and the program quits, returning to XDIVAUX2. This is done in lines 19 to 23. After that there are two cases, $k > 1$ (i.e., K > 2), and $k = 1$ (i.e., K = 2). Lines 24 to 38 cover the first case, placing 999999 into the appropriate place in ∟QUOT in accordance with Lemma A.2.11 and lines 39 to 64 implementing the procedure suggested by Lemma A.2.12. And, of course, the final line calculates the value of V to be used in the next repetition of the loop of XDIVAUX2.

A.2.13. Exercise. *Store* $\{1, 2, 3, 5, 6, 7\}$ *in the variable* ∟NUM *and run* DIVIDE *for the following choices of* ∟DENOM:
i. $\{1, 2, 3, 4\}$
ii. $\{1, 2, 3, 5\}$
iii. $\{1, 2, 3, 6\}$
iv. $\{1, 2, 4, 4\}$.

There is one final task needing to be done to complete our suite of division programs, namely, to perform the cleanup operation to clear RAM of unneeded variables:

```
PROGRAM:XCLEANUP
:DelVar ∟X
:DelVar ∟Y
:DelVar T
:DelVar C
:DelVar W
:DelVar V
:DelVar ∟NUM1
:DelVar Q
:DelVar N
```

:DelVar D
:DelVar A
:DelVar B
:DelVar ∟Z
:DelVar E
:DelVar F
:DelVar K .

There is, of course, no preferred order in which to list the variables to be deleted. Above, in order to ensure that I didn't miss any, I simply went through the various programs and deleted the variables in the order in which they were introduced. Alphabetical order might be more æsthetically pleasing and grouping the real and list variables might lend the program some semblance of organisation, but such touches are not necessary.

The use of auxiliary programs in defining DIVIDE has primarily served the purpose of handling different cases. The expert programmer will notice that I have been wasteful of space in repeating certain tasks, albeit applied to different variables, in these auxiliaries and have not created auxiliaries to handle common tasks—as I did in creating XPREP to code the common steps taken by SUM and DIFF. I leave such a more efficient organisation of the suite of division programs to the more energetic reader as a project/exercise. I now turn my attention, in the coming section, to utilising the programs we have at hand to carry out some computations that would be rather painful to perform by hand.

3. Extended Calculations on the *TI-83*

It is natural to assume at this point that, having discussed the Chinese-Horner method in section 1 and exact calculations in section 2, I would now perform some very precise calculations of roots of polynomials automating the Chinese-Horner algorithm on the *TI-83 Plus*. I will not do so for several reasons. First, rational numbers get involved. Consider Newton's polynomial,

$$x^3 - 2x - 5 = 0. \tag{138}$$

An approximation, say, 2.09455148 to its root is not an integer amenable to exact computations using the programs of section 2, and must therefore be replaced either by the integer 209455148 or the fraction 209455148/100000000, i.e., by the pair of integers 209455148 and 100000000. If we consider the first alternative as the simpler, we run up against the problem that 209455148 is nowhere near a root of (138), but is an approximation to the root of

$$\left(\frac{x}{10^8}\right)^3 - 3\frac{x}{10^8} - 5 = 0,$$

i.e.,

$$x^3 - 2 \cdot 10^{16} x - 5 \cdot 10^{24} = 0,$$

which has two large coefficients. The fact is that, no matter which route we follow, the Chinese-Horner method will quickly lead to polynomials with large multidigit coefficients, which will have to be treated as lists. The general polynomial will thus be a *list of lists*.

Now, the calculator does have a means of handling lists of lists as *matrices*. A matrix is a rectangular array of numbers and can be thought of as a list of its columns. The calculator even has built-in procedures for extracting specific columns as lists and for constructing matrices out of a number of lists of a common length. But I don't particularly feel like carrying out the details of such an implementation of the Chinese-Horner method. After all the work of the preceding section we ought to relax with some quick applications and not immediately launch into a new programming project of any depth. We should consider a fixed polynomial, like (138), where we have a fixed number of coefficients (in this case four: 1, 0, −2, −5) and hence a fixed number of lists that would need to be dealt with.

The second reason for not programming the Chinese-Horner algorithm applies as much to individual problems like solving (138) as to solving the general problem. This is its inherent inefficiency. The traditional schoolroom approach to Horner's Method of securing one digit at a time, testing 0, 1, 2, ..., 9 successively for the next digit, is fine for hand-calculation, but can include many unnecessary steps. Qín's estimation of the next digit by dividing two coefficients of the current polynomial eliminates many of these steps. Choosing only a single digit from this quotient means less work in dividing, again fine for hand computation, but it offers no advantage on the calculator which can quickly calculate several digits. Throwing away the extras just means they have to be recalculated. Keeping them all results in Newton's originally formulated method, which simplifies to Raphson's formulation which doesn't require the repeated calculation of new coefficients.

Hence I propose to use our extra precision in combination with the Newton-Raphson Method to solve a polynomial equation or two to some excessive degree of accuracy. As for the choice of equation, there are two natural candidates:

$$x^2 - 2 = 0 \quad \text{and} \quad x^3 - 2x - 5 = 0.$$

The "2" in the first equation can be replaced by any constant. The point is that finding square roots is historically rooted and one of the traditional algorithms applied to this task is just the Newton-Raphson Method in disguise. The second equation is Newton's and it became customary to illustrate a new technique for approximating the root of a polynomial with this example. Joseph Louis Lagrange did so in 1769 when he devised his technique of expanding the root of a polynomial into an infinite continued fraction.[8] And Augustus de Morgan informs us:

> Another instance of computation carried to a paradoxical length, in order to illustrate a method is the solution of $x^3 - 2x = 5$, the example given of Newton's method, on which all improvements have been tested. In 1831, Fourier's posthumous work on equations showed 33 figures of solution got with enormous labour. Thinking this a good opportunity to illustrate the superiority of the method

---

[8] *Cf.* Jean-Luc Chabert (ed.), *A History of Algorithms; From the Pebble to the Microchip*, Springer-Verlag, Heidelberg, 1999, pp. 228 – 230, for details.

3. EXTENDED CALCULATIONS ON THE *TI-83*

of W.G. Horner[9], not yet known in France, and not much known in England, I proposed to one of my classes, in 1841, to beat Fourier on this point, as a Christmas exercise. I received several answers, agreeing with each other, to 50 places of decimals. In 1848, I repeated the proposal, requesting that 50 places might be exceeded: I obtained answers of 75, 65, 63, 58, 57 and 52 places. But one answer, by Mr. W. Harris Johnston, of Dundalk, and of the Excise Office, went to 101 decimal places... The results are published in the *Mathematician*, Vol. III. p. 290. In 1851, another pupil of mine, Mr. J. Power Hicks, carried the result to 152 decimal places without knowing what Mr. Johnston had done. The result is in the *English Cyclopædia*, article INVOLUTION AND EVOLUTION.[10]

This certainly throws down the gauntlet.

Let us begin by taking up de Morgan's initial challenge of bettering Fourier, which we can do by finding the solution to 6 micresimal places, i.e., to 36 decimal places.

We could start with $a = 2$ as our initial estimate for the root of Newton's equation (138). However, as we already have the programs HORNER and HORNER2 which yield 2.094551 quickly enough, it would not be cheating to use this number as our initial estimate.

Another reason for eschewing the general problem and sticking to a specific one is that we can modify the procedure to fit the problem. In Chapter 2, section 10, we applied the Mean Value Theorem of the Second Order to Newton's polynomial to conclude that the number of secured decimals doubles at each step. Starting with 2.094551 with 6 decimals secured, one application will give us 12 decimals, then 24, then 48—beating Fourier and almost matching de Morgan's students of 1841.

Now $P(x) = x^3 - 2x - 5$ and $P'(x) = 3x^2 - 2$. Evaluating P at 2.094551, i.e., $\{2,94551\}/\{1,0\}$ results in

$$\frac{\{2,94551\}^3}{\{1,0\}^3} - 2 \cdot \frac{\{2,94551\}}{\{1,0\}} - 5,$$

i.e.,

$$\frac{\{2,94551\}^3 - 2 \cdot \{1,0\}^2 \cdot \{2,94551\} - 5 \cdot \{1,0\}^3}{\{1,0\}^3}.$$

Likewise, $P'(2.094551)$ is

$$\frac{3 \cdot \{2,94551\}^2 - 2 \cdot \{1,0\}^2}{\{1,0\}^2}.$$

Thus we get the next estimate by evaluating

$$\frac{\{2,94551\}^3 - 2 \cdot \{1,0\}^2 \cdot \{2,94551\} - 5 \cdot \{1,0\}^3}{3 \cdot \{2,94551\}^2 \cdot \{1,0\} - 2 \cdot \{1,0\}^3}.$$

---

[9] By which is meant not the Chinese-Horner method, but a more sophisticated refinement to be discussed in the next section.

[10] Augustus de Morgan (David Eugene Smith, ed.), *A Budget of Paradoxes*, 2 vols., 2nd ed., Open Court Publishing Company, Chicago and London, 1915; volume II, pp.66 –67 .

Applying MULT twice yields
$$\frac{\{9,189096,625296,766151\} - 2\cdot\{1,0\}^2\cdot\{2,94551\} - 5\cdot\{1,0\}^3}{3\cdot\{4,387143,891601\}\cdot\{1,0\} - 2\cdot\{1,0\}^3}.$$
Multiplying by the constants and applying XCARRY transforms this into
$$\frac{\{9,189096,625296,766151\} - \{4,189102\}\cdot\{1,0\}^2 - 5\cdot\{1,0\}^3}{\{13,161431,674803\}\cdot\{1,0\} - 2\cdot\{1,0\}^3}.$$
Multiplication by $\{1,0\}$ simply adds a 0 to the end of the list:
$$\frac{\{9,189096,225296,766151\} - \{4,189102,0,0\} - \{5,0,0,0\}}{\{13,161431,674803,0\} - \{2,0,0,0\}}.$$
All these lists have length 4, so we can use the ordinary list subtraction and XBORROW to get
$$-\frac{\{5,374703,233849\}}{\{11,1161431,674803,0\}}.$$
We want 12 decimals of this, so we multiply by 10^{12} putting
$$\{5,374703,233849,0,0\}$$
into ∟NUM and $\{11,1161431,674803\}$ into ∟DENOM. (Actually, to possibly save some time, we might drop the final 0 in each list.) Then we run DIVIDE. ∟QUOT is $\{481542\}$, giving us 2.094551481542 after tacking it on to the end of $\{2,94551\}$.

We could now repeat the above steps using $\{2,94551,481542\}/\{1,0,0\}$ in P and P', but multiplying by 10^{24} before applying DIVIDE to determine the next two micresimals.

A.3.1. Exercise. *Carry this out two more steps to exceed Fourier and obtain the root of* (138) *correct to* 48 *decimals.*

Working through a problem using calculator and paper was a common procedure in the early days (1960s) of pocket calculators as they initially had no memory, then stored only one number in memory, and eventually only stored a single list. Today we would not proceed as above for too many iterations. There are a lot of lists to keep track of and a lot of repetitive steps, whence one would probably write a program to perform the computations of the Exercise, and maybe even that portion of the computation I have just presented. Probably the hardest part of writing such a program, since it depends on all the programs developed in the preceding section, is to avoid clashes of variables with those earlier programs.

The program NEWTPOLY will start with the approximation $\{2,94551\}$ and will assume a number of iterations of the above procedure to be stored in the as yet unused variable P. It will then repeatedly call a program NEXTNEWT to calculate the next approximation to a root of (138). These iterations can be time consuming, so it is recommended one stick to low values for the variable P or move the project over to a computer with greater speed and memory.

3. EXTENDED CALCULATIONS ON THE *TI-83*

PROGRAM:NEWTPOLY
:{2,94551}→L$_1$
:For(S,1,P)
:prgmNEXTNEWT
:End
:L$_1$→∟ROOT.

After completing NEXTNEWT we will know what cleanup commands to append to this program.

NEXTNEWT itself simply codifies the steps we took in the above calculation.

PROGRAM:NEXTNEWT
:L$_1$→∟X
:L$_1$→∟Y
:prgmMULT
:∟Z→∟L$_2$
:∟Z→∟Y
:prgmMULT
:∟Z→∟L$_3$.

L$_1$, L$_2$, L$_3$ will contain $\{2, 94551\}$, $\{2, 94551\}^2$, and $\{2, 94551\}^3$, respectively, on its first call, $\{2, 94551, 481542\}$, $\{2, 94551, 481542\}^2$, and $\{2, 94551, 481542\}^3$, respectively, on its second call, etc. The next step is to multiply L$_1$ by 2 and L$_2$ by 3.

:2∗L$_1$→∟Z
:prgmXCARRY
:∟Z→L$_4$
:3∗L$_2$→∟Z
:prgmXCARRY
:∟Z→L$_5$.

Then we have to multiply various things by appropriate powers of $\{1, 0\}$ or of $\{1, 0, 0\}$ etc., the list depending on the current value of S. Multiplication by these lists consists of appending an appropriate number of 0's.

:2^(S−1)→N
:seq(0,X,1,N)→∟Z1
:augment(∟Z1,∟Z1)→∟Z2
:augment(∟Z2,∟Z1)→∟Z3
:augment(L$_4$,∟Z2)→L$_4$
:augment({5},∟Z3)→∟NCON
:augment({2},∟Z2)→∟DCON.

The lists ∟NCON and ∟DCON are the constant terms of the numerator and denominator, respectively. It is now time to perform the subtractions. The denominator is easy:

:L$_5$→∟X
:∟DCON→∟Y
:prgmDIFF
:∟Z→∟DENOM.

The numerator requires two subtractions, basically subtracting $2x$ from x^3 and then 5 from the result of the first subtraction. Now the last decimal digit in the last micresimal of L_1 is rounded. This means that L_1 could represent a number less than the root and the result of the second subtraction will be negative, or it could represent a number larger than the root and the result of the subtraction is positive. With the limited calculations I have performed, the former has always been the case, but to be on the safe side our program will incorporate a test and subtract the smaller from the larger number. It really doesn't matter where addition and subtraction are concerned[11], but DIVIDE requires positive ∟NUM and ∟DENOM to work right.

    :L_3→∟X
    :L_4→∟Y
    :prgmDIFF
    :∟Z→∟Y
    :∟NCON→∟X
    :"WEAK"→Str1
    :prgmXLESS
    :If T
    :Then
    :prgmDIFF
    :Else
    :∟Z→∟X
    :∟NCON→∟Y
    :prgmDIFF
    :End.

Before performing the long division, we must multiply the numerator by an appropriate power of 10^6.

    :augment(∟Z,∟Z1)→∟NUM
    :prgmDIVIDE.

It is now time to add or subtract ∟QUOT to L_1. To do so, we must first append the correct number of 0's to the end of L_1 and then perform the addition or subtraction.

    :augment(L_1,∟Z1)→∟X
    :∟QUOT→∟Y
    :If T
    :Then
    :prgmSUM
    :Else
    :prgmDIFF
    :End
    :∟Z→L_1.

---

[11]Though the result of subtracting a larger from a smaller will look strange: {1,2} minus {2,3} is {−2,999999} not −{1,1}.

This completes **NEXTNEWT**. When P is given a large value like 4 the program takes a few minutes. On my calculator I finished with a DISP L_1 command. This fairly quickly prints one line and after a while a second, which reassures the user that some progress is indeed being made. Because **NEWTPOLY** ends with a command producing ∟ROOT, the final answer will appear again on the screen, this time in scrollable form.

The cleaning up commands to add to the program **NEWTPOLY** can now be listed:

:ClrList L_1,L_2,L_3,L_4,L_5
:DelVar M
:DelVar N
:DelVar P
:DelVar S
:DelVar T
:DelVar ∟X
:DelVar ∟Y
:DelVar ∟Z
:DelVar ∟Z1
:DelVar ∟Z2
:DelVar ∟Z3
:DelVar ∟DCON
:DelVar ∟NCON
:DelVar ∟NUM
:DelVar ∟DENOM
:DelVar ∟QUOT
:DelVar ∟REM
:DelVar Str1.

After these are added, a scrollable listing of ∟ROOT will no longer appear. A convenient listing can be had by tacking on one more command at the end of **NEWTPOLY**, namely

:SetUpEditor ∟ROOT.

Then one can enter the list editor after running **NEWTPOLY** and see the micresimal digits nicely presented.

Storing 3 in P, I ran the program and in a couple of minutes came up with

2.094551 481542 326591 482386 540579 302963 857306 105628,

thus bettering Fourier as requested by de Morgan. I then ran the program with 4 stored in P and acquired 96 figures beyond the decimal point, almost matching Johnston's achievement. The computation took noticeably longer than that for P = 3. Running the program for P = 5 took several minutes, producing 192 decimal places, thus surpassing even Hicks's performance:

2.094551 481542 326591 482386 540579 302963 857306 105628
239180 304128 529045 312189 983483 667146 267281 777157
757860 839521 189062 963459 845140 398420 812823 701739
655313 940554 761602 258281 889491 443972 226659 155954

Without the *Mathematician* or the *English Cyclopædia* at hand, how do we know that my figures are correct?[12]

A.3.2. Exercise. *Write a program* XCHECKRT *to check if* ∟ROOT *indeed represents a root of* (138). *It should check that* $x^3 - 2 \cdot 10^{2s} x$ *is less than* $5 \cdot 10^{3s}$ *when x is replaced by* ∟ROOT *and that* $x^3 - 2 \cdot 10^{2s} x$ *is greater than* $5 \cdot 10^{3s}$ *when x is replaced by the list resulting by adding 1 to its last entry—where s is the number of micresimals in* ∟ROOT *following the decimal point, i.e., 1 less than the length of* ∟ROOT.

The other polynomial cited as an example to which to apply our exact calculations to was

$$P(x) = x^2 - 2, \qquad (139)$$

the root of which is the square root of 2. Given the long history of algorithms for producing square roots outlined in Chapter 1, this is a natural, if not particularly exciting, choice. In one of my earlier books[13] I programmed the *TI-83 Plus* to find square roots of integers to any desired degree of accuracy using the familiar single-digit-at-a-time algorithm and was disappointed to have to report that it took about 15 minutes to produce 50 decimals of $\sqrt{2}$. The speed with which the program NEWTPOLY solved (138) to 192 decimal places gives me renewed hope of being able to find square roots to many places on the *TI-83 Plus* in a reasonable amount of time.

The analysis of convergence of the iterates for (138) carries over to (139). Letting P be the polynomial of this latter equation, we have

$$P'(x) = 2x \quad \text{and} \quad P''(x) = 2.$$

Defining

$$g(x) = x - \frac{P(x)}{P'(x)} = x - \frac{x^2 - 2}{2x} = \frac{2x^2 - x^2 + 2}{2x} = \frac{x^2 + 2}{2x},$$

we have

$$g'(x) = \frac{(2x)(2x) - (x^2 + 2) \cdot 2}{(2x)^2} = \frac{4x^2 - 2x^2 - 4}{4x^2} = \frac{2x^2 - 4}{4x^2} = \frac{x^2 - 2}{2x^2}$$

$$g''(x) = \frac{(2x^2) \cdot 2x - (x^2 - 2) \cdot 2 \cdot 2x}{(2x^2)^2} = \frac{4x^3 - 4x^3 + 8x}{4x^4} = \frac{8x}{4x^4} = \frac{2}{x^3}.$$

By the Mean Value Theorem of the Second Order, for any approximation x to $\sqrt{2}$,

$$g(x) - \sqrt{2} = \frac{P''(c)}{2P'(x)}(x - \sqrt{2})^2$$

---

[12] In point of fact, the volume of the *Mathematician* in question is available online at the Hathi Trust Digital Library and I can testify that the figures cited agree as far as they are given by de Morgan.

[13] Craig Smoryński, *Chapters in Mathematics*, College Publications, London, 2012.

for some c between x and $\sqrt{2}$. Choosing $\sqrt{2} - .1 < x < \sqrt{2} + .1$ to begin with,

$$|g(x) - \sqrt{2}| = \frac{2}{2 \cdot 2x}(x - \sqrt{2})^2 = \frac{(x - \sqrt{2})^2}{2x} < \frac{(x - \sqrt{2})^2}{2(\sqrt{2} - .1)} < .4(x - \sqrt{2})^2.$$

The accuracy will again double at each iteration.

As the convergence is the same as before, there is no need for a major change of strategy in writing a program to generate the micresimals of $\sqrt{2}$. The differences are occasioned by the simpler nature of the new polynomial P. We only have to perform one multiplication in generating ∟NUM and one multiplication by a constant in finding ∟DENOM. Plus, everything in sight is positive and we don't need to perform any test before calling DIVIDE. Thus, programming SQROOT will be an overall simpler task than programming NEWTPOLY. There is only one complication that I've added, and that is to incorporate a final adjustment. Note that if the error is less than 10^{-n}, the n-th decimal itself need not be correct: it could be the n-th decimal of the root or one greater. We can check this quickly by calculating the square of the estimate and seeing if it is greater or less than 2, i.e., if the leading micresimal digit is 2 or 1.

```
PROGRAM:SQROOT
:{1,414213}→∟L₁
:For(S,1,P)
:prgmNEXTROOT
:End
:L₁→∟X
:L₁→∟Y
:prgmMULT
:If ∟Z(1)=1
:Then
:L₁→∟ROOT
:Else
:dim(L₁)→N
:L₁(N)−1→∟L₁(N)
:L₁→∟Z
:prgmXBORROW
:∟Z→∟ROOT
:End
(clean up commands)
:SetUpEditor ∟ROOT .
```

And, of course:

```
PROGRAM:NEXTROOT
:L₁→∟X
:L₁→∟Y
:prgmMULT
:∟Z→∟X
:2^(S−1)→N
:seq(0,X,1,N)→∟Z1
```

```
:augment(∟Z1,∟Z1)→∟Z2
:augment({2},∟Z2)→∟Y
:prgmSUM
:∟Z→∟NUM
:2*L₁→∟Z
:prgmXCARRY
:augment(∟Z,∟Z1)→∟DENOM
:augment(∟NUM,∟Z2)→∟NUM
:prgmDIVIDE
:∟QUOT→∟L₁ .
```

I was too lazy to get my watch from the other room and time it with any degree of accuracy, but I can report that running the program with 4 stored in P produced $\sqrt{2}$ to 96 places in much less time than the 15 minutes it took my earlier program to produce 50. Putting 5 in P produced the following

> 1.414213 562373 095048 801688 724209 698078 569671 875376
> 948073 176679 737990 732478 462107 038850 387534 327641
> 572735 013846 230912 297024 924836 055850 737212 644121
> 497099 935831 413222 665927 505592 755799 950501 152782

Again I didn't keep track of the time, but I can report starting a bowl of cereal and finishing it before the program completed its execution. I suppose one could run the program for P = 6 to get 384 decimals, but it would take a rather long time.

A.3.3. Exercise. *Write a general program for finding the square roots of integers to up to 192 decimal places.*

Finding the square root of 2 to 192 decimal places, or doing the same for the root of Newton's polynomial, is hardly a matter of great practical importance. However, it does illustrate how far we have come over the centuries. Neither Newton nor Lagrange matched the 14 digits of the built-in equation solver of the *TI-83 Plus*. Fourier exceeded this, but, as de Morgan notes, "with enormous labour". De Morgan later summed up the situation with reference to an improved version of Horner's Method:

> My reason for publishing this account is, to fix a limit which any future proposed method of solving equations to great extent (I do not speak of special methods for a few figures) must pass before it has established itself against Horner's. Fourier, who certainly lent some new power to Newton's method, went to 33 figures, but more than ten youths, in their elementary studies, have carried the result by Horner's process beyond fifty figures; and one to 103 figures[14]. I may add that the complete method—invented (or at least extended from the ordinary evolution of the square root) in some of its main parts two centuries and a half ago by Vieta—thrown by for at least a century and a half, as superseded by shorter methods—suddenly

---

[14]De Morgan's "figures" include the "2" to the left of the decimal point.

restored by a then obscure country schoolmaster, who invented the part which had always been wanting—and not only restored to algebra, but made to take a place in arithmetic[15], from which it will not be driven as long as the common rule of division remains—possesses an historical interest which, as sufficiently appears above, is sufficient to induce beginners to exercise themselves in computation of a more extensive character than it is usually thought reasonable to require of them.[16]

The same reasoning applies today. The student can now use a pocket calculator to obtain results to previously unheard of degrees of accuracy without the many hours of tedious and repetitive arithmetic. All the effort goes into writing a few programs, and here the main difficulty lies in the limitations of TI-BASIC. Keeping track of continually changing non-mnemonically named variables is not easy and I confess to having had to correct so many mistakes that I had to commit to excessive accuracy in my final answers in order to guarantee that the entire process of programming and debugging took less time than calculating the solution by hand. A switch over to the *TI-89 Titanium*, with its longer variable names, ability to program functions, and built-in exact calculations with fractions, eases the programming task somewhat.

4. Extended Calculations on the *TI-89*

The task of performing extended calculations on the *TI-89 Titanium* is much easier than that on the *TI-83 Plus*. The *TI-89* has been preprogrammed to handle integers and fractions exactly. Programming is easier because one has greater freedom naming variables, variables can be local, and one can define functions. And a lot of the symbolic computation has been programmed into the calculator as well. Here is a simple program[17] for solving Newton's polynomial using the Chinese-Horner-Newton algorithm:

```
:newtPoly(n,d)
:Func
:Local p
:Local list1
:newList(n)→list1
:2→list1[1]
:x^3−2*x−5→p
:Local k
:Local a,q,b
:For k,1,n−1
:p|x=y+list1[k]→q
:q|y=0→a
```

---

[15]De Morgan notes in an earlier footnote that his students hadn't take up algebra yet, but were first learning arithmetic.

[16]Augustus de Morgan, "Remark on Horner's method of solving equations", *The Mathematician*, vol. III, (1850), pp. 289 – 291; here: pp. 290 – 291.

[17]As stated in the preface, I assume anyone who has a *TI-89* not to require any explanation.

```
:(q−a)/y→b
:expand(⁻a/(b|y=0))→list1[k+1]
:q|y=x→p
:EndFor
:Local total
:sum(list1)→total
:iPart(10^d*total)
:EndFunc.
```

The program generates a list list1 of rational approximations to the roots of (138) and the subsequent substituted polynomials and then outputs (ignoring the decimal point) the resulting approximation to d decimal places. The choices of n and d must judiciously depend on each other. If $n = 1$, list1 contains only the entry 2 and one should not really take d greater than 0. However, the first digit of the root after the decimal point is 0, so one can take $d = 1$. After that, one can double the number d with each unit increment of n: $d = 2^{n-1}$. Running newtPoly(n,2^(n−1)) for $n = 1, 2, 3, 4, 5, 6$ gives successively

2.0, 2.10 (correct to within .01), 2.0945, 2.09455148, 2.0945514815423265

2.09455148154232659148238654057930,

this last matching Fourier. It took a noticeable amount of time, but not "with enormous labour" (on my part at least). Plugging in $n = 7$ produced 64 digits, all correct. Trying $n = 8$ to produce 128 digits resulted in the message:

Error: Overflow.

Obviously, even the built-in exact calculations of the *TI-89 Titanium* have their limitations. The manual that is packaged with the calculator does not explain this error, but evidently there is some limit to the size of the integers that can be represented in EXACT mode on this calculator. To match what we have done on the *TI-83 Plus*, it looks like we would have to repeat our use of lists to ensure extra precision. I leave this task to the reader, noting only that it is an easier task on the *TI-89* than it was on the *TI-83* and that using functions and carefully chosen variable names, the programs should be more readable than those of section 2.

When one codes in the extra precision, one will again be faced with the problem of long coefficients in the substituted polynomials that arise in the Chinese-Horner algorithm and one will probably want to switch over again to Raphson's formulation of the Newton-Raphson Method.

I next take up my pet problem, namely calculating $\sqrt{2}$ to numerous digits. This is very easy. One defines a function sqRoot(,) taking inputs n and d by replacing the lines

```
:2→list1[1]
:x^3−2*x−5→p
```

of newtPoly by

```
:1→list[1]
:x^2−2→p.
```

4. EXTENDED CALCULATIONS ON THE TI-89

sqRoot(1,4), sqRoot(2,4), sqRoot(3,4) yield

$$10000, 15000, 14166,$$

suggesting one calculate sqRoot(n,2^(n−2)) for $n = 2, 3, 4, \ldots$ One gets (with the decimal point re-introduced)

$$1.5,\ 1.41,\ 1.4142,\ 1.41421356, \text{ etc.}$$

sqRoot(8,2^(8−2)) yields $\sqrt{2}$ to 64 decimal places in a second or two, far better than the 15 minutes required to obtain 50 places in my earlier work. $n = 9$ gave 128 decimal places in a few short seconds and $n = 10$ gave 256 places after a noticeable, but not long period. I confess I did not follow up to see how far I could go before getting an Overflow error message.

At this point I should exercise some restraint and not perform any more calculations to what de Morgan referred to as "paradoxical length", but I shan't as I wish to make one more point. This is Horner's improved version of Horner's Method hinted at a couple of times earlier.

If the reader goes all the way back to page 17 he will be reminded of the comparatively slow convergence of the sequence of approximations to $\sqrt[5]{17}$ listed in (15) and how we sped up the convergence by a little trick found in Horner's original paper. Given an approximation a to a root r of a polynomial P, the original Chinese-Horner and Newton's methods approximated the error $b = r - a$ by attempting to solve

$$Q(y) = P(y + a) = 0.$$

This was done by dropping the terms of Q of degree higher than 1 and solving the resulting linear equation to obtain $b = -P(a)/P'(a)$. Horner takes this as a starting point and then re-introduces the quadratic term, replacing one y by b. For Newton's polynomial $P(x) = x^3 - 2x - 5$, we have

$$Q(y) = y^3 + 3ay^2 + 3a^2y - 2y + a^3 - 2a - 5,$$

which yields

$$b = -\frac{a^3 - 2a - 5}{3a^2 - 2} \tag{140}$$

when $y^3 + 3ay^2$ is dropped and one solves the linear equation. Re-introducing the quadratic term gives

$$Q(y) \approx 3ay^2 + 3a^2y - 2y + a^3 - 2a - 5$$
$$\approx (3ay + 3a^2 - 2)y + a^3 - 2a - 5$$
$$\approx (3ab + 3a^2 - 2)y + a^3 - 2 - 5,$$

on making the substitution. We again have a linear equation. Setting this equal to 0, plugging in the value (140) of b, and solving the equation yields a new estimate for b:

$$b' = \frac{3a^5 + 2a^3 + 30a^2 - 10}{6a^4 - 6a^2 + 15a + 4},$$

as the reader can readily verify after applying the requisite amount of algebra.[18]

In general, one can "raphsonise" this, defining
$$h(x) = x - \frac{2f(x)f''(x)}{2(f'(x))^2 - f(x)f''(x)},$$
and considering the sequence
$$x_0 = \text{initial approximation to a root } r$$
$$x_{n+1} = h(x_n).$$
And one can again analyse the rapidity of convergence of the sequence of approximants.

Now, the algebra involved can be horrendous in the general case, but for the function f at hand it is not too bad. We have
$$h(x) = \frac{3x^5 + 2x^3 + 30x - 10}{6x^4 - 6x^2 + 15x + 4}.$$
In theory differentiating this is a simple matter. In practice it is a bit of a pain, and this is where the built-in symbolic calculation on the *Ti-89 Titanium* comes in handy. Enter

(3x^5+2x^3+30x^2−10)/(6x^4-6x^2+15x+4)

for y1 in the equation editor. Then on the homescreen enter

d(y1(x),x).

The calculator yields
$$\frac{6 \cdot (3x^8 - 11x^6 - 30x^5 + 8x^4 + 50x^3 + 79x^2 + 20x + 25)}{(6x^4 - 6x^2 + 15x + 4)^2}.$$
Store this in y2:

ans(1)→y2(x).

And factor the numerator:

factor(getNum(ans(2))).

The result is slightly more pleasant than the unfactored form:
$$h'(x) = \frac{6(x^3 - 2x - 5)^2 (3x^2 + 1)}{(6x^4 - 6x^2 + 15x + 4)^2}.$$
The denominator does not factor over the rationals, which means that $x^3 - 2x - 5$ is not a factor, whence the denominator is nonzero at the root r of $x^3 - 2x - 5$.[19] It follows that $h'(r) = 0$.

One can now differentiate y2 to get $h''(x)$, the numerator of which is

$$12(30x^9 + 405x^8 + 18x^7 - 990x^6 - 2229x^5$$
$$- 450x^4 + 313x^3 + 480x^2 + 466x - 335)$$

---

[18]Using the built-in algebraic operations of the *TI-89 Titanium*, this can be done fairly effortlessly.

[19]A simpler way of seeing this is to note that, for $x > 1$, $6x^4 - 6x^2 = 6x^2(x^2 - 1) > 0$, whence the denominator has no root > 1.

and the denominator of which is
$$(6x^4 - 6x^2 + 15x + 4)^3.$$

Store ans(1) in y3(x). The numerator again contains $x^3 - 2x - 5$ as a factor, whence $h''(r) = 0$.

One more differentiation yields $h'''(x)$ with an absolutely hideous numerator

$$-12\big(540x^{12} + 9720x^{11} + 1080x^{10} - 33480x^9 - 124965x^8$$
$$- 35640x^7 + 74322x^6 + 124830x^5 + 76452x^4$$
$$- 28440x^3 - 10536x^2 + 22200x - 16939\big)$$

and the pleasant denominator
$$(6x^4 - 6x^2 + 15x + 4)^4.$$

I have no idea if the numerator factors. It is not divisible by $x^3 - 2x - 5$. What is important here is a bound. If one stores the result in y4(x) and graphs the curve using the window

xmin=2.
xmax=2.2
xscl=.1
ymin=⁻1.
ymax=2.,

one will see that h''' is monotone decreasing on the interval $[2, 2.2]$ with a maximum value of $h'''(2) \approx 1.97 < 2$.

Applying the Mean Value Theorem of the Third Order,

$$h(x) = h(r) + h'(r)(x-r) + \frac{h''(r)}{2}(x-r)^2 + \frac{h'''(c)}{6}(x-r)^3$$
$$= r + 0 + 0 + \frac{h'''(c)}{6}(x-r)^3 = r + \frac{h'''(c)}{6}(x-r)^3,$$

for some c between x and r, whence

$$|h(x) - r| = \left|\frac{h'''(c)}{6}\right| \cdot |x-r|^3 < \frac{2}{6}|x-r|^3 = \frac{1}{3}|x-r|^3.$$

Thus
$$|x_{n+1} - r| = |h(x_n) - r| < \frac{1}{3}|x_n - r|^3.$$

If we start at $x_0 = 2$, with $|x_0 - r| < .1 = 10^{-1}$, we get successively

$$|x_1 - r| < \frac{1}{3}(10^{-1})^3 < 10^{-3}$$
$$|x_2 - r| < \frac{1}{3}(10^{-3})^3 < 10^{-9}$$
$$|x_3 - r| < \frac{1}{3}(10^{-9})^3 < 10^{-27},$$

etc., tripling the accuracy at each iteration.[20]

Programming the procedure is straightforward. Bearing in mind the rapid growth in the numbers of digits involved and the failure of newtPoly(,) to handle too many iterations, I have incorporated the practice of truncating each approximant to its secure digits before using it to find the next approximant.

```
:newtHorn(n)
:Func
:Local h
:(3*x^5+2*x^3+30*x^2−10)/(6*x^4−6*x^2+15*x+4)→h(x)
:Local a
:2→a
:Local i,pows
:newList(n)→pows
:1000→pows[1]
:For i,1,n−1
:pows[i]^3→pows[i+1]
:EndFor
:For i,1,n
:iPart(h(a)*pows[i])/pows[i]→a
:EndFor
:iPart(a*pows[n])
:EndFunc.
```

Running the program for $n = 1, 2, 3, 4$ yielded values of the root correct to $3, 9, 27, 81$ decimal places, respectively, the correctness of which was verified by comparison with the 192 decimal places given on page 325, above. Running the program for $n = 5$ gave 243 decimal places in about half a minute. I could not, of course, check all of these digits against the earlier result, but they agreed in all 192 decimal places of this earlier result. Running the program for $n = 6$ resulted in an Overflow message.

The obvious thing to do here, of course, is to use lists to increase the precision of the *TI-89 Titanium* beyond its built-in limits. After all the work in section 2, this may at first strike one as an unpleasant task, rife with the possibilities of making small programming errors. However, the ability to use mnemonic names for variables on the *TI-89* should make for more readable programs and a reduced likelihood for error. All-in-all, porting the programs over to the *TI-89* from the *TI-83 Plus* ought not to be an odious task. Possibly the most perplexing part of the undertaking is the choice of base. 10^6 was a good choice for the *TI-83 Plus* because the result of multiplying two micresimal digits, being a 12-digit number, was guaranteed correct. On the *TI-89*, one needn't restrict oneself to so small a base. For the sake of variety, I would suggest 10^{10} or 10^{12}, though one could conceivably even use 10^{100}.[21] I leave the consideration of

---

[20] For those not enamoured of this calculation, I note that a less computational analysis of this convergence can be found in my first history book: Smoryński, *History...*, *op. cit.*, pp. 193–196.

[21] The larger the base, the greater the reliance on the calculator's speedier built-in functions, and thus the quicker the overall execution.

this problem, as well as the implementation of the programs, to the reader—together with the proposal that he also program and carry out the application of, say, the improved Horner iteration at least one step farther than I was able to do above.

On the matter of implementing the iterative procedures for finding roots, I am sure the reader will have noticed that, unlike the case with the programs HORNER1 and HORNER2 of section 1, the programs NEWTPOLY, SQROOT, and newtPoly of these last two sections were written for specific polynomials, not general ones. The reason is simply that each polynomial was accompanied by an analysis of convergence that told us how many digits to keep after n iterations. This varies from polynomial to polynomial, a fact not revealed by the samples chosen. We already know from our discussion of Newton's Method in Chapter 2 that the iteration of that procedure can be erratic and thus the procedure should not be applied without prior analysis of convergence. The same applies to Horner's Method. The following example was brought to my attention.

A.4.1. Example. *At one point Qín considers the following polynomial:*

$$P(x) = x^2 + 82655x - 2269810000.$$

The root lies between 21742 *and* 21743. *Starting at* $x_0 = 21742$, *the Newton iteration converges very quickly:*

$$x_1 = \frac{210963428}{9703} = 21742 + \frac{802}{9703}$$
$$\approx 21742.082654849016$$
$$x_2 = \frac{258204114223801184}{11875776498663}$$
$$= 21742 + \frac{981589870238}{11875776498663}$$
$$\approx 21742.082654794854,$$

correct to 12 *decimals. Iterating the raphsonised Horner Method, when starting with* $x_0 = 21742$, *yields*

$$x_1 = \frac{6927590712989}{318627114} = 21742 + \frac{401}{318627114}$$
$$\approx 21742.000001258524$$
$$x_2 = \frac{23205783681793194901673307378868217865 6290617841}{1067325162317482522620124125744122896 6110153}$$
$$\approx 21742.0000025$$
$$x_3 \approx 21742.0000038$$
$$x_4 \approx 21742.000005$$
$$x_5 \approx 21742.0000063,$$

etc.

A.4.2. Exercise. *Approximately how many iterations of the procedure will it take for x_n to be correct to two decimal places?*

Index

$\cos(x)$, 85, 86, 92–94, 104–105, 114, 128
$\cos^{-1}(x)$, 114
d/dx, 84
Δx, 83
e, 86
E_h, 124
e^x, see $\exp(x)$
$\exp(x)$, 85–88, 92, 112, 113, 167, 194, 227, 246–249, 286
f', 83, 84, 280
f'', f''', 84
$f^{(n)}$, 84
I, 124
$\int f(x)dx$, 91
$\int_a^b f(x)dx$, 79
$\lim_{n\to\infty}$, 173
$\overline{\lim}$, 216
$\ln(x)$, 85–88, 92, 103, 107–108, 112, 113, 116–118, 128, 194, 227, 246–249, 286
$_nC_k$, 27
$n!$, 28
$_nP_k$, 27
π, 116
$\sin(x)$, 85, 86, 92–94, 104–105, 112, 113, 128, 140, 253
$\sin^{-1}(x)$, 108–110, 113, 128
$\tan^{-1}(x)$, 105, 107–108, 113, 128

Abel's Summation Formula, 257
Abel, Niels Henrik, 15, 153, 161, 192–196, 212, 224–226, 244, 250–279, 284, 285
Abu l-Wafā', 18, 19
Aepinus, Franz, 147
Alembert, Jean le Rond d', 170–172, 177, 181–185, 189–191, 193–195, 199, 255, 280
Algebra (al-Khayyāmī), 18
Alsdorf, Ludwig, 34
Ampère, André Marie, 283
The Analyst (Berkeley), 169, 170

antiderivative, 90
Apian, Peter, 54, 56
Apollonius of Perga, 54
Archimedes, 54, 98
arithmetic-geometric mean, 245
Arithmetica infinitorum (Wallis), 63, 120
Arithmetica integra (Stifel), 54
Arithmetica Logarithmica (Briggs), 107
Arithmetical Triangle, 23
Âryabhaṭa, 19, 20, 36
Âryabhaṭiya (Âryabhaṭa), 19, 21

Bag, Amulya Kumar, 34, 35
Barrow, Isaac, 90
Barrow-Green, June, 138
Bennett, Albert A., 253
Berggren, J.L., 18, 19
Berkeley, George, 169–171
Bernal, John Desmond, 32
Bernoulli, Jakob, viii, 61, 120, 121
Bernoulli, Johann, 96, 118, 258
Berthelot, Marcellin, 32
Biernatzki, Karl Leonhard, 40
binomial coefficients, 1, 23
    combinatorial definition, 23, 27–29
    Horizontal Recursion, 23, 29, 36, 53
    properties, 58–61
    recursion, 23, 26–27
biquadratic formula, 15
Birkhoff, Garrett, 253, 264
bisection method, 6
Boethius, 54
Bolzano, Bernard, 96, 97, 137, 147, 153, 161, 179, 191–226, 229, 232, 234, 235, 237–239, 243, 244, 246, 253, 254, 259, 270, 272, 276, 284
Bombelli, Rafael, 56
Bottazzini, Umberto, 193, 194, 212, 227, 250, 251, 265
Bradley, Robert E., 228, 238, 244, 247
Brahmagupta, 36, 61, 118

Briggs, Henry, 53, 54, 61, 69, 107, 119
Browne, E.J., 226
Budan, François, 47
Busse, 147
Bynum, W.F., 226

Calcul différentiel (Cauchy), 137, 143
Calcul infinitésimal (Cauchy), 137, 229, 251, 284, 285
Cardano, Girolamo, 52, 53, 56, 61
Casorati, Felice, 265
Castillon, Johann, 147
Cauchy completeness, 179
Cauchy product, 153, 248, 272–276
Cauchy, Augustin Louis, 49, 78, 129, 137, 143, 153, 161, 167, 171, 179, 193–196, 199, 205, 212, 222–255, 265–268, 272, 274, 276, 282–286
Ch'in Chiu-shao, *see* Qín Jiǔsháo
Chabert, Jean-Luc, 47, 129, 133, 134, 282, 320
Chain Rule, 87
Chakravarti, G., 34
Chandaḥsūtra (Piṅgala), 33, 34
Chia Hsien, *see* Jiǎ Xiàn
China, mathematics in, 10, 36–51
Chu Shih-chieh, *see* Zhū Shìjié
Clairaut, Alexis Claude, 147
Colebrooke, Henry Thomas, 30
Colson, John, 147
combinations, 27
completeness of the real numbers, 177
completing the square, 2
continuity, 199, 205, 212, 223–236
  left, 257
  on a set, 205
  uniform, 204, 205, 223–232, 236, 239, 243
continuous differentiability, 211, 213, 280
continuous function, 78
convergence, 173–191, 223–226
  absolute, 179, 181
  Cauchy, 179, 215
  pointwise, 211
  uniform, 210–216, 223–226
  uniform Cauchy, 215–216
convergence tests, 175–181
  alternating convergence test, 175
  comparison test, 177, 182, 188
  monotone convergence test, 176
  ratio test, 177, 180, 191, 248, 255
  root test, 216
Cours d'analyse algébrique (Cauchy), 137, 192, 224, 226, 227, 229, 234, 244, 248, 250, 254, 266, 267, 276, 285
Crelle's Journal, 40, 193, 252, 266, 275
Crelle, August Leopold, 147, 252
Crossley, John N., 38, 61, 120, 121
cubic formula, 15
Cunha, José Anastácio da, 179

Datta, Bibhutibhushan, 19, 21, 33
Dauben, Joseph W., 19, 30, 31, 34, 38, 40, 42
Der binomische Lehrsatz (Bolzano), 147, 192, 222, 238
derivative, 84
  table of derivatives, 85
A Detailed Analysis of the Mathematical Methods (Yáng Huī), 38
Deutsche Arithmetica (Stifel), 54
Diderot, Denis, 171
Dieudonné, Jean, 266
difference quotient, 83
differentiation, 84
Dijkstra, Edsger W., 293
Dirichlet, Peter Gustav Lejeune, 261, 262, 270
directrix, 81
Discourse on the Residual Analysis (Landen), 149
divergence, 173
du Bois-Reymond, Paul, 266–271
Dù Shírán, 38, 40, 43
Dugac, Pierre, 265
Dumbleton, *see* John of Dumbleton

Edwards, A.W.F., 30, 35, 36, 51–55, 61
Edwards, C.H., 109, 227
Edwards, Harold, x
Encyclopédie (Diderot), 171, 172
Epistola posterior, 64
Epistola prior, 64
epsilontics, 199
equicontinuity, 230–231, 237, 240, 243
Ersch, J.S., 193
Essays on Mathematics (Simpson), 134
Euclid, 30, 36, 54
Euler, Leonhard, 96, 147, 148, 151, 153, 170, 175, 179, 185, 192, 193, 195, 201, 217, 218, 222, 243
Extreme Value Theorem, 138, 282
Ezra, Abraham ben Meir ibn, 51

factorial, 28
false lemmas
  in Abel, 263–266
  in Bolzano, 208–210

in Cauchy, 195, 212, 224–225, 229, 230, 232, 244, 252, 253, 263, 266
Fermat, Pierre de, 56, 61, 78, 79, 121, 138
Ferrar, W.L., 153, 262
Ferraro, Giovanni, 102, 107, 118, 184, 190, 191, 194
figurate numbers, 53
Finite Binomial Theorem, 23
Fisch, Menachem, 167
Fischer, Johann Karl, 147
focus, 81
Fourier series, 193
Fourier, Jean Baptiste Joseph, 137, 192, 193, 320–322, 325, 328, 330
Frend, William, 161
Freudenthal, Hans, 61, 223, 225, 232
Fundamental Theorem of Calculus, 90

Gauss, Carl Friedrich, 192, 194, 245, 253
General trattato di numeri et misure (Tartaglia), 52
geometric progression, 94
Gillispie, Charles Coulston, 56
Goldbach, Christian, 96
Goldstine, Herman H., 107
Goriely, Alain, viii
Grabiner, Judith V., 223, 224, 228, 229
Grandi series, 162
Grandi, Guido, 258
Grant, Edward, 98–100
Grattan-Guinness, Ivor, 222
Gregory series, *see* Madhava-Gregory series
Gregory, James, 107, 117–119
Gruber, J.G., 193
Guō Shŏujìng, 118
Gupta, Radha Charan, 31

Hahn, Hans, 197
Halāyudha, 34, 35
Halley, Edmond, 117
Hamilton, James R., vii
Hankel, Hermann, 163, 168, 191, 193, 195, 207, 222, 251, 253
Hansteen, Christoffer, 250
Harmonicorum libri XII (Mersenne), 54, 55
Harmonie universelle (Mersenne), 54, 55
Harriot, Thomas, 54, 119
Heine, Heinrich Eduard, 270
Heine-Borel Theorem, 270
Hérigone, Pierre, 62
Heytesbury, William, 98
Hicks, J. Power, 321, 325
Hindenburg, Carl Friedrich, 147

Hipparchus, 118
Höfler, Alois, 197
Holmboe, Bernt Michael, 250–252
Holmes, Sherlock, vii, viii
Horizontal Recursion, 60
Horner's Method, 40–51, 287–299, 319–321, 328–331
  Improved, 331–336
Horner, William George, 40–42, 47, 321, 328, 331
Horsley, Samuel, 147
Hutton, Charles, 148, 150, 153–155, 157, 159, 161, 201, 215, 217
Huzler, C.L.B., 228
hypergeometric function, 194
hyperreal numbers, 226

India, mathematics in, 18–19, 21–22, 30–36, 137
infinitesimals, 200, 226–230
integral
  definite, 79
  indefinite, 91
  linearity of the integral, 92
  table of integrals, 92
integrand, 93
integration by parts, 93
Intermediate Value Theorem, 234, 282
interpolation, 118–125
Introduction to Arithmetic (Nicomachus), 53
Islam, mathematics in, 18–19, 36
Itzigsohn, Carl, 228

Jiǎ Xiàn, 37, 38, 40
John of Dumbleton, 98
Johnson, Jimmie, xi
Johnston, William Harris, 321, 325
Joseph, George Gheverghese, 31, 33, 43, 104, 105, 107
Jungius, Joachim, 147
Jyesthadeva, 104, 105

Karajī, Abū Bakr ibn Muhammand ibn al Ḥusayn al-, 36
Karsten, Johann Gustav, 147
Kāshī, Ghiyāth al-Dīn Jamshīd al-, 19, 36
Kästner, Abraham Gotthelf, 147
Kaussler, 147
Kaye, George Rusby, 31, 32
The Key to Mathematics (Jiǎ Xiàn), 37, 38
Khayyāmī, 'Umar al-, 18, 19, 33, 36
Klein, Felix, 125
Kline, Morris, 96

Klügel, Georg Simon, 147
Krause, 147
Kronecker, Leopold, 265
Kuo Shou-ching, *see* Guō Shŏujìing

L'Huillier, Simon Antoine Jean, 147
Lagrange, Joseph Louis, 133, 137, 145, 147, 170, 192, 193, 251, 279–284, 320, 328
Landen, John, 147, 148, 150, 151, 153, 159, 170, 195, 201, 215, 217
Laplace, Pierre Simon de, 55
Laugwitz, Detlef, 224–228, 230
Leçons sur le calcul des fonctions (Lagrange), 281
Leçons sur le calcul différentiel (Cauchy), 284
Leibniz, Gottfried Wilhelm, 64, 71, 78, 80, 84, 90, 106, 116, 118, 175, 177, 185, 195, 258
Levi ben Gerson, 51
Lǐ Yǎn, 38, 40, 43
Libbrecht, Ulrich, 38, 40–43, 46, 48
Liber de triplici motu (Thomaz), 100
Lie, Marius Sophus, 252, 271
limit, 84, 171–174, 198
limit supremum, 216
lim sup, 216
Liouville's Journal, 261
Liouville, Joseph, 261
Liu Ch'uo, *see* Liú Zhuó
Liú Zhuó, 118
locally uniform boundedness, 269
Logarithmotechnica (Mercator), 107
Loria, Gino, 41, 42
Luckey, Paul, 33–35
Lun, Anthony W.-C., 38

Maclaurin series, 106
Maclaurin, Colin, 106, 126–128, 151, 171, 215
Madhava of Sangamagramma, 104–106
Madhava-Gregory series, 107
Markov, Andrei Andreevich, xi
Maseres, Francis, 153, 155, 157, 159, 161
Mathematical Treatise in Nine Sections (Qín), 40, 42
matrices, 320
Matthew Effect, 119
maxima/minima, 3–5
McKern, Leo, viii
Mean Value Theorem, 137–145, 213, 229, 237–238, 280–285
  Extended, 280
  of the Second Order, 143, 281, 282, 321
  of the Third Order, 281, 282, 333

Mengoli, Pietro, 103
Menzler-Trott, Eckart, xi
Mercator's series, 107
Mercator, Gerardus, 107
Mercator, Nicolaus, 107, 116
Mersenne, Marin, 54, 55
Mertens, Franz, 275
Merton scholars, 98, 100
meru-prastāra, 34
Meschkowski, Herbert, x
Methodenreinheit, 238
Methodus incrementorum (Taylor), 118
micresimal digit, 299
Mikami Yoshio, 38, 41–43
modulus of continuity, 205
modulus of convergence, 174
Moivre, Abraham de, 118
Morgan, Augustus de, 55, 121, 320, 321, 325, 326, 328, 331
Moriarty, James, vii, viii
Moulton, Derek, viii
Mourraille, Jean Raymond Pierre, 129, 134, 136
multinomial coefficients, 154
Multinomial Theorem, xi, 147
multiple precision, 298

Nahin, Paul J., 5
Needham, Joseph, 34, 35, 38–40
Nested Interval Property, 176
Neuenschwander, Erwin, 265
Newton Forward Difference Formula, 119, 120
Newton's cubic equation, 129–130, 136, 141–144, 294, 319–326, 329, 331–334
Newton's Method, 49, 128–137, 141–145, 320–331, 335
Newton, Isaac, viii, 46, 49, 62–64, 68–71, 73, 78, 80, 84, 90, 92, 97, 106–109, 112, 116, 118, 119, 127–133, 147, 148, 169, 171, 172, 195, 294, 298, 320, 328
Newton-Gregory Interpolation Formula, 119
Newton-Raphson Method, *see* Newton's Method
Nicomachus of Gerasa, 52, 53
Nilakantha, 104
Nine Chapters on the Mathematical Art, 37
Nordmann, 147
Novæ quadraturæ arithmeticæ (Mengoli), 103

odd function, 234
Oldenburg, Henry, 64, 71, 116

INDEX 341

On Obtaining Cube and Fourth Roots
 (Abu l-Wafā'), 18
operators, 124
Opus geometricum (Saint-Vincent), 102
Opus novum (Cardano), 53
Opuscules mathématique (d'Alembert),
 182
Ore, Oystein, 251
Oresme, Nicole, 63, 98–101, 103, 163
Oughtred, William, 56

Pacioli, Luca, 56
Paget, Sydney, vii
Paradoxien des Unendlichen (Bolzano),
 197
Paramesvara, 104
partial sum, 174
Pascal's triangle, 24
Pascal, Blaise, 1, 23, 24, 36, 51, 53–55,
 57–62, 80, 120, 121
Pascal, Etienne, 54, 55
Pasquich, Johann, 147
Peacock, George, 161–169, 196
Peletier, Jacques, 56
permutations, 27
Pfaff, Johann Friedrich, 147
Piṅgala, 33, 35
Pinyin transliteration, 37
Porter, Roy, 226
power series, 106
Precious Mirror of the Four Elements,
 38
Příhonský, Franz, 197
Principle of Mathematical Induction, 60
Principle of the Permanence of
 Equivalent Forms, 161–169
Principle of the Permanence of Formal
 Laws, 163
The Principles of Analytical Calculation
 (Woodhouse), 167, 170
problem of points, 56
Problemgeschichte, x
programs
 CONVERT, 299
 DALEM2, 189
 DALEM, 189
 DIFF, 301
 DIVIDE, 305
 HORNAUX, 297
 HORNER2, 295
 HORNER, 291
 NEWTPOLY, 323
 NEWTROOT, 76
 NEXTNEWT, 323
 NEXTROOT, 327

 POLYSUB2, 289
 POLYSUB, 288
 POWDIFF, 114
 POWPROD, 116
 POWQUOT, 116
 POWREVRT, 113
 POWSUM, 114, 115
 SQROOT, 327
 SUM, 301
 SYNTHDIV, 287
 XBORROW, 301
 XCARRY, 301
 XCHECKDV, 314
 XCHECKRT, 326
 XCLEANUP, 318
 XCOUNTER, 300
 XDIVA2A, 312
 XDIVA2B, 312
 XDIVA2C, 316
 XDIVAUX1, 306
 XDIVAUX2, 311
 XLESS, 304
 XPAD, 300
 XPREP, 301
 XUNPAD, 300
 newtHorn, 334
 newtPoly, 329
 nextPair, 13
 powDiff, 114
 powProd, 116
 powQuot, 116
 powRevrt, 113
 powSum, 114, 115
 sqRoot, 330
purity of method, 238

Qín Jiǔsháo, 37, 40–42, 44, 46, 49–51,
 130, 291, 294, 295, 320, 335
quadratic equation, 1
quadratic formula, 2
Quadrature of the Parabola
 (Archimedes), 98

radius of convergence, 181, 182, 214
Rajagopal, C.T., 104
Ramasubramanian, K., 105
Rangachari, M.S., 104
Raphson, Joseph, 128, 131–133, 320, 330
*Reclassification of the Mathematical
 Methods* (Yáng Huī), 38
*Réflexions sur les suites et sur les
 racines imaginaires* (d'Alembert),
 182–188, 191
Reiff, Richard, 98, 118
Rein analytischer Beweis (Bolzano),
 179, 222

Residual Analysis (Landen), 149
Riemann, Bernhard, 78, 79, 285
Robertson, Abram, 147
Robson, Eleanor, 138
Rolle's Theorem, 139, 143, 280, 281
Rolle, Michel, 138
root extraction, 6–23, 43–51, 291–298, 319–336
Rösling, 147
Rothe, 147
Rudin, Walter, 262
Ruffini, Paolo, 40
Ruffini-Horner Method, *see* Horner's Method
Russ, Steve, 147, 196, 197, 201, 204, 206, 223

Saint-Vincent, Grégoire, 102, 107
Samaw'al ben Yaḥyā ben Yahūda al-Maghribī, al-, 36
Sandifer, C. Edward, 228, 238, 244, 247
Sarasa, Alphonse Antonio de, 102, 107
Sarasvati, T.A., 105
Sarkar, Benoy Kumar, 31
Sarma, K.V., 105
Scherfer, Karl, 147
Scheubel, Johann, 56
Schultz, Johann Friedrich, 147
Scriba, Christoph J., 19, 30, 31, 34, 38, 40, 42
Secant Method, 132, 133
Segner, Johann Andreas von, 147
sequence, 94
series, 94
  alternating harmonic, 263
  harmonic, 99, 103, 263
  reversion, 112
Sewell, William, 147
Shùshū jiǔzhāng, *see Mathematical Treatise in Nine Sections*
Sī shū suǒ suàn, *see The Key to Mathematics*
Simon, Heinrich, 194
Simpson, Thomas, 133, 147
Singh, Avadesh Narayan, 19, 21, 33, 34
Smith, David Eugene, 42, 55, 56, 58, 61, 62, 67, 71, 169, 252, 253, 257, 264, 321
Smoryński, Craig, ix, 8, 12, 15, 42, 83, 153, 180, 282, 287, 326, 334
Sørensen, Henrik Kragh, xi, 266, 268, 269
Spalt, Detlef, 226
Srinivas, M.D., 105
Srivam, M.S., 105

Stedall, Jacqueline, 107, 120, 127, 138, 149, 169, 171, 172, 227, 254, 257, 264
Stifel, Michael, 54–56
Stokes, George Gabriel, 224
Struik, Dirk J., 55, 61, 120, 125, 127, 149, 169
sum of a series, 174
summation by parts, 257
Swineshead, Richard, 98
Sylow, Ludvig, 252, 271
synthetic division, 43–51, 287

Tangent Method, 128, 133
Tannery, Paul, 31
Tartaglia, Niccolò, 52, 54–56
Taton, René, 56
Taylor series, 106
Taylor's Theorem, 106, 118–128, 279–286
  with the Cauchy Form of the Remainder, 284
  with the Lagrange Form of the Remainder, 280
Taylor, Brook, 106, 118, 120, 121, 125–127, 145
Théorie des fonctions analytiques (Lagrange), 279, 281
Thomas, Alvarus, *see* Thomaz, Alvaro
Thomaz, Alvaro, 98, 100–103, 117
Thulin, Fred, xi
TI-83 Plus, ix, 8, 12, 13, 15, 17, 51, 73, 113, 114, 116, 117, 129, 135, 141, 189, 287, 290, 292, 295, 298, 299, 319, 326, 328–330, 334
TI-84, ix, 8, 12
TI-89 Titanium, ix, 8, 10, 12, 13, 15, 17, 20, 22, 48, 51, 73, 74, 77, 94, 107, 113, 114, 116, 117, 129, 130, 135, 287, 290, 299, 329, 330, 332, 334
Todhunter, Isaac, 55
Tómas, Alvaro, *see* Thomaz, Alvaro
Tracts on Mathematical and Philosophical Subjects (Hutton), 148, 153
Traité de la Résolution (Mourraille), 134
Traité du triangle arithmétique (Pascal), 55, 56, 61, 80, 121
A Treatise on Algebra (Peacock), 161–168
Treatise on fluxions (Maclaurin), 127
Trenchant, Jean, 56
Trigonometrica Britannica (Briggs), 53, 69
Tūsī, Naṣir al-Din al-, 56

uniform continuity, *see* continuity, uniform
uniform convergence, *see* convergence, uniform

Vandermonde's Theorem, 152, 218, 245
Vandermonde, Alexandre Theophile, 152
Varahamihara, 35
variable quantities, 200
vertex of a parabola, 3
Viète, François, 102, 328
Volkert, Klaus, 151
Vorlesungen über die complexen Zahlen (Hankel), 168

Wade-Giles transliteration, 37
Wallis, John, 63, 64, 78, 80, 81, 92, 120, 121, 171
Wang Hsun, *see* Wáng Xún
Wáng Xún, 118
Watson, John, vii
Weierstrass, Karl, 138, 196, 199, 224, 239, 285
Weintraub, Sol, 29
Wieleitner, Heinrich, 100, 101
Wilde, C.B., 226
Wolf, Christian, 258
Woodhouse, Robert, 167, 170
Wylie, Alexander, 40

Yáng Huī, 37, 38
Yuktibhasa (Jyesthadeva), 105

Zeno of Elea, 100
Zhū Shìjié, 37–39, 56, 118